Enzyme Applications in Fiber Processing

ACS SYMPOSIUM SERIES **687**

Enzyme Applications in Fiber Processing

Karl-Erik L. Eriksson, EDITOR
University of Georgia

Artur Cavaco-Paulo, EDITOR
Universidade do Minho

American Chemical Society, Washington, DC

Library of Congress Cataloging-in-Publication Data

Enzyme applications in fiber processing / Karl-Erik L. Eriksson, editor, Artur Cavaco-Paulo, editor.

p. cm.—(ACS symposium series, ISSN 0097–6156; 687)

"This book presents papers from the ACS Symposium "Enzyme Applications in Fiber Processing" held in San Francisco, April 1997"—Pref.

Includes bibliographical references and indexes.

ISBN 0–8412–3547–3

1. Wood-pulp—Biotechnology—Congresses. 2. Enzymes—Biotechnology—Congresses. 3. Textile fibers—Congresses. 4. Microbial enzymes—Industrial applications—Congresses.

I. Eriksson, Karl-Erik. II. Cavaco-Paulo, Artur. III. ACS Symposium Enzyme Applications in Fiber Processing (1997 : San Franciso, Calif.) IV. Series.

TS1176.6.B56E58 1998
676′.12—dc21

98–10370
CIP

This book is printed on acid-free, recycled paper.

Distributed by Oxford University Press

PRINTED IN THE UNITED STATES OF AMERICA

Advisory Board

ACS Symposium Series

Foreword

THE ACS SYMPOSIUM SERIES was first published in 1974 to provide a mechanism for publishing symposia quickly in book form. The purpose of the series is to publish timely, comprehensive books developed from ACS sponsored symposia based on current scientific research. Occasionally, books are developed from symposia sponsored by other organizations when the topic is of keen interest to the chemistry audience.

Before agreeing to publish a book, the proposed table of contents is reviewed for appropriate and comprehensive coverage and for interest to the audience. Some papers may be excluded in order to better focus the book; others may be added to provide comprehensiveness. When appropriate, overview or introductory chapters are added. Drafts of chapters are peer-reviewed prior to final acceptance or rejection, and manuscripts are prepared in camera-ready format.

As a rule, only original research papers and original review papers are included in the volumes. Verbatim reproductions of previously published papers are not accepted.

ACS BOOKS DEPARTMENT

Contents

TEXTILES

Preface

THE WORLDWIDE EFFORTS devoted over the past few decades to research for a better understanding of the enzyme mechanisms involved in the degradation of cellulose, hemicelluloses, pectins, lignins, and other natural polymers has provided a solid base for a successful use of enzymes in fiber processing. However, it is not only progress in this research but also progress in the biosciences in general, i.e., in biochemistry, microbiology, molecular biology, and genetics that has made the use of enzymes in fiber processing possible and even economically feasible for both the pulp and paper and the textile industries. The progress in these biosciences allows, for example, for cloning of enzymes from extreme thermophiles or from extreme alkalophiles into efficient production systems which means that enzymes, suitable and efficient in fiber processing, be it in bleaching, deinking or other paper making processes, scouring of cotton fibers or treatment of wool fibers in fabric making processes, can be produced at costs you could only dream of 5 to 10 years ago.

This book presents papers from the ACS symposium, "Enzyme Applications in Fiber Processing" held in San Francisco, April 1997. The first half of the book is devoted to enzymes in pulp and paper fiber processing and the second half to enzymes in textile fiber processing. The pulp and paper section starts with an overview of biotechnology in the pulp and paper industry which involves the use of both microorganisms, white-rot fungi in particular, and enzymes in bleaching, pitch degradation, deinking, etc. One chapter is fully devoted to biomechanical and biochemical pulping, another to biodegradation of resin acids in the chip pile by microorganisms and in the white-water systems by enzymes. Enzymes are also applied for the modification of kraft pulp fibers, to increase freeness in recycled fiber slurries, and for use in deinking of papers. Laccases have been used to catalyze the bonding of wood fibers together in the production of composites. Enzymes involved in lignin degradation and modification such as laccases, peroxidases, and oxidoreductases are described and characterized. These enzymes emanate from white-rot fungi. Finally in the pulp and paper section, expression of genes structuring for hemicellulases from extremely thermophilic culturable and unculturable bacteria are described in two different chapters.

The textile section starts with an overview relating to processing of textile fibers with enzymes. Most of the interest in textile fiber processing is devoted to enzyme treatment of cotton fibers and bioscouring of cotton fabrics with cellulases. Pectin degrading enzymes have also been used for scouring of cotton and for enzymatic treatment of linen fabrics and for retting of flax. Also, dissolving pulp fibers have been modified using cellulases. Finally, various types of proteases are described for use in modification of wool fibers.

While patents have been issued in many of the areas described in this book, this is to our knowledge the first book where up-to-date knowledge is put together for enzyme treatment of both pulp and paper and textile fibers. We therefore hope that this book will be of value to those in both the paper and the textile industries.

This book was developed from a symposium sponsored by the Divisions of Cellulose, Paper, and Textile and Biochemical Technology at the 213th National Meeting of the American Chemical Society in San Francisco, CA on April 13-17, 1997.

KARL-ERIK L. ERIKSSON
Professor of Biochemistry
Eminent Scholar of Biotechnology
University of Georgia
Department of Biochemistry and Molecular Biology
Athens, GA 30602-7229

ARTUR CAVACO-PAULO
Professor Auxiliar
Departmento de Engenharia Textil
Universidade do Minho
4800 Guimaraes Portugal

Pulp and Paper

Chapter 1

Biotechnology in the Pulp and Paper Industry: An Overview

Karl-Erik L. Eriksson

Department of Biochemistry and Molecular Biology, Center for Biological Resource Recovery, University of Georgia, Athens, GA 30602–7229

Development of biotechnology for the pulp and paper industry was probably started prematurely, i.e., during the 1970's, the cradle period of biotechnology. At that time the knowledge about the enzyme mechanisms involved in the degradation of wood and wood components was in it's infancy. Also, production of crucial, thermostable and alkaphilic enzymes in commercial scale and at acceptable costs has only been possible in recent years. However, with these hurdles overcome, it seems very likely that the pulp and paper industry will develop to be a very biotechnology dependent industry. In this chapter the status of biotechnology, use of enzymes in particular, for pulping, pitch removal, bleaching and enzymatic deinking is summarized.

The oil crisis during the 1970's turned interest towards the utilization of renewable resources and towards lignocellulosic materials in particular. The 1970's were also the cradle period of biotechnology and the years when biotechnical utilization of lignocellulosic wastes from agriculture and forestry gained priority. This was a logical step to take since one of natures most important biological reactions is the conversion of wood and other lignocellulosic materials to carbon dioxide, water, and humic substances. Traditionally there has been a conflict between the microorganisms and the forest industries. Fungi, in particular, have stained and rotted wood and lumber, bacteria have caused slime problems in pulp and paper mill's production systems. However, since biotechnology, by definition, is the technical utilization of biological reactions, we are now trying to join the microorganisms instead of defeating them as we have done in the past. Since wood and other lignocellulosic plants constitute the raw material for the forest industries, there should be ample opportunities to use biotechnology for manipulation and improving both the production and the use of these resources.

There are, however, difficulties for implementation of such technologies in the pulp and paper industry. While biotechnology in areas like medicine and pharmacology concerns production of expensive products on a small scale, biotechnical utilization and conversion of lignocellulosic materials means production of inexpensive products on a large scale. Biotechnological utilization of lignocellulosic materials is thus a very difficult task, and the commercial utilization of this technology has therefore only recently gained momentum. There are several reasons for this. One has been the lack of people in the pulp and paper industry with an educational background in the biosciences. Another reason was the lack of basic knowledge about the enzyme mechanisms involved in the degradation and conversion of wood, other lignocellulosics and their individual components. However, the world-wide efforts devoted over the past few decades to research for a better understanding of these mechanisms, now provide a solid base for successful development of biotechnology for the pulp and paper industry. For somebody who has been deeply involved in these investigations, it is obvious that the investments in this research have been a dazzling success. Two recent reviews summarize our knowledge of how microorganisms and their enzymes degrade wood and other plant fiber cell walls (*1,2*).

However, it is not only progress in this research, but also progress in the biosciences in general, i.e., in biochemistry, microbiology, molecular biology and genetics that has made the use of biotechnology interesting for the pulp and paper industry. The progress in the biosciences allows, for example, for cloning of enzymes from extreme thermophiles or from extreme alkalophiles into efficient production systems which means that enzymes, suitable and efficient in pulping, bleaching, and paper-making processes, can be produced at costs, you could only hope for 5-10 years ago. Microorganisms can now also be tailor-made to fit into certain processes by addition or deletion of genes and enzymes can be designed to better catalyze certain important reactions. We certainly also see the paths for how to improve on the starting resource - the forest trees, through genetic engineering, even if we not yet can do it efficiently enough to produce such trees in a cost-effective way. An overview of where in the production chain of timber, pulp and paper, biotechnology is now being introduced or already used is presented in Figure 1.

Biotechnology in the Chip Pile

Lignin Removal. Mechanical pulping of wood is done with a very high input of electrical energy while chemical pulp is produced at high temperatures and pressures, at extreme pH-levels, and with input of large amounts of chemicals. Pretreatment of wood chips, using white-rot fungi, can give rise to delignification which saves energy in mechanical pulping and reduces chemical consumption in chemical pulping (*3, 4*). White-rot fungi degrade wood using different modes of attack. While most attack all the wood components simultaneously, other remarkably have the ability to selectively attack lignin (*1*). The most studied white-rot fungus, for its ability to degrade cellulose and lignin, *Phanerochaete chrysosporium*, causes simultaneous decay. In spite of this, Eriksson and co-workers at the Swedish Pulp and Paper Research Institute, (STFI), Stockholm, Sweden, used this fungus for delignification of wood chips (*5*). However, to avoid cellulose degradation, cellulase-less mutants of the fungus, obtained by UV radiation of asexual conidial spores, were used (*6*). Several patents were obtained on the use of such mutants for biopulping (*7, 8*).

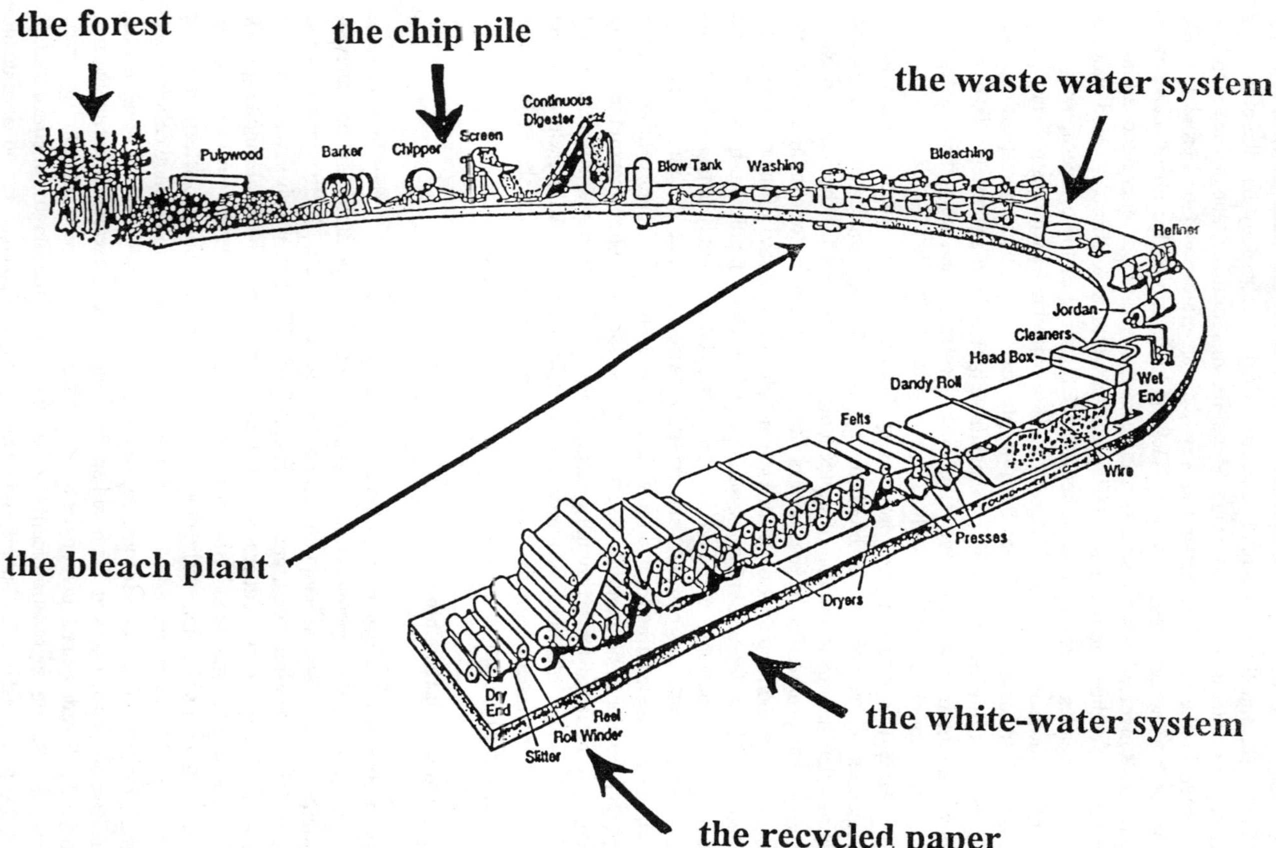

Figure 1. Applications of biotechnology in the production chain of timber pulp and paper.

The first generation Cel⁻-mutants obtained gave rise to little lignin removal and small savings in refining energy. However, a second generation Cel⁻-mutants of *P. chrysosporium* had a higher production of both phenoloxidases and xylanase and was also shown to degrade lignin better than earlier strains. Cel⁻-mutants with even greater lignin degrading capacity than all earlier cellulase-less mutants were obtained from *P. chrysosporium* by selection, mutation, and intercrossing of homokaryotic strains (*9*). However, the overall results obtained with Cel⁻-mutants from *P. chrysosporium* were not convincing and the savings obtained were not considered to be enough to cover the extra costs for the fungal pretreatment.

Comprehensive studies of the feasibility of biopulping was started in 1987 at the Forest Products Laboratory (FPL) in Madison, WI. Prior to the start of this work, a Biopulping Consortium had been established which involved FPL, the University of Wisconsin Biotechnology Center, The University of Minnesota, Department of Plant Pathology, Pulp and Paper and other related companies. In these feasibility studies, two fungi were examined in more detail, namely *P. chrysosporium* and *Ceriporiopsis subvermispora*. Various techniques were used to test the overall feasibility. Simons Stain, which had previously been used to investigate the degree of fibrillation in beating of pulp fibers, turned out to be a valuable tool for evaluation of the effectiveness of a fungal treatment (*10, 11*). Pulp obtained from chips pretreated with *C. subvermispora* turned out to have extensively fibrillated fibers that stained deep orange with Simons Stain. The intensity of yellow staining of biopulped fibers obtained under different experimental conditions, correlated well with the energy savings earned in the refining (*12*). Based on energy savings and strength improvement, *C. subvermispora* was identified as the best fungus (*13*). Different strains of this fungus were found to be effective on both aspen and loblolly pine and a patent was issued on the use of *C. subvermispora* for biomechanical pulping (*14*). After years of work, a 50 ton outdoor chip pile experiment was conducted at FPL in the fall of 1996. An economic analysis of the feasibility of biopulping based on the results obtained in this large-scale trial makes it look very attractive. The FPL researchers claim that savings of more than 40 US dollars per ton may be realized with a 30% increase in mill throughout. This should make the payback period for necessary investments shorter than one year. If an increased production cannot be realized because of other process limitations, savings of up to 20 US dollars per ton may be realized based on the 30% reduction in electric energy consumption that is always obtained. These analysis do not reflect other benefits such as increased paper strengths, reduced pitch content, and reduced environmental impact that is obtained as a bonus of the biopulping process. Several patents have been issued on the process. The Energy Center of Wisconsin has sponsored the larger trials.

Pitch Removal. Pitch problems in pulp and paper mills are caused by resinous materials, pitch, in wood. The pitch content varies from about 2 to 8%, depending upon wood species. Pine species, but also some hardwood species, particularly from tropical hardwoods, are very rich in resins.

Traditional techniques to keep pitch problems at a minimum include seasoning of round-wood, i.e., outdoor storage for several months, or seasoning of chips in the piles which takes at least 3 to 4 weeks. Seasoning of wood or wood chips has the drawback of decreased yield due to fungal decay which also causes decreased brightness. In Japan,

Hata et al., (*15-17*), found that pitch troubles were caused by the triglycerides in wood resins. To overcome pitch problems in pulp and paper mill systems a pitch control method dependent upon the use of the enzyme lipase was developed (*15-17*).

In the U.S., Farrell and co-workers also studied biotechnological solutions to pitch problems (*18, 19*). Their approach was to reduce the pitch problems already in the chip pile, inoculating the chips with the fungus *Ophiostoma piliferum* (*20, 7*). The fungus not only decreased pitch problems but also increased pulp brightness. The fungal technique is now commercially utilized and the inoculum marketed as Cartapip.

Biotechnology in Bleaching of Wood Pulp

Conventional techniques for bleaching wood pulp has involved the use of chlorine and chlorine derivatives. It is widely accepted that the effluents from such bleach plants are harmful to the environment (*21*). Particularly after dioxin was discovered both in bleach plant effluents and in some paper products, bleached with molecular chlorine, the pulp and paper industry has devoted enormous amounts of work and money to development of more environmentally benign bleaching processes. The bleaching of the wood pulp is therefore another area where biotechnology could be used in efficient and environmentally benign processes. Various biotechnology approaches have been made, the most important ones are listed in Table I.

Table I. Biotechnical Approaches to Pulp Bleaching

Use of White-rot Fungi
Use of Lignin Peroxidase (LiP)
Use of Xylanase
Use of Laccase plus Redox-mediators
Use of Manganese Peroxidase (MnP)
Use of Cellobiose-dehydrogenase (CDH)

Use of White-Rot Fungi. The first approach, to use white-rot fungi for bleaching of kraft pulp, has been studied particularly at the Pulp and Paper Research Institute of Canada (PAPRICAN). It was found that the white-rot fungus, *Trametes versicolor*, when grown in a pulp slurry, decreases the pulp kappa number and increases brightness of pulp without diminishing pulp strengths or yield. However, the direct use of white-rot fungi is not a feasible bleaching process since the process takes too long. The highest reported brightness, accomplished by fungal treatment, is 80% ISO. This was accomplished after a 2-stage treatment of hardwood pulp with a white-rot fungus named YK-624 (*22*).

Use of Lignin Peroxidase. The discovery of lignin peroxidase (LiP) (*23, 24*) generated a great deal of optimism and it was generally considered that this enzyme would give rise to a real breakthrough for biotechnology in the pulp and paper industry in general and for biobleaching in particular. Several laboratories tried to develop techniques for use of LiP in pulp bleaching, however, with limited or no success. While its high redox-potential allows it to oxidize both phenolic- and non-phenolic lignin structures these reactions give rise to phenoxyradicals which spontaneously repolymerize. The net reaction was therefore, not really a depolymerization of the lignin.

Use of Xylanase. Since xylanases or other hemicellulases, do not directly attack the lignin, the use of xylanases is an indirect bleaching method. Different hypothesis for the mechanisms by which xylanases support pulp bleaching have been introduced (*25*). In kraft pulping, most of the substituents of the xylan backbone are stripped off. The remaining mainly unsubstituted backbone can therefore be degraded with only one enzyme, i.e., an endo-xylanase. In kraft cooking, the pH level decreases due to formation of acetic acid from acetyl groups on the hemicelluloses. At a certain point, xylan reprecipitates on the fiber surfaces. One hypothesis is that the action of xylanase is able to remove this reprecipated xylan and consequently open pores in the fibers, making the remaining lignin more accessible to bleaching chemicals in subsequent stages. Another hypothesis is that the xylan backbone, with lignin attached to it, is hydrolyzed and made water soluble by xylanases. This lignin carbohydrate complex diffuses easier from the pulp fibers (*26*). The various hypotheses for the action of xylanases in pulp bleaching have been thoroughly discussed by Viikari et al., (*25*).

At the University of Georgia, we have developed a totally chlorine-free bleaching process (TCF). This process, "The EnZone Process," combines oxygen and enzymatic delignification of hardwood pulps with ozone treatments and a final peroxide bleaching stage to produce pulps of equal brightness as compared to conventional chlorine and chlorine dioxide bleached pulps. An alkaline extraction stage is inserted between the enzyme and the ozone stages to obtain fully bleached softwood kraft pulp. This process allows for a brightness of hardwood kraft pulps between 90-92% ISO and for softwood pulp, 86-88% ISO. Both hardwood and softwood kraft pulps, bleached with the EnZone process, have fully acceptable strength values (*27,28*). In the development of this process we have used various xylanases in laboratory scale. However, to test the process at a pilot plant scale, we have on the campus of the University of Georgia which allows for bleaching of up to 5 to 10 tons of pulp per day. In this process we are presently using a very thermostable xylanase from *Thermotoga maritima* cloned in *E. coli* (*29*). This xylanase is thermostable at 92°C for the time we use in the xylanase bleaching stage, i.e., 90 min. (Fig. 2). At 92°C, it loses only 20% of its activity during 24 hours. The xylanase is now being cloned in a better production system to allow for mass production. However, we can already produce enough of the enzyme in *E. coli* to allow for its use in the pilot plant. In the EnZone Process, the xylanase stage typically enhances brightness of the total bleaching sequence with 5-8 points ISO.

Use of Laccase Plus Redox-Mediators. Laccase, a 4-copper phenoloxidase, is one of several phenoloxidases produced by white rot fungi (*1*). Lignin degrading activity by *Phanerochaete chrysosporium* has been associated with its production of lignin peroxidase (LiP) and manganese peroxidase (MnP) (*30, 31*). It was found, only very recently, that *P. chrysosporium* also seems to produce a laccase-like activity (*32, 33*). A combination of laccase with either LiP and/or MnP seems to be a more common pattern of phenoloxidases than the LiP/MnP pattern found in *P. chrysosporium* (*34*). It has generally been considered that laccase was not very important for lignin degradation since it could not oxidize non-phenolic lignin structures. However, Bourbonnais and

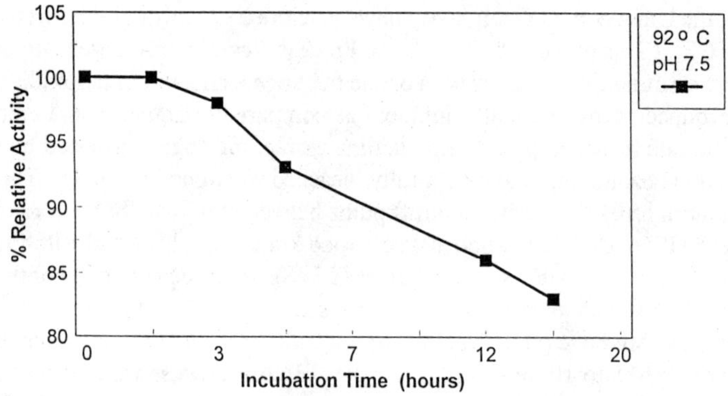

Figure 2. Temperature stability of *Thermotoga* xylanase.

Paice (*35*) showed that two artificial laccase substrates, ABTS (2,2'-azinobis-[3-ethylbenzoline-6-sulfonate]) and Remazol blue, could act as redox-mediators which allowed laccase to oxidize even non-phenolic lignin model compounds. When the same authors (*36*) later demonstrated that laccase from *Trametes versicolor,* in combination with ABTS, could delignify and demethylate kraft pulp, the interest for laccase plus redox-mediators in pulp bleaching was considerably enhanced. Call and Mücke in Germany, have explored the biotechnological possibilities to use laccase plus redox-mediators for bleaching of wood pulp (*37*). To really investigate the importance of laccase in lignin degradation, we have worked with the white-rot fungus *Pycnoporus cinnabarinus,* which was found, after an extensive screening process to be an ideal model organism for our purpose. This fungus produces only one iso-form of laccase and only trace amounts of a phenol oxidase which is neither LiP nor MnP. Since *P. cinnabarinus* could extensively degrade lignin, it was natural to believe that the laccase we produced had a higher redox potential than other laccases. After purification and characterization of the *P. cinnabarinus* laccase, we could demonstrate that this hypothesis did not hold true (*38*). Instead we found that the fungus produces its own natural redox-mediator which we isolated and identified as 3-hydroxyanthranilic acid (3-HAA). While laccase alone could not attack non-phenolic lignin model substances, it could do so in the presence of 3-HAA. We could furthermore demonstrate that the laccase/3-HAA redox-system also was able to depolymerize synthetic lignin (*39*).

To further demonstrate the importance of laccase for lignin degradation, we decided to produce Lac^- mutants of *P. cinnabarinus*. Cultures of these Lac^- mutants, grown in the presence of ^{14}C-ring labeled synthetic lignin, could not metabolize this lignin as evidenced by the lack of $^{14}CO_2$ evolution. However, addition to the cultures of laccase purified from the wild type, allowed the mutants to degrade lignin as equally well as the wild type. Also, while the wild type caused a considerable weight loss in wood blocks of both hardwood and softwood species, about 50% in 12 weeks, it was difficult for the Lac^- mutants to even establish themselves in the wood blocks and the weight loss was minimal (*40*). These findings demonstrate that laccase is of the utmost importance for lignin degradation by the white-rot fungus *P. cinnabarinus* and reveal a new mechanism for lignin degradation, different from that proposed in *P. chrysosporium*. It also suggests that white-rot fungi may go about lignin degradation in different ways.

Laccase plus redox-mediators may be a more efficient way for lignin removal in bleaching of wood pulp since these reactions do not seem to give rise to the same amount of repolymerization as do other phenoloxidase mechanisms. We are now screening for various laccases and redox-mediators and hope to scale up the use of laccase and redox-mediators for pulp bleaching to pilot plant scale in the near future (*41*).

Use of Manganese-Peroxidase. Manganese-peroxidase (MnP) is another phenoloxidase produced by many white-rot fungi during wood degradation. Iso-enzymes of MnP from *Trametes versicolor* have been used in delignification studies, particularly at PAPRICAN and Kondo et al., (42) have used MnP from *Phanerochaete sordida*YK-624. Results from their studies indicate that MnP is a necessary enzyme activity for pulp bleaching by both fungi. The enzyme oxidizes Mn(II) to Mn(III), and chelated Mn(III) is the most likely species for lignin oxidation. Addleman et al., (*43*) were able to demonstrate than an MnP negative mutant would not brighten or delignify kraft pulp, while this ability was at least partially restored by the addition of exogenous MnP. *T. versicolor* also produces laccase, however, a Lac⁻ mutants from this fungus, to which laccase was added, could not cause bleaching of pulp. In conclusion, in *T. versicolor* MnP plays an important role in bleaching of pulp. This enzyme is therefore certainly also of the greatest interest in our attempts to use enzymes in bleaching of wood pulp.

Use of Cellobiose Dehydrogenase. The first cellobiose dehydrogenase was discovered by Westermark and Eriksson (*44, 45*). It was found to reduce quinones produced by oxidation of phenols or lignins by phenoloxidases. This enzyme was therefore named cellobiose quinone:oxidoreductase. Later, Ayers et al., (*46*), discovered what was thought to be a cellobiose oxidase. It has now been renamed cellobiose dehydrogenase (*47*). It has later been demonstrated that cellobiose quinone:oxidoreductase (CBQ) is a proteolytic cleavage product of CDH (*48*). CDH has a molecular weight of around 95 kDa, while CBQ has a molecular weight of around 65 kDa. CDH contains two domains, one FAD and one heme domain. CBQ is the FAD domain of CDH. The two enzymes can be distinguished by their absorption spectra, but also by the fact that only CDH can reduce cytochrome c. Already Westermark and Eriksson (*44*) found that CBQ reduces quinones produced by oxidation of lignin or lignin degradation products by phenoloxidases in the presence of cellobiose, a cellulose degradation product. CBQ was therefore suggested to be a link between cellulose and lignin degradation. The same goes for CDH. In addition to the ability to reduce quinones, we have also demonstrated that CDH and CBQ can reduce phenoxyradicals and also cation radicals (*49*). This again points to the possibility that cellobiose dehydrogenases could play a role in lignin degradation by preventing repolymerization reactions.

The bleaching ability of CBH/CBQ is definitely worth investigating and has been done so at PAPRICAN (*50*) and now also in my laboratory. The use of these enzymes for bleaching of wood pulp is certainly of the highest interest and is presently pursued in my laboratory.

Enzymatic Deinking of Recycled Papers. The traditional deinking processes using caustic, silicate, and peroxide have been useful techniques for deinking of oil-based printing materials such as newspapers and magazines. However, with the growing use of coating and new types of inks containing synthetic polymers in laser and xerographic printing, the conventional deinking methods are inadequate for the production of high-quality pulps. Recycling mills are therefore increasingly dependent upon mechanical devices to break down the larger non-impact ink particles to allow for removal by flotation or washing.

Wood degrading microorganisms, fungi in particular, produce a variety of enzymes which in white-rot fungi are able to degrade all the wood components, cellulose, hemicelluloses and the lignins. While paper produced from bleached chemical pulp mainly has a cellulosic surface unless it is coated, newsprint and magazine papers containing mainly mechanical fibers also have lignin and hemicellulose surface components. In printing, the ink adheres to these exposed surfaces and can be readily removed if treated by specifically formulated enzymes which attack the surface components.

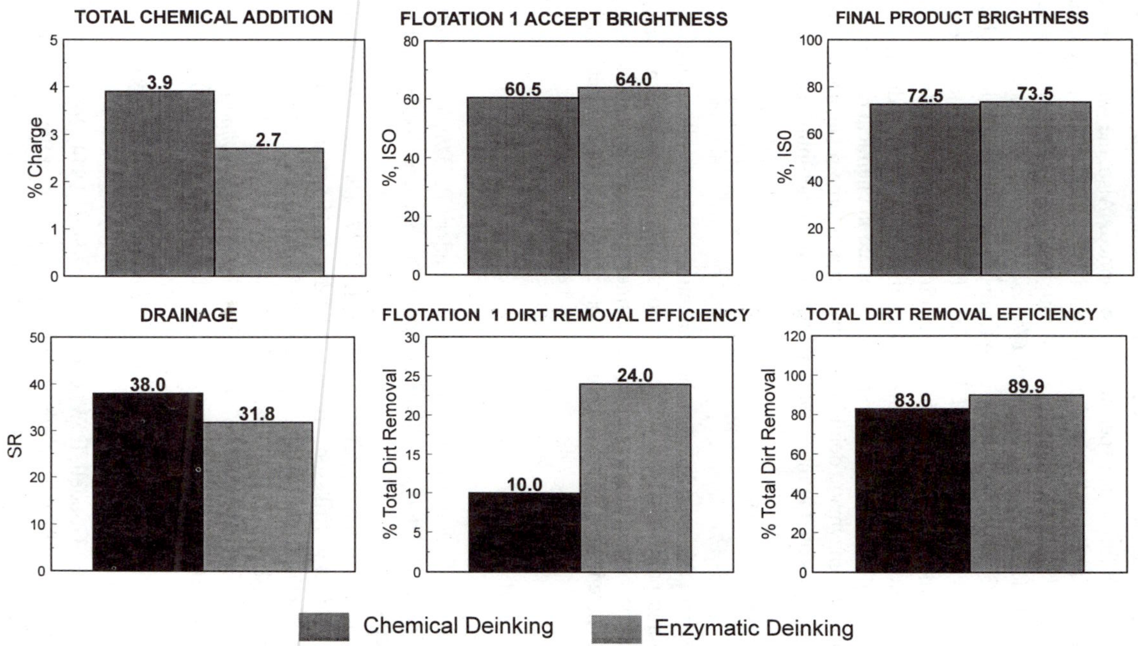

Figure 3. Comparison of chemical and enzymatic deinking of mixed office waste paper (MOW).

We have, at the University of Georgia, developed an enzymatic technique that allows for deinking of all kinds of recycled papers (*51*). This technique is particularly superior for deinking of laser and xerox printed mixed office waste. The furnish, in every recycling mill, varies in it's composition. Not one mill has a uniform starting material. It is therefore necessary to do incisive studies of the furnish mixture in the specific mill until a full scale experiment is performed. Also, the water composition and other conditions in the mill are of the utmost importance. Our enzymatic deinking process is now being commercialized through the company Enzymatic Deinking Technologies, (EDT) located in Atlanta, GA.

In Figure 3, a comparison has been made of the results obtained with enzymatic deinking and conventional chemical deinking of mixed office waste. It can be seen in this figure that although a much lower addition of chemicals is used in enzymatic deinking, the results obtained with enzymatic deinking in terms of drainage, brightness, dirt count are in all cases superior to results obtained in chemical deinking. Tests with both chemical and enzymatic deinking can be run at a large scale, about 3 tons per day, in our pilot plant located in the same building as our bleach pilot plant.

Conclusion

It is obvious from this overview that there are numerous possibilities for biotechnology in the pulp and paper industry. Yes, even to the extent that it seems likely that this industry will be very dependent upon biotechnology based processes or as somebody phrased it, "There is no doubt that there is a way ahead - only if we have the head for the way."

Literature Cited

1. Eriksson, K.-E. L.; Blanchette, R. A.; Ander, P. *Microbial and Enzymatic Degradation of Wood and Wood Components*. Springer Verlag, Berlin, Germany **1990,** pp 416.
2. Kuhad, R. C.; Singh, A.; Eriksson, K.-E. L. **1997**. *Biotechnology in the Pulp and Paper Industry* Eriksson, K.-E. L., Ed.; In Special Issue - Adv. Chem. Eng./Biotechnol., *57*, 45-125, Springer Verlag.
3. Eriksson, K.-E. *TAPPI J.,* **1985,** *68*, pp 46-55.
4. Akhtar, M.; Blanchette, R. A.; Kirk, T. K. **1997**. *Biotechnology in the Pulp and Paper Industry* Eriksson, K.-E. L., Ed.; In Special Issue - Adv. Chem. Eng./Biotechnol., 57, 159-195, Springer Verlag.
5. Eriksson, K.-E. *The Marcus Wallenberg Prize Symposium,* **1985**, Falun, Sweden. pp 72.
6. Eriksson, K.-E.; Goodell, E. W. *Can. J. Microbiol.,* **1974**, *20*, pp 371-378.
7. Method for producing cellulose pulp, (patented in several countries, US pat 3962033).
8. Method of obtaining cellulase deficient strains of white-rot fungi. Intl. Patent PCT/SE85/00358.
8. Johnsrud, S. C.; Eriksson, K.-E. *Appl. Microbiol. Biotechnol,* **1985**, *21*, pp 320-327.
9. Simons F. L. *TAPPI J.,* **1950**, *33*, pp 312-314.

10. Wurz, O. *The Paper Maker*, **1969**, *158(6)*, p 59.
11. Akhtar, M.; Blanchette, R. A.; Burnes, T. *Wood Fiber Sci.,* **1995**, *27*, pp 258-264.
12. Akhtar, M.; Attridge, M. C.; Myers, G. C.; Blanchette, R. A. *Holzforschung,* **1993** *47*, pp 36-40.
13. Blanchette, R. A.; Leatham, G. F.; Attridge, M. C.; Akhtar, M.; Myers, G. C. **1991,** U.S. Patent no. 5,055,159.
14. Irie, Y.; Hata, K.; *Proceedings of Papermakers Conference,* **1990,** Atlanta, GA.
15. Fujita Y.; Hata K. *Japan TAPPI J,* **1991**, *45(8)*, pp 905-921
16. Fujita, Y.; Hata, K., *TAPPI J,* **1992**, *74(4)* pp 112-117.
17. Blanchette, R. A.; Farrell, R. L.; Burnes, T. A.; Wendler, P. A.; Zimmerman, W.; Bursh, T. S.; Snyder, R. A. *TAPPI J.,* **1992**, *75(12)* pp 102-106.
18. Farrell, R. L.; Hadar, Y.; Wendler, P. A.; Zimmerman, W. **1992,** European Patent No. 470929.
19. Brush, T. S.; Farrell, R. L.; Ho, C. *Tappi J.,* **1994**, *77*, p 155.
20. Bajpai, P.; Bajpai, P. K. *Biotechnology in the Pulp and Paper Industry* ***1997,*** Eriksson, K.-E. L., Ed.; In Special Issue - Adv. Chem. Eng./Biotechnol., *57*, 213-259, Springer Verlag.
21. Tsuchikawa, K.; Kondo, R.; Sakai, K. *Japan TAPPI J.*, **1995,** *49*, pp 1332-1338.
22. Tien, M.; Kirk, T. K *Science,* **1983,** *221*, pp 661-663.
23. Glenn, J. K.; Morgan, M. A.; Mayfield, M. B.; Kuwahara, M.; Gold, M. H. *Biochem. Biophys. Res. Comm.,* **1983**, *114*, pp 1077-1083.
24. Viikari, L.; Suurnäkki, A.; Buchert, J. *Enzymes for Pulp and Paper Processing;* American Chem. Society, T. W. Jeffries and L. Viikari, Eds.; In ACS Symposium Series 655; Washington, D.C.; **1996,** pp. 15-25.
25. Yang, J.; Eriksson, K.-E. L. *Holzforschung,* **1992**, *46(6)*, pp 481-488.
26. Yang, J. L.; Sacon, V.; Lowe, E.; Eriksson, K.-E. L. *TAPPI J.*, **1993**, *76(7)*, pp 91-96.
27. Yang, J. L.; Cates, D. H.; Law, S. E.; Eriksson, K.-E. L. *TAPPI J.*, **1994**, *77(3)*, pp 243-250.
28. Chen, C.-C.; Adolphson, R.; Dean, J. F. D.; Eriksson, K.-E. L.; Adams, M. W. W.; Westpheling, J. *Enzyme Microb. Technol.,* **1996,** *20*, pp 39-45.
29. Tien, M.; Kirk, T. K. *Science,* **1983**, *221*, pp 661-663.
30. Glenn J. K.; Morgan, M. A.; Mayfield, M. B.; Kuwahara, M.; Gold, M. H. *Biochem. Biophys. Res. Comm.,* **1983**, *114,* pp 247-250.
31. Srinivasan, C.; D'Souza, T. M.; Boominathan, K.; and Reddy, C. A. *Appl. Environ. Microbiol.,* **1995**, *61*, pp 4274-4277.
32. Perez, J.; Martinez J.; Dela Rubia, T. *Appl. Environ. Microbiol.,* **1997**, *62*, pp 4263-4267.
33. Eggert, C.; Temp, U.; Eriksson, K.-E. L. *Enzymes for Pulp and Paper Processing;* American Chem. Society, T. W. Jeffries and L. Viikari, Eds.; In ACS Symposium Series 655; Washington, D.C.; **1996,** pp 130-150.
34. Bourbonnais, R.; Paice, M. G. *Biotechnol. Lett.*, **1993**, *15(3)*, pp 289-294.
35. Bourbonnais, R.; Paice, M. G. *FEBS Lett.,* **1990**, *267*, pp 99-102.
36. Call, H. P.; Mücke, I. *Conference Proceedings*; In TAPPI Pulping Conference Proceedings, San Diego, CA; TAPPI : San Diego, CA **1994**, p 83.

37. Eggert, C.; Temp, U.; Eriksson, K.-E. L. *Appl. Env. Microbiol.,* **1996**, *62(4)*, pp 1151-1158.
38. Eggert, C.; Temp, U.; Dean, J. F. D.; Eriksson, K.-E. L. *FEBS Lett.,* **1996**, *391*, pp 144-148.
39. Eggert, C.; Temp, U.; and Eriksson, K.-E. L. *FEBS Lett.* **1997**, (In Press).
40. For Information on our pilot plants see our web site at: http://www.bae.uga.edu/outreach/pulp_plant/
41. Kondo, R.; Harazono, K.; Sakai, K. *Appl. Environ. Microbiol.* **1994**, *60*, 4559-4363.
42. Addleman, K.; Dumonceaux, T.; Paice, M. G.; Bourbonnais, R.; Archibald, F. S. *Appl. Environ. Microbiol.,* **1995**, *61*, pp 3687-3694.
43. Westermark, U.; Eriksson, K.-E. *Acta Chem. Scand.*, **1974,** *B28*, pp 204-208.
44. Westermark, U.; Eriksson, K.-E. *Acta Chem. Scand.*, **1974,** *B28*, pp 209-214.
45. Ayers, A. R.; Ayers, S. B.; Eriksson, K.-E. *Eur. J. Biochem.,* **1978**, *90*, pp 171-181.
46. Wenjun, G.; Usha, S. N.; Renganathan, V. *Arch. Biochem. Biophys.,* **1974**, *B28*, pp 209-214.
47. Habu, N.; Samejima, M.; Dean, J. F. D.; Eriksson, K.-E. L. *FEBS Lett.,* **1993**, *327(2)*, pp 161-164.
48. Samejima, M.; Eriksson, K.-E. L. *Eur. J. Biochem.,* **1992**, *207*, pp 103-107.
49. Paice, M. G.; Archibald, F. S.; Bourbonnais, R.; Jurasek, L.; Reid, I. D.; Charles, T.; Dumonceaux, T. *Enzymes for Pulp and Paper Processing;* American Chem. Society, T. W. Jeffries and L. Viikari, Eds.; In ACS Symposium Series 655; Washington, D.C.; **1996,** pp. 151-164.
50. Yang, J. L.; Ma, J.; Eriksson, K.-E. L. *Biotechnology in the Pulp and Paper Industry, Recent Advances in Applied and Fundamental Research;* E. Srebotnik; K. Messner, Eds.; Proceedings of the Sixth International Conference on Biotechnology in the Pulp and Paper Industry. Vienna, Austria, **1996** pp 157-162.

Chapter 2

Overview of Biomechanical and Biochemical Pulping Research

M. Akhtar[1-3,6], G. M. Scott[3], R. E. Swaney[4], and T. K. Kirk[5]

[1]Biopulping International, Inc., Madison, WI 53705
[2]University of Wisconsin Biotechnology Center, Madison, WI 53706
[3] Forest Products Laboratory, U.S. Department of Agriculture Forest Service, Madison, WI 53705
[4]Department of Chemical Engineering, University of Wisconsin, Madison, WI 53706
[5]Department of Bacteriology, University of Wisconsin, Madison, WI 53706

The fungal pretreatment of wood chips prior to mechanical pulping saves electrical energy, improves paper strength properties, and reduces pitch content. Fungal pretreatment is also effective on nonwoody plants and benefits sulfite, organosolv, and dissolving pulp production. Preliminary results with biokraft pulping are encouraging, but an in-depth evaluation is required. The results of two pilot-scale outdoor chip pile experiments recently conducted at the Forest Products Laboratory were comparable to those obtained using laboratory-scale bioreactors. The economics of the process are attractive, and the process appears to fit into existing pulp mill facilities with some modifications.

The pulp and paper industry utilizes mechanical or chemical pulping methods or a combination of these to produce pulps with desired characteristics. Mechanical pulping involves the use of mechanical force to separate the wood fibers. Mechanical processes are high yield (up to 95%) and produce paper with high bulk, good opacity, and excellent printability. However, these processes are electrical energy-intensive and produce paper with lower strength, higher pitch content, and higher color reversion rate (tendency to turn yellow with time) compared to chemical processes. Chemical pulping involves the use of chemicals to degrade and dissolve lignin from the wood cell walls, releasing high-cellulose fibers. Chemical pulping processes yield pulps with higher strength; however, these processes are low yield (about 40% to 50%) and are very capital-intensive.

Biopulping, which is defined as the treatment of lignocellulosic materials with lignin-degrading fungi prior to pulping, appears to have the potential to overcome some problems associated with conventional mechanical and chemical pulping methods. The following text summarizes biomechanical and biochemical pulping research and describes key findings.

[6]Current address: Forest Products Laboratory, U.S. Department of Agriculture Forest Service, Madison, WI 53705.

Biomechanical Pulping of Wood

Background. The use of white-rot fungi for the biological delignification of wood was perhaps first seriously considered by Lawson and Still (*1*) at the West Virginia Pulp and Paper Company (now Westvaco Corporation). These researchers published a survey of the literature (72 lignin-degrading fungi), which pointed to the dearth of knowledge about the fungal degradation of lignin. In the 1970s, Eriksson at the Swedish Forest Products Laboratory (STFI) launched a fairly comprehensive investigation that demonstrated that fungal pretreatment could result in significant energy savings and strength improvements for mechanical pulping (*2-5*). That research also resulted in a U.S. patent that described a "method for producing cellulose pulp" (*6*). Although this research met with limited success-encountering difficulties in scale up-it provided valuable insights (*7,8*). Treatment of coarse mechanical pulp with enzymes such as cellulases and hemicellulases prior to secondary refining has also resulted in both electrical energy savings and improvements in paper strength properties (*9*). However, the information is scanty; details can be found in several review articles and the literature cited therein (*10-13*).

The literature cited in these review articles indicates that fungal pretreatment of wood chips for mechanical pulping can have certain benefits. Accordingly, a comprehensive evaluation of biomechanical pulping was launched in 1987 at the USDA Forest Service, Forest Products Laboratory (FPL), under the auspices of a Biopulping Consortium. The Consortium involved the FPL, the Universities of Wisconsin and Minnesota, and up to 22 pulp and paper and related companies. The overall goal was to comprehensively evaluate the technical and economical feasibility of fungal pretreatment with mechanical pulping to save energy and/or improve paper strength. The consortium research was conducted by several research teams (*12*). In this chapter, we focus only on the work conducted by the fungal, pulp and paper, and engineering scale-up teams. Key findings are described in the following sections.

Selection of Lignin-Degrading Fungi. The screening of white-rot fungi followed procedures for selecting fungi that preferentially degrade lignin. One of the most appropriate methods appeared to be an assessment of decay (chemical analyses of lignin and wood sugar content) using wood blocks in accelerated decay chambers (*14*). Fast-growing species that rapidly colonize wood with a preference for degrading lignin were selected. These fungi with selective lignin-degrading ability showed promise for biomechanical pulping. Recently, a new and relatively easier and faster method was developed using Simons stain that predicts the efficacy of fungal pretreatments during biomechanical pulping by staining selectively for fibrillated fibers in coarse pulps (*15*). This method is now being used routinely to help optimize the biopulping process as well as to select new species and strains. After screening more than 400 species and strains. The best fungi found to date are *Ceriporiopsis subvermispora, Phanerochaete chrysosporium, Phlebia tremellosa, Phlebia subserialis, Phlebia brevispora, Dichomitus squalens, and Hyphodontia setulosa.*

Evaluation of Selected Fungi for Mechanical Pulping. White-rot fungi screened as described were evaluated for their ability to decrease the electrical energy requirement and to improve paper strength properties during refiner mechanical pulping. The process involved treatment of wood chips with fungi in bioreactors on a bench scale at appropriate temperature and humidity, followed by mechanical pulping of control and fungus-treated chips in a single-disk atmospheric refiner, preparation of paper, and testing of paper for physical properties (*12*).

Energy Savings and Physical Properties. The test fungi saved significant amounts of electrical energy and improved paper strength (*12*). Unfortunately, the most effective fungi on aspen (hardwood) were found to be relatively ineffective on loblolly pine (softwood). However, we discovered that a relatively uncommon fungus, *C. subvermispora,* was effective on both wood species (*16*). Our results indicate that *C. subvermispora* pretreatment of different wood species saves energy during both initial fiberization and subsequent refining (*17*). A U.S. patent was issued on the use of C *subvermispora* for biomechanical pulping (*18*).

We recently discovered that *Phlebia subserialis* gives results comparable to those obtained with *C. subvermispora* . Importantly, *P. subserialis* has a broader temperature range (27°C-39°C) than *C. subvermispora* (27°C-32°C). *P. subserialis* produces fewer aerial hyphae than *C. subvermispora* on wood chip surfaces. Also, *P. subserialis-treated* chips are not as soft (and presumably not as compressible) as those treated with *C. subvermispora.* These advantages of *P. subserialis* should reduce the cost of biotreatment significantly during scale up (i.e., decreased need for temperature control and greater ease of ventilation). A U.S. patent has been filed on the use of *P. subserialis* for mechanical pulping.

Optimization Studies. Many variables can affect biopulping. In our initial work, we simply made best guesses based on the literature, knowledge of fungal growth, and past experience. However, as in any industrial microbial process, the opportunity to increase the effectiveness and efficiency of biopulping and decrease its cost through optimization of variables is great. Consequently, we selected certain variables, such as rate of ventilation, addition of nutrients, and wood species, for initial optimization studies; details can be found in a previous publication (*12*). Some of the key variables are described in the following text.

Wood Chip Decontamination. In general, white-rot fungi are not able to outcompete indigenous microorganisms in unsterilized wood chips, which often are quite dirty. Even with the most aggressive white-rot fungi, maintaining reproducible results on a routine basis requires some decontamination. In bench-scale studies, wood chips were sterilized by autoclaving prior to fungal inoculation to maintain control. However, recent experiments have shown that surface decontamination achieved by brief atmospheric steaming (as short as 15 s) is sufficient to give *C. subvermispora and P. subserialis* the competitive advantage, even in heavily contaminated chips. After steaming, the temperature of the chips is near 100°C, at least at the surface.

Inoculum. In any industrial microbial process, the inoculum is of key importance. In early experiments, precolonized wood chips were used as inoculum. In recent experiments, we evaluated a liquid inoculum of *C. subvermispora* on loblolly pine chips. Since this fungus does not produce spores, fragmented mycelium was used (*19*). Three kg/ton of inoculum (dry weight basis) was required to achieve acceptable results, which is probably too high to be practical (*20*). A remarkable reduction was achieved by adding corn steep liquor, a byproduct of the corn wet-milling industry, to the inoculum suspension. Addition of only 0.5% unsterilized corn steep liquor (dry weight basis) reduced the amount of inoculum to ≤ 5 g/ton of wood (dry weight basis) without sacrificing biopulping efficacy of the fungus (Table I). Corn steep liquor is relatively inexpensive ($110/ton corn steep liquor on dry weight basis). This amount of inoculum is probably well within a commercially attractive range.

Since corn steep liquor is produced widely in the United States and its composition (despite the variation) from one source to another or from one batch to another does not seem to affect the biopulping efficacy of the fungus, pulp and paper companies should be able to obtain a regular supply from the nearest location and minimize transportation costs. The component or components of corn steep liquor responsible for the beneficial effect are not known. A U.S. patent has been issued on the use of corn steep liquor in biopulping (*21*).

Table 1. Energy Savings and Tear Index Improvement From Biomechanical Pulping of Aspen Chips With Lignin-Degrading Fungi With and Without Corn Steep Liquor[a]

Fungus	*Com Steep Liquor*	*Savings or Improvement Over Control* (%) *Energy*	*Tear Index*
Ceriporiopsis subvermispora L- 14803 SS-3	-	0	0
	+	33	22
Phlebia brevispora HHB 7099	-	0	0
	+	38	19
Phlebia subserialis RLG 6074-sp	-	0	0
	+	40	0
Phlebia tremellosa FP 102557-sp	-	0	0
	+	27	24
Hyphodontia setulosa FP 106976	-	0	0
	+	36	16

[a]Chips steamed, cooled, and inoculated with 5 g fungus/ton of wood (dry weight basis); ±0.5% unsterilized corn steep liquor (dry weight basis); 2-week incubation.

Bleaching Studies. Fungal pretreatment significantly reduced the brightness of the resulting mechanical pulps. However, experiments showed that the pulps can easily be bleached with either alkaline hydrogen peroxide or sodium hydrosulfite. Biomechanical pulp was readily bleached to about 60% Elrepho brightness with 3% hydrogen peroxide and I% sodium hydrosulfite-a brightness suitable for newsprint (*22*); brightness values approaching 80% were achieved with a two-step bleach sequence. Thus, bleachability of biomechanical pulps does not appear to be a problem.

Analyses of Effluents: Samples of the wastewater from the first refiner passes of fungus-treated aspen chips were analyzed for Microtox toxicity, biochemical oxygen demand (BOD), and chemical oxygen demand (COD). Fungal pretreatment decreased effluent toxicity substantially, decreased or increased BOD values slightly depending on experimental conditions, but somewhat increased COD values compared to that of values for control effluents (*23*). Increased COD values may be due to the release of biorecalcitrant lignin-related products from fungal action.

Microscopic Studies. To gain insight into the mechanism of biopulping, we examined the fungal growth patterns of *P. chrysosporium and C. subvermispora in* aspen wood chips at the microscopic level. *P. chrysosporium* (used in our early experiments) grew well across the chip surfaces and throughout the cell walls. The hyphae penetrated the chips through the lumens of wood vessels and fiber cells as well as through natural wood cell pits and fungal bore holes. Partial degradation of the cell lumen walls was evident. Erosion troughs and localized wall fragmentation or thinning was clearly visible as was generalized swelling and relaxing of the normally rigid wood cell wall structure.

Interestingly, some fungi, including *C. subvermispora,* produced abundant calcium oxalate crystals on their hyphae as well as manganese deposits in localized areas on the wood cell walls. The deposition of calcium and manganese occurred in wood after a relatively short incubation with *C. subvermispora* (*24*).

Our observations suggest that the physical basis for the biopulping efficacy of the fungal treatment is likely to involve an overall softening and swelling of the cell walls as well as thinning and fragmentation in localized areas (*25*). Further studies showed that the biopulped fibers have increased fibrillation compared with that of conventional mechanically pulped fibers (*26*) and that chip size has only a minor effect on biopulping efficacy of the fungus (*27*).

Pitch Reduction. Pitch refers to the mixture of hydrophobic materials that cause a number of problems in pulp and paper manufacture, including downtime for cleaning, breakage of papers on the paper machine, and holes in the paper (*28*). Treatment of wood chips with different fungi has shown promise in reducing significant amounts of pitch (*29*). In our laboratory studies, we have shown that the best biopulping fungus, C *subvermispora,* can remove as much as 30% of the extractives compared to the control (*19*).

Biomechanical Pulping of Nonwoody Plants

The preservation of forests and increasing environmental awareness have focused research on exploration of agro-based resources for papermaking. In particular, developing countries are using such resources for papermaking on a commercial scale. However, in the United States, their use is almost negligible, although nearly 330 million tons of various agricultural biomass is available (*30*).

Standard chemical pulping processes are typically used for making pulp from agro-based fibers; however, these processes are energy-intensive, require large capital investments, and contribute to air and water pollution. Studies in collaboration with the University of Wisconsin Department of Forestry showed that fungal (C. *subvermispora)* pretreatment of kenaf prior to mechanical pulping saved at least 30% of the electrical energy normally used during refining and improved paper strength properties; values approaching those for hardwood kraft pulps were observed (Table II) (*31*). Biopulping of bagasse has also shown very promising results (*4*). These findings suggest that biopulping technology also has potential for papermaking from nonwoody plants.

Table II. Comparison of Paper Strength Properties

Pulping Process	*Burst Index (kNIg)*	*Tear Index (mNm²lg)*
	Aspen[a]	
Stone groundwood	1.00	1.85
Refiner mechanical pulping	0.85	2.38
Thermomechanical pulping	1.18	2.50
Chemithermomechanical pulping	2.35	6.28
Semichemical pulping	4.45	6.00
Kraft pulping	4.60	8.00
	Kenaf[b]	
Control	2.00	6.00
Fungus-treated (2-week)	3.35	9.50

[a]Paper made by various pulping processes without fungal pretreatment.
[b]Paper made by mechanical pulping process (*31*).

Biochemical Pulping

In chemical pulping, the goal is to remove lignin from wood and leave the cellulose and hemicelluloses. However, the chemical processes are not completely selective and they tend to degrade the polysaccharides to some extent. Fungal pretreatment of wood chips removes some of the lignin and modifies other lignin; these changes might make it easier to remove lignin in the subsequent pulping process. As mentioned earlier, fungal treatment causes softening and swelling of wood cells. It is possible that these fungus-induced changes and the removal/modification of lignin may result in improved chemical penetration during pulping operations, which could result in more easily bleachable, lower kappa pulps; reduced cooking times and temperatures; reduced pulping chemical needs; and reduced effluent waste load.

Biosulfite Pulping. Sulfite pulping processes include several that differ in the base used for the pulping chemicals as well as the pH of the pulping liquor. Historically, sulfite pulping has been dependent on calcium-based liquor of high acidity (pH 1-2). However, in the last half-century, the more soluble bases of sodium, magnesium, and ammonium have come into use. The use of these bases has extended the possible pH range to less acidic conditions so that sulfite pulping is now done at pH of 3-5 and neutral sulfite semichemical (NSSC) pulping is done at pH 7-9. In the United States, 14 sulfite mills are in operation (*32,33*). Recent Environmental Protection Agency (EPA) regulations propose dramatic reductions in air and water discharges, and new technologies are required to meet these regulations. Recent studies suggest that biopulping technology can help in complying with these regulations; the results are summarized below.

Magnesium-Based Sulrite Pulping. In an experiment on magnesium-based sulfite pulping, Fischer et al. (*19*) showed that 2-week treatment with C. *subvermispora* of birch and spruce wood led to kappa reduction of approximately 30% with both woods. Longer fungal treatment led to greater kappa reduction. These authors also reported a significant reduction in cooking time with fungal pretreatment.

Sodium- and Calcium-Based Sulfite Pulping. Using loblolly pine and *C. subvermispora,* Scott et al. *(32,33)* obtained results (Table III) similar to those of Fischer et al. (*19*). Scott et al. also observed lower shives content in treated pulp samples compared to the control, indicating a more complete pulping with fungal pretreatment (unpublished results). Effluents from fungus-treated samples showed substantial reduction in toxicity with no increases in BOD and COD values. These results suggest the potential of biopulping technology for sulfite pulping.

Table III. Yield and Kappa Number of Loblolly Pine After Calcium-Acid Sulfite Pulping[a]

Treatment	*Yield* (%)	*Kappa Number*
Control	47.6	26.8
Strain CZ-3[b]	47.7	13.7
Strain SS-3[b]	47.8	21.1

[a]Source: Ref. *32.*

[b]Chips steamed, cooled, and inoculated with either strain of C. *subvermispora;* 2-week incubation. Data are average of three replicates.

Biokraft Pulping. Very few studies have shown that fungal pretreatment is effective for increasing pulp yield and strength and for reducing kappa number and cooking time during kraft pulping *(34,35)*. Recent studies have indicated that fungal pretreatment can reduce the amount of pulping liquor and cooking time, and it can improve paper strength properties with increased brightness (unpublished results). However, an in-depth evaluation is required; all studies to date must be regarded as preliminary.

Fungal Pretreatment for Organosolv Pulping. Technologies are currently in development that would reduce the environmental impact of kraft pulping. The problem of emission of volatile sulfur compounds and high chemical oxygen demand bleaching effluents is yet to be solved. Organosolv pulping circumvents the environmental problems related to sulfur emissions, and it has been found to be effective on several wood species with a broad range of organic solvents in acid or alkaline media *(36,37)*. In addition to the high delignification efficiency of several organic solvent systems, the acidic conditions can also result in carbohydrate degradation. As a result, pulps with low papermaking quality are obtained. Therefore, selective delignification without carbohydrate degradation is required.

With the use of selective lignin-degrading fungi, increases in delignification rates have been noted even at low pulp weight losses in organosolv pulping (38). Alternatively, the same residual lignin contents could be achieved in shorter reaction times with fungus-pretreated samples, thus providing energy savings in the organosolv process.

Fungal Pretreatment for Dissolving Pulp Production. Dissolving pulp is a low yield chemical pulp (30%-35%) with a high cellulose content (95%-98%) and relatively low hemicellulose (1%-10%) and lignin (< 0.05%) contents. Dissolving pulp is manufactured by both prehydrolysis-kraft and acid sulfite methods. The end uses of dissolving pulp include cellophane and rayon, cellulose esters, cellulose ethers, graft derivatives, and cross-linked cellulose derivatives. The bleaching of both sulfite and prehydrolysis-kraft pulps is achieved by removing residual lignin to increase final brightness and (α-cellulose content of the dissolving pulp. Although bleaching chemicals are selective and cost-effective, they cause environmental problems. The use of fungal pretreatment might result in savings of chemicals in pulping and bleaching and improvement of the quality of dissolving pulps.

Biobleaching of sulfite pulp from eucalyptus wood chips with *C. subvermispora* enhanced brightness significantly, but it reduced the cellulose content of the resulting dissolving pulp. A combined fungus and xylanase pretreatment of sulfite pulps led to the production of dissolving pulp using 63% less active chlorine and having a brightness of over 93% ISO (*39*). Subsequent studies showed that *C. subvermispora* pretreatment of eucalyptus wood chips not only increased lignin removal during sulfite pulping and increased brightness of the dissolving pulp, but also improved the selectivity of the bleaching process, thereby increasing the final pulp yield (Christov et al., personal communication).

Table IV. Energy Savings and Improvements in Strength Properties From Biomechanical Pulping of Spruce Chips With *C. subvermispora* L-14807 SS-3 at Laboratory and Pilot Scales[a]

Parameter[b]	*Laboratory Scale (1.5 kg, Dry Weight Basis)*	*Pilot Scale (50-ton, Dry Weight Basis)*
Energy savings	24	38
Burst index	35	22
Tearindex	52	35
Tensileindex	27	9

[a]Two-week incubation.
[b]Percentage of energy savings or strength improvements calculated on basis of untreated control values. Data are average of three replicates.

Engineering and Scale-Up

Our recent work involved engineering and scale-up of the biopulping process towards economic evaluation. Several engineering challenges were met, which involved redesigning a successful laboratory procedure to be practical on a larger scale. We developed methods for decontaminating wood chip surfaces, cooling the chips, and inoculating them with C *subvermispora* on a continuous basis. Methods were also developed to maintain the optimum growth temperature of the fungus (27°C-32°C) and the moisture in the chip pile (50%-60%, wet weight basis) so that the fungus could perform biopulping effectively on a larger scale.

The goal of maintaining proper temperature and moisture in the chip pile was accomplished by ventilating the pile with conditioned and humidified air. We found that ventilation was necessary; when a 1-ton chip pile was biopulped without forced ventilation, the center of the pile reached about 42°C within 48 h as a result of the metabolic heat generated by the fungus. No biopulping action was noted in that region.

In October 1996, we conducted a 50-ton (dry weight basis) outdoor chip pile experiment at FPL to test our large-scale design in a cold climate. In this experiment, the decontamination of chips using low-pressure steam, subsequent cooling, and inoculation were performed sequentially in screw conveyers. Inoculated chips were then incubated for 2 weeks to allow the fungus to grow and "soften" the chips. During incubation, the chip pile was ventilated with conditioned air. During subsequent mechanical pulping, significant energy savings and improvements in paper strength properties were noted (Table IV); variation among different regions of the pile was insignificant. In July 1997, another outdoor chip pile experiment was conducted at FPL to test our large-scale design in a hot climate. Similar results were obtained. The fungus-treated chips from this trial were also refined at a thermomechanical pulp (TMP) mill, and a 30% savings in electrical energy and significant improvements on paper strength properties were realized.

Process Economics

The economic attractiveness of the biopulping process was evaluated on the basis of the process studies and engineering data obtained to date. The economic benefits gained through the use of the process result from the following effects.

Refiner Energy Savings. For a 2-week process, the savings should be a minimum of 25% under the worst-case conditions, whereas up to nearly 40% can be achieved under optimized conditions.

Process Debottlenecking. The reduction in the specific refiner power is coupled to higher refiner throughput. Mills that are currently limited by refiner capacity may assign substantial value to the debottlenecking effect that the fungal pretreatment can provide.

Furnish Blend Advantages. The biopulping process results in pulps with improved paper strength. This is advantageous in situations where the product is a blend of mechanical pulps and more expensive kraft pulps. The improved strength of the biomechanical pulps may allow the required strength of the blend to be achieved with a lower percentage of the kraft pulp.

An economic evaluation was performed for a 600 ton/day TMP mill. The storage time was 2 weeks in a flat-pile geometry. Capital costs to incorporate biopulping technology into this paper mill are estimated to be between $5 and $7 million. Savings of about $10/ton pulp may be realized with 30% savings in electrical energy. This is equivalent to an annual savings of $2.1 million, which, compared to the estimated capital costs, results in a simple payback period of 2 to 3 years. Mills that are refiner-limited can experience increases of more than 30% from the reduction in energy by refining to a constant total power load. A 20% increase in throughput results in savings of about $55/ton pulp or $11.5 million annually. With this improvement, the payback period for this technology is approximately 6 months. Additionally, if 5% of the kraft pulp is substituted by biomechanical pulps in a blend, a savings of over $13/ton pulp might be realized. This preliminary analysis is subject to appropriate qualifications. The capital costs are subject to some variability, in particular the cost associated with integrating a new facility into an existing site. The additional advantages of biopulping, such as environmental benefits and pitch reduction, were not quantified, but we did quantify the cost of additional bleach chemicals. Much of the analysis will be site specific, depending on the operating conditions at a particular mill.

Industrial-Scale Process Flowsheet

The fungal treatment process fits well into a mill's woodyard operation. Wood is debarked, chipped, and screened (normal mill operation). Chips are briefly steamed to reduce surface contaminants (natural chip microorganisms), cooled with forced air to the appropriate temperature, and inoculated with a water suspension of the biopulping fungus augmented with unsterilized corn steep liquor. The inoculated chips are piled and ventilated with filtered and humidified air for 2 weeks prior to processing.

Acknowledgments

The authors wish to thank the many people who played a major role in this research, including Michael Lentz, Eric Horn, Rick Bergman, and the staffs of the Engineering Mechanics Laboratory, Research Facilities Engineering, and Pilot Plant at the Forest Products Laboratory. This work was financially supported by 22 pulp and paper and related companies, the Energy Center of Wisconsin, the University of Wisconsin, and the Wisconsin Alumni Research Foundation.

Literature Cited

1. Lawson, L. R.;Still, C. N. *TAPPI J.* **1957,** *40*, 56A-80A
2. Ander, P.; Eriksson, K.-E. *Svensk Papperstidning,* **1975,** *18*, 641-642.
3. Eriksson, K.-E.; Vallander, L. *Svensk Papperstidning,* **1982,** *85*, R33-R38.
4. Johnsrud, S. C.; Fernandez, N.; Lopez, P.; Guitieffez, I.; Saez, A; Eriksson K.-E. *Nordic Pulp & Paper Res. J.* (Special Issue) **1987,** *2*, 47-52.
5. Samuelsson, L.; Mjoberg, P. J.; Hartler, N.; Vallander, L.; Eriksson, K.-E. *Svensk Papperstidning* **1980,** *8,* 221-225
6. Eriksson, K.-E.; Ander, P.; Henningsson, B.; Nilsson, T.; Goodell, B. *U.S. Patent* 3, 962,033, 1976.
7. Eriksson K.-E. *Wood Sci. Technol..* **1990,** *24*, 79-101.
8. Eriksson, K.-E.; Johnsrud, S. C.; Vallander, L. *Arch. Microbiol.* **1983,** *135,* 161-168.
9. Pere, J.; Siika-aho, M.; Viikari, L. In *Proceedings of 213th ACS National Meeting;* San Francisco, CA, April 13-17, 1997, Abstract CELL 04 1.
10. Akhtar, M.; Blanchette, R. A.; Kirk, T. K. *Advances in Biochemical Engineering/Biotechnology,* Springer-Verlag: Berlin, Heidelberg, 1997, *57*, pp. 159-195.
11. Kirk, T. K.; Burgess, R. R.; Koning **J.** W. Jr. In *Frontiers in Industrial Mycology. Proceedings of Industrial Mycology Symposium;* Leatham, G. F. Ed.; June 25-26, 1990, Madison, WI. Routledge, Chapman & Hall: New York, 1992, Chapter 7, P. 99-111.
12. Kirk, T. K.; Koning, J. W. Jr.; Burgess, R. R.; Akhtar, M. et al. Res. Rep. FPL-RP-523, U.S. Department of Agriculture, Forest Service, Forest Products Laboratory, Madison, Wisconsin, 1993, 74 p.
13. Messner, K.; Srebotnik, E. *FEMS Microbiology Reviews* **1994,** *13,* pp. 351-364.
14. Otjen, L.; Blanchette, R. A.; Effland, M.; Leatham, G. *Holzforschung* **1987,** 41, 343.
15. Akhtar, M.; Blanchette, R. A.; Burnes, T. A. *Wood Fiber Sci.* **1995**, *27*, pp. 258-264.
16. Akhtar, M.; Attridge, M. C.; Blanchette, R. A.; Myers, G. C. et al. In *Biotechnology in Pulp and Paper Industry;* Kuwahara, M.; Shimada, M., Eds.; UNI Publishers Co., Ltd.: Kyoto, Japan, 1992, pp. 3-8.
17. Akhtar, M. *Holzforschung* **1994,** *48,* pp. 199-202.
18. Blanchette, R. A.; Leatham, G. F.; Attridge, M. C.; Akhtar, M.; Myers, G. C. *U.S. Patent* 5,055,159, October 1991.

19. Fischer, K.; Akhtar, M.; Blanchette, R. A.; Burnes, T. A.; Messner, M.; Kirk, T. K. *Holzforschung* **1994,** *48*, pp. 285-290.
20. Akhtar, M.; Lentz, M. J.; Blanchette, R. A.; Kirk, T. K. *TAPPI J.* **1997,** *80*, pp. 161-164.
21. Akhtar, M. *US patent* 5,620,5 64, April, 1997.
22. Sykes, M. S. *TAPPI J.* **1993,** *76*, pp.121-126.
23. Sykes, M. S. *TAPPI J.* **1994,** *77*, pp. 160-166.
24. Blanchette, R. A.; Krueger, E. W.; Haight, J. E.; Akhtar, M.; Akin, D. E. *J. Biotechnol.* **1997**, *53*, pp. 203-213.
25. Sachs, I. B.; Leatham, G. F.; Myers, G. C. *Wood Fiber Sci.* **1989,** *21*, pp.331-342.
26. Sachs, I. B.; Leatham, G. F.; Myers, G. C.; Wegner, T. H. *TAPPI J.* **1990,** *73*, pp. 249-254.
27. Sachs, I. B.; Blanchette, R. A.; Cease, K. R.; Lithium, G. F. *Wood Fiber Sci.* **1991,** *23*, 363-375.
28. Hata, K.; Matsukura, M.; Taneda, H.; Fujita, Y. In *Enzymes for Pulp and Paper Processing;* Jeffries, T. W.; Viikari, L., Eds.; ACS Symposium Series 655, Washington, D. C., 1996, pp. 280-296.
29. Breuil, C.; Iverson, S.; Gao, Y. *In: Environmentally Friendly Technologies for the Pulp and Paper Industry;* Young, R. A.; Akhtar, M., Eds.; John Wiley & Sons, Inc.: New York, 1998, pp. 541-565.
30. Atchison, J. E. *TAPPI J.* **1996,** *79*, 87-95.
31. Sabharwal, H. S.; Akhtar, M.; Blanchette, R. A.; Young, R. A. *TAPPI J.* **1994,** *77*, 105-112.
32. Scott, G. M.; Akhtar, M.; Lentz, M. J. In *Proceedings of 1995 TAPPI Pulping Conference.* TAPPI: Atlanta, Georgia, 1995, Book 1, pp. 355-362.
33. Scott, G. M.; Akhtar, M.; Lentz, M. J.; Sykes, M.; Abubakr, S. In: *Proceedings of 1995 International Environmental Conf.* TAPPI: Atlanta, Georgia, 1995, pp. 1155-1161.
34. Oriaran, T. P.; Labosky, P. Jr.; Blankenhom, P. R. *TAPPI J.* **1990**, *73*, 147-152.
35. Oriaran, T. P.; Labosky, P. Jr.; Blankenhom, P. R. *Wood Fiber Sci.,* **1991,** *23,* 316-327.
36. Johansson, A.; Aaltonen, O.; Ylinen, P. *Biomass* **1987**, *13*, 45.
37. McDonough, T. *TAPPI J.* **1993,** *76*, 186-193.
38. Ferraz, A.; Mendonga, R.; Cotrim, A. R.; da Silva, F. T. In *Proceedings of 6th International Conf. Biotechnol. Pulp and Paper Industry;* Vienna, Austria, June 1996, pp. 221-224.
39. Christov, L. P.; Akhtar, M.; Prior, B. A. *Holzforschung,* **1996**, *50*, 579-582.

Chapter 3

Biodegradation of Resin Acids in Pulp and Paper Industry: Application of Microorganisms and Their Enzymes

Kunio Hata, Motoo Matsukura, Yuko Fujita, Kazumasa Toyota, and Hidetaka Taneda

Central Research Laboratory, Nippon Paper Industries Company, Ltd., 5–21–1 Oji Kita-ku, Tokyo 114, Japan

Resin acids are compounds which come from the mechanical pulping process and are of major concern in papermaking because they cause pitch problems. To discover ways to prevent these problems, biological treatment of pulp was studied. Several soil inhabiting microorganisms were selected, and had a strong degradation ability. By the microorganism treatment of groundwood pulp, the contents of resin compounds were apparently decreased. The quality of the treated pulp was also improved.

Wood resins exist widely in pine tree species. Rosin, an extracted product from resins, generally consists of three types. One is gum rosin which is produced from raw resin. The second type is wood rosin which is extracted from pine stumps by petroleum solvents. The third type is tall oil rosin which is separated from black liquor in the kraft pulping process. Oleoresin, which is collected from pine wood by tapping, contains pimaric acids, neoabietic acids and levo pimaric acid. On the other hand a purified rosin contains a great amount of abietic acid (ABA) and dehydroabietic acid (DHA) because an isomerization or an oxidation occurs during the production process. These acids are generally called resin acids.

Rosin compounds are useful materials for industrial applications because they are comparably less expensive. The several derivatives are used for such things as printing inks, paints and hot melt adhesives. One of the most important usage of such derivatives is as sizing agent for paper. A large amount of resin acid can be observed in a papermaking process. Some are added as a sizing agent, and the others come from the wood through the mechanical pulping process. In the case of chemical pulp, especially in the kraft pulping process, resinous compounds are removed at the cooking stage. But in the case of mechanical pulp, especially in groundwood pulp of red pine, it contains a large amount of resin acids, fatty acids and triglycerides. Resinous compounds cause pitch problems in the papermaking process. The resinous compounds, which are called pitch, are likely to stick to the surfaces of paper machine rolls. As a result, web breaks and damage to the paper product occur frequently. In the paper industry, several methods to avoid pitch problems have been used for a long period of time.

There are several ways to prevent pitch problems. A seasoning of logs can make the content of resinous compounds decrease, and by the addition of chemicals such as detergents, or a powdered clay, resinous compounds are absorbed on them and retained on the surface of fibers as stable small particles. However, these ways have not completely solved pitch problems. Therefore we have successfully developed the enzymatic pitch control method to solve them. Lipase, which can hydrolyze triglycerides, has been added to the early stage of groundwood pulping(GP) or refiner groundwood pulping (RGP) for the last 10 years (*1-4*). Since then the method to use lipase has improved(*5-9*). This enzymatic method can decrease the damage of products and increase the productivity of paper machines. Now, it has been applied at several paper mills in Japan on a large scale (*10*). On the other hand, the Cartapip method uses a living microorganism, *Ophiostoma piliferum*, to prevent pitch accumulation. Many reports and reviews have been published in this field of research (*11-15*).

Recently, alkaline papermaking is increasing rapidly, even for wood containing papers. In acidic conditions, resin acids are effective as a sizing agent, but tend to be dissolved in the alkaline conditions. It is reasonable since pKa of the resin acids is about 6. Therefore, the alkaline papermaking process may suffer from the accumulation of resin acids in the process water. When lipase is applied, the amount of fatty acids increases due to the enzymatic hydrolysis of triglycerides. In the acidic papermaking process, the enzymatic action of lipase is accelerated by aluminum ions. Resulting resin acids and fatty acids are fixed on the fibers as aluminum salts. However, the effect of alum in the alkaline papermaking process decreases as compared with the acidic process. These acids accumulate in the white water of paper machines and cause pitch problems.

Both ABA and DHA have a poisonous effect on fish even in extremely low concentrations. It would be a serious problem to discharge the effluent containing these acids to the environment (*16-17*). Several papers report that resin acids which exist in softwood can be degraded with microorganisms such as molds or bacteria (*18-25*). In this paper we will discuss the biodegradation of resin acids by microorganisms or enzymes. This consists of two parts. The first part is a screening of microbial candidates from nature and an evaluation of their activities for the degradation of resin acids. The second part is a treatment of groundwood pulp (GP) with microorganisms under actual papermaking conditions, and determine the effect of this treatment on paper characteristics.

Screening of Microorganism to Degrade Wood Resin Acids.

Screening. Several strains were collected by screening with the agar medium or the liquid shaking culture which contained ABA. One was found in the soil of a suburb in Tokyo, and the others from the soil of a pine forest in Karuizawa. In order to compare the activities, several *Pseudomonous* type cultures were used. The screened strains and type cultures were characterized by the following methods: Gram stain types, shapes of

Table 1. Bacteria degrading abietic acid

	1	2	3	4	5	6	7	8	9	10
Gram stain	Negative	Negative	Negative	Negative	Negative	Negative	Negative	Negative	Negative	Negative
Flagellum	Tail type	Tail type	Tail type	Tail type	Tail type	Tail type	Tail type	Tail type	Tail type	Tail type
Cell chromophore (yellow)	–	–	–	–	–	–	–	–	–	–
Oxidase	+	+	+	+	+	+	+	+	+	+
Accumulation of PHB	+ –	+ –	+ –	+ –	+ –	+ –	+	+	+	–
Degradation of starch	–	–	–	–	–	–	–	–	–	–
Lipase (TW80)	+	–	–	+	+	+	–	+	+	d
Proliferation (>41C)	–	–	–	–	+ –	+ –	–	–	–	–
Cleavage of PCA	o	o	o	o	o	o	o	m	m	o
Quinone type	Q9	Q9	Q9	Q9	Q9	Q9	Q8	Q8	Q8	Q9
Species	1	1	1	1	1	1	2	3	3	1
Carbon source fixation										
Glucose	+	+	+	+	+	+	–	–	–	+
Fructose	+	+	+	+	+	+	+	–	+	+
Mannose	–	+	+	+	–	–	–	–	–	d
Galactose	–	+	+	–	–	–	–	–	–	d
Arabinose	+	+	+	–	–	–	–	–	–	+
Mannitol	–	+	+	+	–	–	–	–	+	d
Glycerol	+	+	+	+	–	–	–	–	+	+
Nicotic acid	+	–	–	–	–	–	+	–	+	–
Benzoic acid	+	+	+	+	+	+	+	+	–	d
o-OH-benzoic acid	–	–	–	–	–	+	–	–	–	+
Testosterone	–	+	+	+	–	+	+	+	–	
Abietic acid	–	+	+	+	+	+	–	–	–	

all +: acetic acid, lactic acid, hydroxy lactic acid, citrical acid, p-OH-benzoic acid

all –:xylose, rhamnose, cellobiose, trehalose, starch, lactose, sucrose, salicin, sorbitol, inositol

1 JCM-6157 (*Pseudonomonas putida*,Biotype A)
2 AB-1
3 AB-2
4 AB-3
5 AB-4
6 ATCC-14235 (*Pseudomonas resinovorans*)
7 ATCC-17697 (*Alcaligenes eutrophus*)
8 JCM-5832 (*Pseudomonas teststeroni*)
9 JCM-5833 (*Pseudomonas acidovorans*)
10 *Pseudomonas putida* (Biotyupe B) (d;10 – 90% Positive)

cell, chromatophore, the activity of oxidase or lipase, the degradation of poly--hydroxybutyrate, starch, protocatechuic acid, ubiquinone, and the growth rate at 41, and so forth. The results of these characterizations are shown in Table I. Among them, NAB-3 and NAB-4 showed a very strong ABA degrading activity.

Besides ABA, tall rosin was used as the carbon source of a basic medium.(Table II). ABA and tall rosin were added to the medium to 0.5g/l concentrations. After a 7-day incubation with shaking, the contents of resin acids were determined with gas chromatography. The data showed that all kinds of resin acids were completely degraded(Figure 1). Further study revealed that the degradation could be completed in a day even at a 1.0 g/l concentration. To simplify the measuring method, the content of ABA in the medium was determined by HPLC(Figure 2).

ABA was dissolved in ethyl alcohol and the solution was neutralized with sodium hydroxide. The concentration was adjusted at 70%(v/v). It was added into the medium after pasteurization as a form of emulsion.

Table II. The constitution of basic medium in 1000ml

ABA	1.0g	$ZnSO_4/7H_2O$	70μg
NH_4NO_3	0.5	$CuSO_4/5H_2O$	5
KH_2PO_4	1.0	$MnSO_4/5H_2O$	10
NaCl	0.5	$NiSO_4/7H_2O$	10
$MgSO_4/7H_2O$	50mg	$CoSO_4/7H_2O$	10
$FeSO_4/7H_2O$	25	H_3BO_4	10
$CaCl_2/2H_2O$	25	Na_2MoO_4	10
pH 7.2			

Evaluation of Degradation Activity. In order to compare the degradation activity of each strain, NAB-1, NAB-2, NAB-3, NAB-4, ATCC-14235(*Ps. resinovorans*), and ATCC-17697(*Alcaligenes eutrophus*) were incubated with shaking. When the initial concentration of ABA was 1.0 g/l, it disappeared within 24 hours at 30. These strains, except ATCC-17695, showed a strong degradation ability, and the order of degradation rate was as follows:

ATCC-14235=NAB-4>NAB-2>NAB-1=NAB-3

As shown in Figure 3, an induction period was necessary before starting the degradation of ABA. In order to increase the degradation rate, NAB-1 and ATCC-14235 were cultured thoroughly in a nutrient medium, and excess living cells were added to the ABA containing medium. After a two hour induction phase, ABA was completely degraded within 4 hours(Figure 4). No differences in the degradation ability were observed between the two strains.

Effect of Glucose. The effect of glucose on the degradation of ABA is important because the relationship between growth rate and degradation could be a significant factor for the industrialization of this bio-system. The effect of glucose was studied with an agar plate and the result is shown in Table III. The addition of glucose inhibited the

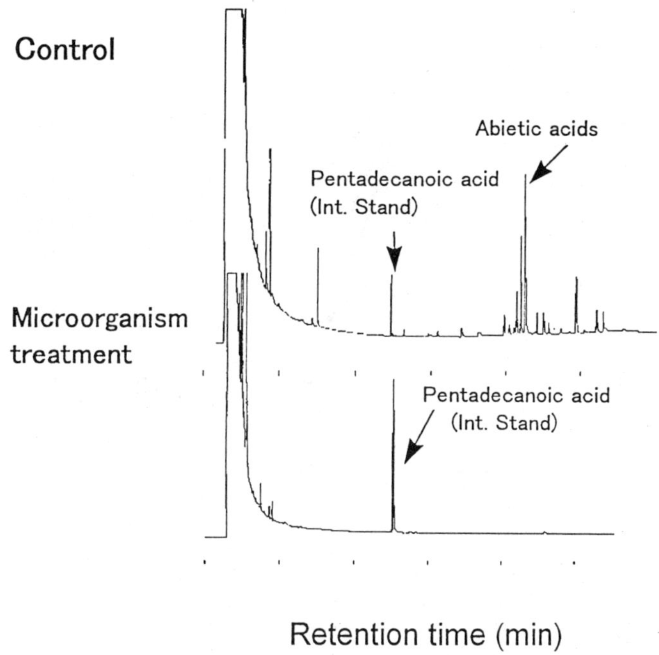

Figure 1. Degradation of abietic acid analyzed by gas chromatography.

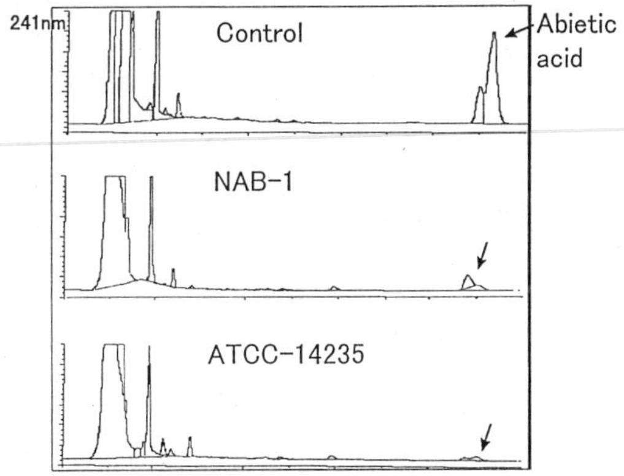

Figure 2. Degradation of abietic acid in GP analyzed by HPLC.

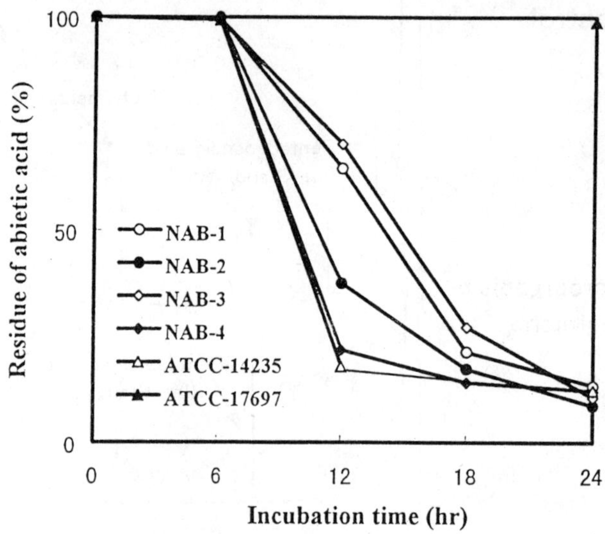

Figure 3. Effect of incubation time on abietic acid degradation.

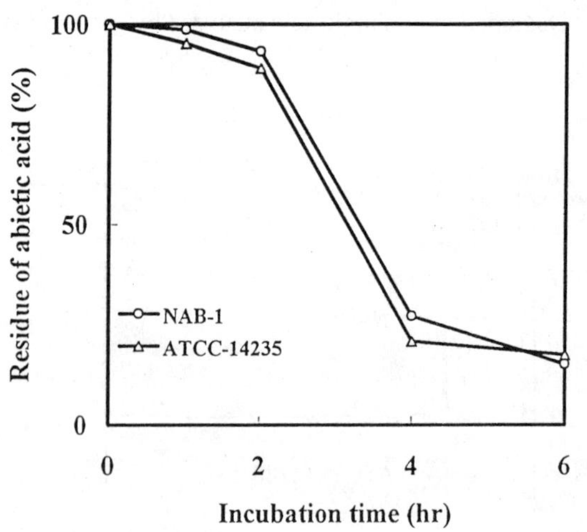

Figure 4. Effect of incubation time on abietic acid degradation (with excess amount of microorganism body).

degradation of ABA suggesting that a catabolite repression occurred during the reaction. Therefore, in terms of industrialization, a sufficient amount of cell body could be obtained from on-site cultivation equipment if the microorganisms were cultured in a glucose enriched medium.

Table III. Effect of glucose on abietic acid degradation

	Concentration of glucose (g/L)				
	0	1	2	4	8
NAB-1	+	+	-	-	-
NAB-2	+	+	-	-	-

+:Clear zone was formed (abietic acid was decreased)
-:Clear zone was not formed (abietic acid was not decreased)

Trial to Isolate Enzymes. The above experiments only show the case when living cells used. A long reaction time would be needed to grow and start-up degradation of ABA. With application of this bio-system in the actual papermaking process, the most significant operational factor is reaction time. In addition, living cells may be troublesome in papermaking. Therefore, perhaps a better way to reduce resinous substances in pulp is to utilize enzymes from the strains.

Cells of NAB-1 and ATCC-14235 were ground with fine glass beads and enzyme isolation was attempted using conventional methods. However, activity was not observed even with the addition of hydrogen peroxide which would promote peroxidase activity. It could be considered then that the pathway to degrade ABA consists of many redox steps(Figure 5). P-450 is a well known redox enzyme, which can catalyze the oxidation reaction in the presence of a co-factor such as NAD. Generally it is rather difficult to use such an enzyme in an industrial application because it needs an electron transfer system to activate the enzyme.

GP treatment with Microorganism. As already mentioned, the other way to utilize this system would be a direct application of living microorganisms to a GP slurry. In the laboratory experiments, GP was incubated while shaking with an excess amount of living cells for 8 hours. ABA in GP completely disappeared(Figure 6) and the water absorbency of the GP was greatly reduced (Table IV). This is attributed to the decrease of resinous extractives. When the same experiment was carried out under less aerobic and stationary conditions, degradation of ABA was not observed. It is concluded that it will be necessary to maintain aerobic conditions to degrade ABA in a GP slurry.

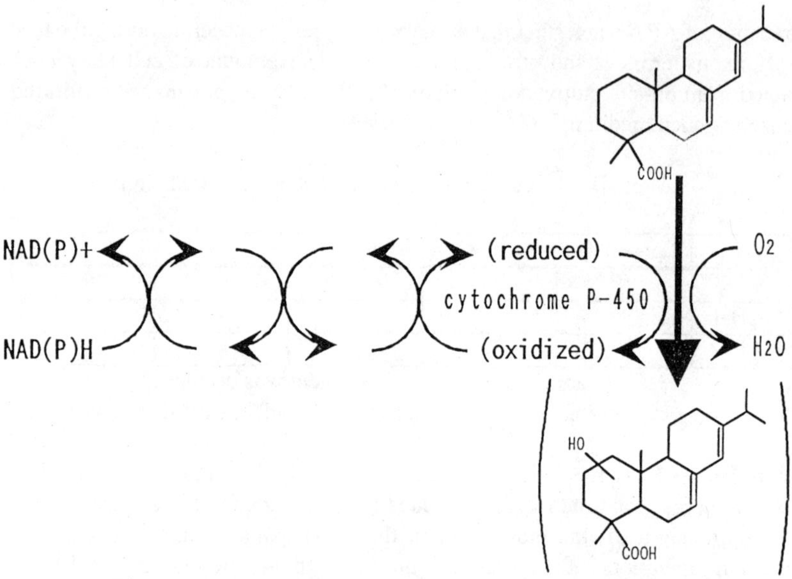

Figure 5. Mechanism of hydroxylation of abietic acid.

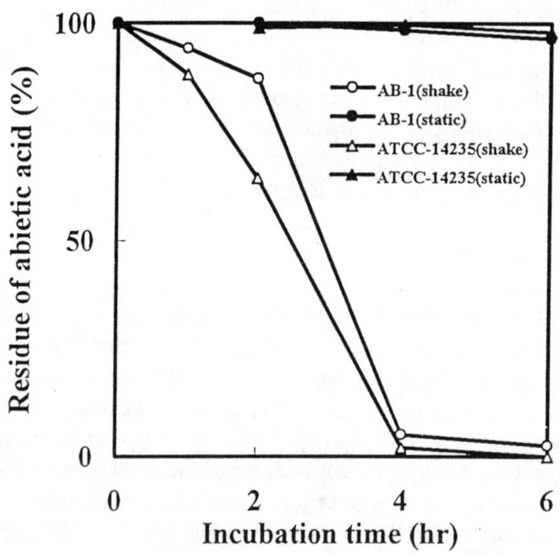

Figure 6. Effect of incubation time on abietic acid degradation in pine GP.

Table IV. Change of water absorbency of GP

	Water absorbency
Control	3.9
NAB-1	0.6
ATCC-14235 *(Ps. resinovorans)*	0.4

Characterization of Biologically Treated GP

In this part of the paper, the changes in characteristics of GP and the results of pitch deposition tests are discussed. The following strains were employed in these experiments.

1. NAB-1 (wild type)
2. ATCC-14235 (*Pseudomonas resinovorans*)
3. JCM-6157 (*Pseudomonas putida*)

Experimental

Biological Treatment of GP. 50 g of red pine GP was dispersed in 5000 ml of water, and then pH of the pulp slurry was adjusted to 7.0 with 1N-sodium hydroxide solution. After the addition of 500 ml of microorganism suspension(1.5g dry cells) which was previously incubated in a nutrient medium, the mixture was shaken for 6 hours.

Solvent Extraction. The samples were extracted in a soxhlet extractor with an ethanol and benzene (1:2) mixture for 6 hours. Fatty and resin acids were determined by GC and triglycerides (TG) were determined with TLC.

Determination of Fatty Acids and Resin Acids. The completely dried sample was dissolved in N,N-dimethylformamide. Penta-decanoic acid was added as an internal standard. The mixture was ethylated by diethyl acetal for 15 minutes at 60. Then it was analyzed by gas chromatography at the following conditions:

Column: Shimadzu HiCapCBP 1-S25-050
Temperature: 150-280(5/min)
Injection: Split, 300
Detector: FID, 300

Determination of Triglycerides (TG). 100 g of 1% pulp slurry was mixed with 5 ml of the internal standard solution which contained 1% choresterol acetate. The slurry was extracted with 200 ml of n-hexane, and then filtered to remove fibers. The hexane solution was dried over anhydrous sodium sulfate. After removing the solvent by evaporation, a part of the sample was dissolved in ethanol and analyzed by TLC with the solvent mixture (hexane: diethyl ether: 25% ammonium solution = 60:8:0.2). TG was determined with TLC-FID detector.

Determination of ABA. 50ml of pulp slurry was mixed with 50ml of ethanol and the mixture was treated with a supersonic disperser for 5 minutes. After millipore filtration, the 20 μl solution was analyzed by HPLC at the following conditions.

Column: Gaskuro-Kogyo Inartosil 2 (4.6 x 250mm)
Solvent: acetonitrile :water: acetic acid=75:25:0.1
Detector: UV(241nm)
Flow rate: 1ml/min

Pitch Deposition Test. This test was based on the TAPPI method No.UM223, and partially modified for a GP sample. In order to conduct a pitch deposition test, a model pitch was added to the pulp slurry, i.e., triolein, ABA and oleic acid (2:2:1) in 10% ethanol solution. Previously the solvent mixture, isopropyl alcohol : acetone : water (100:60:5), was used according to Hassler's method(*26*), but isopropyl alcohol can not be used because it is harmful to the growth of microorganisms. Only ethanol was used as a solvent for the model pitch. In this experiment one native strain NAB-1 and one type culture ATCC-14235 were used as a resin degrading microorganisms. Eight ml of model pitch solution (0.8g model pitch) was added to the 400 ml of GP slurry which contained 5g of mill produced red pine pulp. After the pH was adjusted with 1N-sodium hydroxide, 100ml of culture mixture (ca. 0.3g cells) was added and incubated while stirring for 5 to 7 hours at 30. After incubation, the GP slurry was diluted to 0.5% with water and was kept at 60. After addition of 3% aluminum sulfate, pH of the slurry was adjusted with 1N-sodium hydroxide or 10% sulfuric acid solution. The pitch deposition test was conducted as described above and the weight of the deposited pitch was measured after drying at 105°C.

Results and Discussion. Some species of microorganisms had both ABA degradation activity and lipase activity (Table V). Before and after the biological treatment with the microorganisms, the contents of extractives from GP, such as resin acids, fatty acids and triglycerides were Table VI . Type culture ATCC-14235 had both ABA degradation and lipase activities and could therefore decrease the amount of all resinous compounds. NAB-1 strain could decrease the resin acids content but not the TG content (Table VII). JCM-6157 could decrease none of them.

Table V. Resin acid degradation and lipase production by microorganisms

Activity	NAB-1	ATCC-14235 (*Ps. Resinovorans)*	JCM-6157 (*Ps. Putida)*
Resin acid Degradation	+	+	-
Lipase activity	-	+	+

Table VI. Alcohol-benzene extractives of untreated and treated GP

	Control	NAB-1	ATCC-14235 (*Ps. Resinovorans)*	JCM-6157 (*Ps. Putida)*
Alcohol-benzene extractives(%)	4.24	3.38	2.48	4.21
Wood resin(%)	1.16	0.30	0.24	0.96
Fatty acid(%)	0.36	0.28	0.14	0.23
TG(%)	0.48	0.36	0.03	0.45

Table VII. Extractive contents in untreated and treated GP

	Control	NAB-1	ATCC-14235 *(Ps. Resinovorans)*
Hexane extractives (%)	15.0	8.37(56%)	2.09(14%)
Wood resin (%)	2.34	0.47(20%)	0.10(4%)
Fatty acid (%)	1.75	0.23(13%)	0.20(11%)
TG (%)	3.82	3.49(91%)	0.36(9%)

The biological treatment of GP lowered the brightness of pulp slightly, 2% and a slight increase in pulp strength was also observed (Table VIII). It is considered that internal bonds between fibers become stronger by removal of extractives. When the extractives content of GP was also decreased by biological treatment, water absorbency was decreased because the self sizing was lowered by the removal of resinous extractives. The dynamic friction of paper is thought to be related to the runnablity of paper rolls on web offset printing presses and the dynamic friction coefficient (DFC) is regarded as a quality control parameter in Japan. The treatment with NAB-1 decreased DFC and that of ATCC-14235 increased DFC. The previous paper reports that ABA exists on the surface of fibers and lipase treatment of GP also increases DFC(*3*). ATCC-14235 produced both ABA degradation enzymes and lipase. The degradation of ABA causes a decrease of DFC, while the lipase treatment usually increases DFC. Therefore ATCC-14235 treatment indicated that the effect of lipase would be greater than that of ABA degradation.

The results of the pitch deposition test showed both treatments could clearly decrease the weight of pitch deposition at both acidic and neutral conditions with or without alum (Table IX). NAB-1 had rapid growth capability but produced no lipase activity. A better result was obtained when the combined system of NAB-1 and lipase was applied. Table X shows NAB-1 with lipase could degrade all resinous contents within 24 hours.

Conclusion

Microorganisms with the ability to degrade resin acids are abundant in the soil of pine forests. However it is not easy to utilize their enzyme systems for industrial applications. One reason is that there are many steps to degrade ABA, and many enzymes and cofactors are needed in each step. It is difficult to isolate all the necessary enzymes from the microorganisms and set up the proper conditions for them to act. The only way to utilize biochemical or microbiological degraded resinous substances in the papermaking industry might be to cultivate the microorganisms with the pulp in the pulping process. Actually the resinous extractives in GP were decreased by the treatment by selected microorganisms.

Table VIII. Properties of untreated and treated GP

	Control	NAB-1	ATCC-14235 *(Ps. Resinovorans)*	JCM-6157 *(Ps. Putida)*
CSF (ml)	92	93	94	90
Basis weight (g/m2)	102.1	104.9	105.1	103.3
Caliper (μ)	247.6	235.1	237.4	238.3
Density (g/cm3)	0.41	0.45	0.44	0.43
Burst strength (kg/cm2)	1	1	1.2	0.8
Burst factor	1	1	1.1	0.8
Tear strength (g)	26.8	31.9	32.4	30.4
Tear factor	26.3	30.4	30.8	29.4
Tensile strength (kg)	3	3.6	4.1	3.4
Breaking length (km)	1.9	2.3	2.6	2.2
Stretch (%)	1.9	1.9	1.9	1.9
Folding endurance	0	1	1	0
Brightness (%)	57.6	55.2	55.8	55.9
Water drop test (sec)	4.7	6.2	7.9	5.5
Oil drop test (sec)	1.6	1	1	2.3
Water drop test after aging* (sec)	33.6	9.4	9.5	54.3
Dynamic friction coefficient	0.48	0.44	0.54	0.48

*aging condition: 18 hours at 105°C

Table IX. Results of the pitch deposition test of NAB-1 treated GP(5 hours)

pH	Alum addition (3%)	Deposition (mg)	
		Control	NAB-1 treated
4	+	50.0	2.4(12%)
7	+	27.9	3.6(12%)
7	-	18.2	6.4(35%)

Table X. Contents of extractives in treated GP

	Control	NAB-1	NAB-1+ Lipase* 5 hours	NAB-1+ Lipase* 24 hours
Hexane extractives(%)	15.0	8.37(56%)	6.89(46%)	0.68(5%)
Wood resin(%)	2.34	0.47(20%)	1.81(77%)	0.12(9%)
Fatty acid(%)	1.75	0.23(13%)	0.14(8%)	0.11(6%)
TG(%)	3.82	3.49(91%)	0.28(7%)	0.02(0.5%)

*Lipase : "Resinase A 2X" produced by Novo Nordisk, 200 ppm addition based on dry pulp

Such a treatment had no undesirable effects on the strength and quality of the products. However, it takes approximately 4 hours to treat pulp in the process and it is therefore desirable to shorten the reaction time. We are now investigating thermophilic microorganisms with the hope to find strong enzymatic activity for degradation of resin acids. Progress in this field of research will hopefully give us better pitch control tools for use in the papermaking industry.

Acknowledgments

We would like to thank Mr. Inoue, Mr. Satake, Mr. Tabei, and Ms. Sonoyama for fruitful discussions and experimental assistance. The contribution of selected microorganisms used in these experiments by Toyo Ink Company is greatly acknowledged.

Literature Cited

1. Hata, K.; Irie, Y.; Matsukura, M.; Usui, M. *Tappi Papermakers Conference Proceedings*; TAPPI PRESS: Atlanta, GA, **1990**; pp 1-10.
2. Hata, K; Fujita, Y.; Awaji, H.; Matsukura, M. *Japan Tappi J.,* **1991**, *45*(8), pp 905-921.
3. Hata, K.; Fujita, Y.; Taneda, H.; Matsukura, M.; Shimoto, K.; Sharyo, M.; Sakaguchi, H.; Gibson, K. *Tappi J.,* **1992**, *74*(4), pp 117-122.
4. Hata, K.; Fujita, Y.; Awaji, H.; Taneda, H.; Matsukura, M.; Shimoto, H.; Sharyo,; Abo, M.; Sakaguchi, H. *Biotechnology in the pulp and paper industry*; Kuwabara, M.; Shimada, M., Ed; Uni publishers: Tokyo, **1992**; pp 163-168.
5. Fisher, K.; Messner, K. *Biotechnology in pulp and paper industry;* Kuwabara, M.; Shimada, M., Ed; Uni publishers: Tokyo, **1992**, pp 169-174.
6. Fisher, K.; Messner, K. *Enzyme Microb. Technol.* **1992**, *14*, pp 470.
7. Fisher, K. ; Duchinger, S. K.; Kreiner, N.; Messner, K. *J. Biotechnology*. **1993**, *27*, p 341.
8. Mustranta, A.; Fagernas, L.; Viikari, L. *Tappi J.,* **1995**, *78*(2), pp 140-146
9. Mustranta, A.; Fagernas. L.; Jokela, H.; Viikari, L. *Pre-symposium to the 8th ISWPC, Wood extractives in Pulping and papermaking.* **1995.**
10. Hata, K.; Matsukura, M.; Taneda, H.; Fujita, Y. *Enzymes for Pulp and Paper Proceedings ACS Symposium Series 655,* **1996**; pp 280.
11. Farrell, R. L.; Wender, P. A.; Brush, T. S. *6th Proceedings International Symposium on Wood and Pulping Chemistry*, Melbourne, Australia, **1991**, pp 501-508.
12. Farrell, R. L.; Blanchette, R. A.; Brush, T. S.; Gysin, B.; Hadar, Y.; Perollaz, J. J.; Wendler, P. A.; Zimmerman, W. *Biotechnology in the Pulp and Paper Industry*; Kuwabara, M.; Shimada, M., Ed; Uni publishers: Tokyo, **1992**; pp 27-32.
13. Farrell, R. L.; Blanchette, R. A.; Burnes, T. A.; Wender, P.A.; Brush, T. S.; Snyder, R. A. *Tappi J.* **1992**, *75*(12), pp 102-106.
14. Farrell, R. L.; Brush, T. S.; Ho, C. *Tappi J.,* **1994**, *77*(1), pp 155-159.
15. Kohler, L. J. F.; Dinus, R. J.; Malcolm, E. W.; Rudie, A. W.; Farrell, R. L.; Brush, T. S. *TAPPI journal* **1997**, *80*(3), pp 135-140.

16. Samejima, K.: Shigematsu, A.; Takamura, N. *Mokuzai Gakkaishi*, **1986**, *32*(5), pp 344-350.
17. Leach, J. M.; Thakore, A. N. *Prog. Wat. Tech*, **1977**, *9*, pp 787-798.
18. Biellmann, J. F.; Wenning, R. *Chem. Comm.*, **1970**, pp 346-347.
19. Biellmann, J. F.; Branlant, G. *Tetrahedron*, **1973**, *29*, pp 1227-1236.
20. Biellmann, J. F.; Branlant, G. *Tetrahedron*, **1973**, *29*, pp 1237-1241.
21. Ekman, R.; Sjoholm, R. *Acta hem. Scand.*, **1979**, *33*, pp 76-78.
22. Kutney, J. P.; Singh, H.; Hewitt, G.; Salisbury, P. J.; Brian, R. W.; Servizi, J. A.; Martens, D. W.; Gordon, R.W. *Can. J. Chem.*,**1981**, *59*, pp 2334-2341.
23. Kutney, J. P.; Lewis, S. L. C.; Hewitt, G. M.; Salisbury, P. J.; Singa, M. *Appl. Environ. Microbiol.*,**1985**, *49*, pp 96-100.
24. Nakajima, R.; Nishii, S.; Ishida, T.; Yamamoto, T. *J. Jpn. Oil Chem. Soc.*,**1990**, *39*, pp 191-195.
25. Yano, S.; Nakamura, T.; Uehara, T.; Furuno, T.; Takahashi, A. *Mokuzai Gakkaishi*, **1994**, *40*, pp 1226-1232.
26. Hassler, T.; *Tappi J.,* **1988**, *71*(6), pp 195-201.

Chapter 4

Freeness Improvement of Recycled Fibers Using Enzymes with Refining

Lori A. Eriksson, John A. Heitmann, Jr., and Richard A. Venditti

Department of wood and Paper Science, North Carolina State University, Raleigh, NC 27695

As recycled fiber use and content has increased in paper and board manufacture, the need to improve the drainage rate on the paper machine while maintaining mechanical properties through refining has become more important. This work is focused on the application of an enzyme (Pergalase A40) to blended colored ledger (BCL), varying the level of refining, point of addition of the enzyme, and enzyme dose for application in the manufacture of board from 100% recycled fiber. These results will be compared with those for flexographic newsprint (ONP) and old corrugated container (OCC) fiber types. The objective of the study is to determine the conditions which give the best drainage improvement while maintaining strength properties. The use of enzymes to improve the production of paper and board from recycled fibers is discussed and the most important properties of each fiber type are compared.

The manufacture of paper and paperboard from 100% recycled fiber presents some difficulties relative to virgin fiber. The main obstacles are fiber strength and drainage rate on the paper machine. The differences between recycled and virgin fibers are especially evident in chemical pulp fibers. Most of this can be attributed to hornification, which results from initial drying of the fiber and causes a loss of flexibility and rewetability of pulp fibers. Hornified fibers are found to be stiffer and form bulkier sheets with smaller and weaker interfiber bonds. Horinification is also thought to be one of the reasons that recycled fibers fracture and create fines easily during refining and thus create pulps that drain slowly.

The typical means of recovering strength properties of recycled fibers is refining. Which often dramatically decreases the freeness of the pulp, thus decreasing the machine drainage rate and production rate. This problem has lead to increased interest in the pulp and paper industry to find a means for improving the drainage rate of recycled fibers.

Many research groups have investigated the use of hydrolase enzymes for improving the drainage rate of recycled pulps (*1-5*). Some of the first work was performed by Pommier et al. (*4*) on old corrugated container (OCC) pulp. Their results showed significant drainage improvement with a 30 minute treatment time being optimal with Liftase A40. Bhardwaj et al. (*1*) followed up this work by comparing several different types of enzymes. Pergalase A40 added at 0.2 wt/wt % for 30 minutes provided a significant freeness improvement without compromising the strength properties of the final product. These studies and those by Bhat et al. (*2*) on a bleached softwood kraft pulp showed a decreasing improvement in freeness as enzyme dosage or freeness was increased. This is very important when considering the economic optimization of this type of technology.

Mechanical pulp fibers are known to behave differently from the chemical fibers, but freeness loss due to refining is still a problem when processing old newsprint (ONP). Stork et al. (*6*) studied the effects of enzymes on both virgin and recycled mechanical pulps. They found that ONP had a slight decrease in tensile strength which was attributed to the degradation of fines material by the enzymes. The tear strength of ONP containing chemical fibers decreased slightly, while that without remained virtually unchanged by the enzyme treatment. Most other studies involving ONP treatment with enzymes have dealt with deinking (*7, 8*). The deinking process also improved the freeness of the pulp and strength gains were seen for pulps treated with high xylanase content enzymes. More work is needed in this area to determine the proper enzyme mixture to attain the best results on mechanical fibers.

Another grade of recycled paper which is becoming more available for reuse is mixed office waste (MOW) consisting mainly of ledger paper. The ledger paper is typically divided up into two categories, white and colored. The white ledger is more expensive since it can be used in writing and printing grade papers. Colored ledger (BCL) is a cheaper and underutilized source of chemical fibers which could be used in paperboard grades with minimal processing costs. The main problem with BCL is removal of the dyes for use in other grades of paper and board. Freeness improvements for ledger pulps have mainly been investigated as an additional benefit during research on enzymatic deinking. Rutledge-Cropsy et al. (*5*) showed that their deinking process improved paper machine drainage rate while maintaining the strength properties of the final sheet as compared to the control. Prasad et al. (*9*) had similar results at the laboratory scale. All of the research has been performed on white ledger, therefore the possible effects of dyes and other materials in BCL on the performance of enzymes should be investigated and optimal treatments for drainage improvement should be explored.

Jackson et al. (*3*) showed that cellulases and hemicellulases are preferentially attracted to fines when they are present in kraft softwood pulp. The preferential attraction protects long fibers from excessive degradation by cellulase enzymes. This is the mechanism that allows for freeness improvement without compromising the strength properties of the paper. According to Jackson et al., and others(*1-9*), enzyme treatment of recycled fibers should be a reasonable means of improving pulp freeness.

One topic not covered in the previous studies was the effect of refining in combination with enzyme treatment on drainage and strength. The combination of enzymes with refining was the main objective of this study along with determining the conditions that would provide the greatest freeness improvement without compromising the physical properties of the BCL pulp.

Materials and Methods

Experimental Design. The variables chosen (Table I) were three levels of refining, three types of pulp treatment, and the order of treatment. These variables were used to develop a 3x3x2 factorial design to allow for statistical analysis of the results. Controls were also set to define a control curve for each pulp type and the zero refining values for each treatment. The control curve treatments consisted of the same pretreatments as the other samples followed by the appropriate amount of refining as seen in Table 1. The completion of this design required the collection of 25 sets of data. The pulp was prepared according to the procedure outlined in Figure 1.

Table I. Variables Studied

	1	2	3
Refining	1500 revolutions	2500 revolutions	4000 revolutions
Treatment	Wash	0.2 wt/wt % Enzyme	0.4 wt/wt % Enzyme
Order	Pre-Refining	Post-Refining	

Fiber Sources. The BCL consisted of Springhill® Relay® Plus Canary 20 lb. ledger paper. The paper was torn into approximately 2 in. squares in preparation for addition to the bench-top hydrapulper. The pulp was prepared according to the procedures outlined in Figure 1. This began with a presoak in distilled water at 50°C for 10 minutes. The paper was then pulped at 12% consistency for 30 minutes at 50°C in the bench-top hydrapulper. The pulp was washed in the hyperwasher at 0.5% consistency with two dilutions of distilled water. This removed fines and filler that are removed in a typical washing or thickening system found in recycled fiber processes. This pulp was then concentrated and fluffed for cold room storage and use in the following treatments.

Enzyme Treatment. The enzyme solution used for this study was Pergalase A40, produced by Ciba in Greensboro, NC. It has high cellulase and xylanase activities as seen in Table II, determined by the production of reducing sugar. The xylanase activity was determined using birch xylan as the substrate (*10*). The cellulase activities were determined using carboxymethylcellulose, avicel, and filter paper as the substrates (*11, 12*). The stopping and developing reagent for all of the above reactions is dinitrosalicylic acid (DNSA) (*11*).

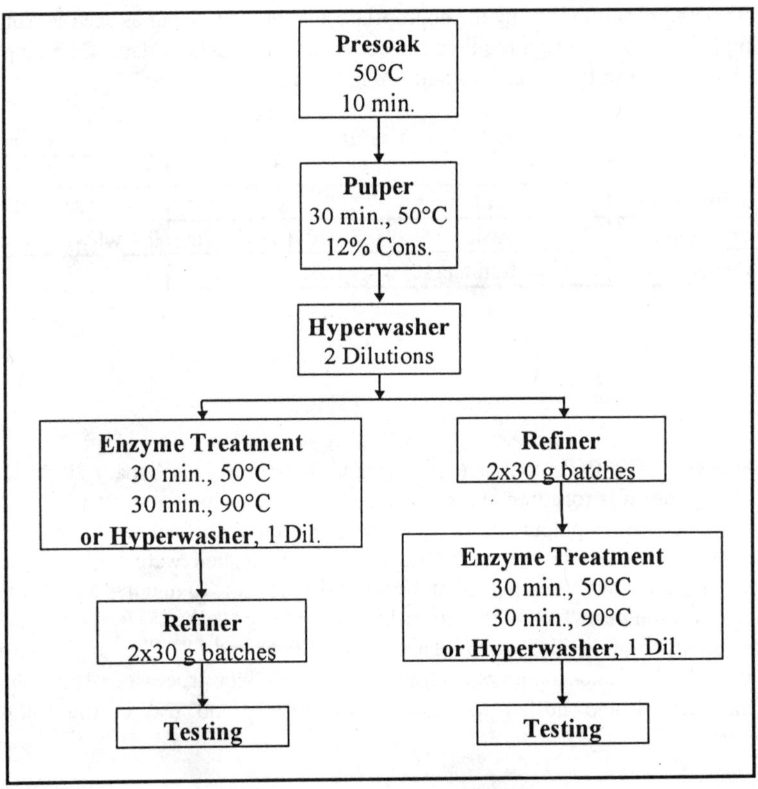

Figure 1. Experimental Procedure

The Pergalase was added to 10% consistency pulp buffered to pH 5.0 with 0.05M citrate at 0.2 wt/wt % or 0.4 wt/wt %. This corresponded to 0.35 mg protein/g pulp or 0.70 mg protein/g pulp respectively. The first was a typical charge known to be effective for drainage improvement in some fiber types, while the second was a high charge. The pulp-enzyme mixture was mixed for 30 min. at 50°C and then treated at 90°C for 30 min. to denature the enzyme and therefore limit the time that it was active on the fibers.

Table II. Pergalase Activities

Assay	Activity (U/ml)
Xylanase (birch xylan)	407 ± 38
CMC	1124 ± 159
Avicell	135 ± 19
Filter Paper	60 ± 6.8
Protein	175 mg/ml

Hyperwasher Treatment. The hyperwasher was used to wash out fines and filler materials from recycled fiber. It is a tank with baffles and a 200 mesh screen in the bottom. A variable speed mixer was used to stir the 0.5% consistency solution as the water and fine material drained out the bottom. The drainage rate could be regulated by adjusting a valve attached to the drain tube at the bottom of the apparatus. The design was based upon the Britt Jar (Tappi method T261). As a comparison to enzyme treatment (Table 1) pulp was treated with a single dilution wash.

Refining. The pulp was treated in a PFI Mill (TAPPI method T248) to four levels of refining; zero, 1500 revs., 2500 revs., and 4000 revs. These levels were chosen to produce a range of freenesses which included a typical production level and a very low freeness.

Pulp Property Evaluation. Drainage was evaluated using Canadian Standard Freeness (TAPPI method T227), and handsheets were made (TAPPI method T205) with a target basis weight of 127 g/m2. Strength properties were tested according to TAPPI Test Methods. Yield loss was determined by weight loss for wash treatments and by sugar production (indicated by the DNSA method) for the enzyme treatments. Fiber length distributions were analyzed using the Kajaani FS-200.

Results

Freeness. In understanding the effect of the different treatments on freeness, it was important to compare the results at different levels of refining. The initial treatment without refining showed significant freeness improvement for all treatments as compared to the control, based on the 95% confidence interval. This trend does not hold true for all of the treatments upon refining. Treating the BCL with enzymes after refining gave the best results at each level of refining with the post-refining wash treatment being the next best (Figure 2). The pulp freeness dropped below the control even at the low level of refining in the case of enzyme treatment prior to refining (Figure 3). As expected, the 0.2 % pre-refining enzyme treatment had less of a detrimental effect on freeness than the 0.4% enzyme treatment. This loss of freeness was probably due to the refining effect of the enzymes themselves on the fibers (*13*).

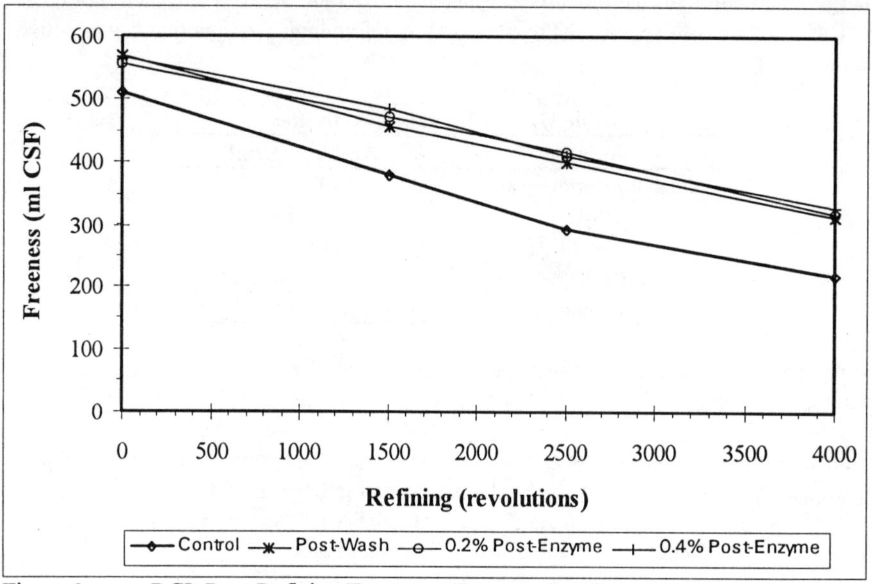

Figure 2. BCL Post-Refining Treatment Freeness Results

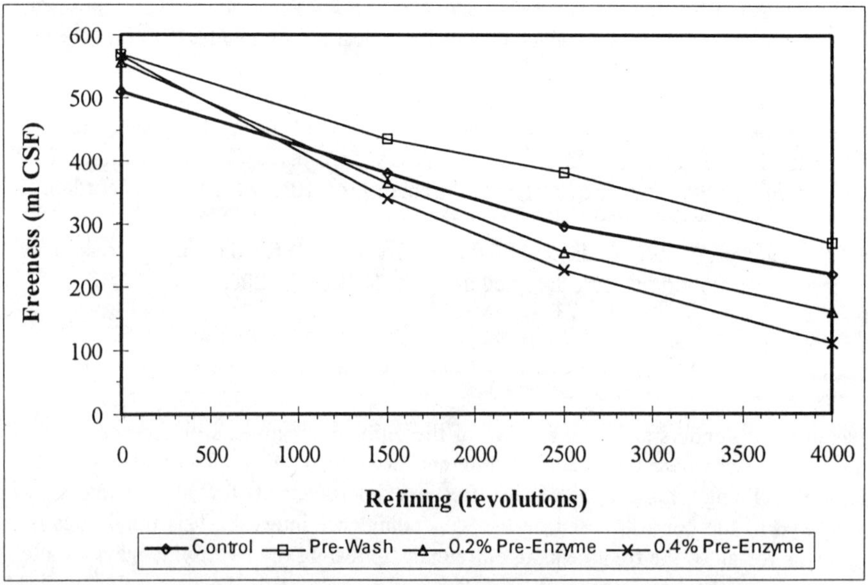

Figure 3. BCL Pre-Refining Treatment Freeness Results

Fiber Length. The average fiber length of each pulp as determined by the Kajaani FS-200 (Figure 4) showed an initial increase due to washing without any refining, compared to the control. Once the washed fiber had been refined, the average fiber length was not different from the control. The pulps treated with enzymes after refining did not have significantly different average fiber lengths from the control except at the highest level of refining. At this point their average fiber length dropped below the control, but there is a very large freeness improvement for this small drop in fiber length. The dosage level of enzyme did not appear to make a difference. Treating the BCL pulp with enzyme prior to refining was detrimental to the average fiber length.

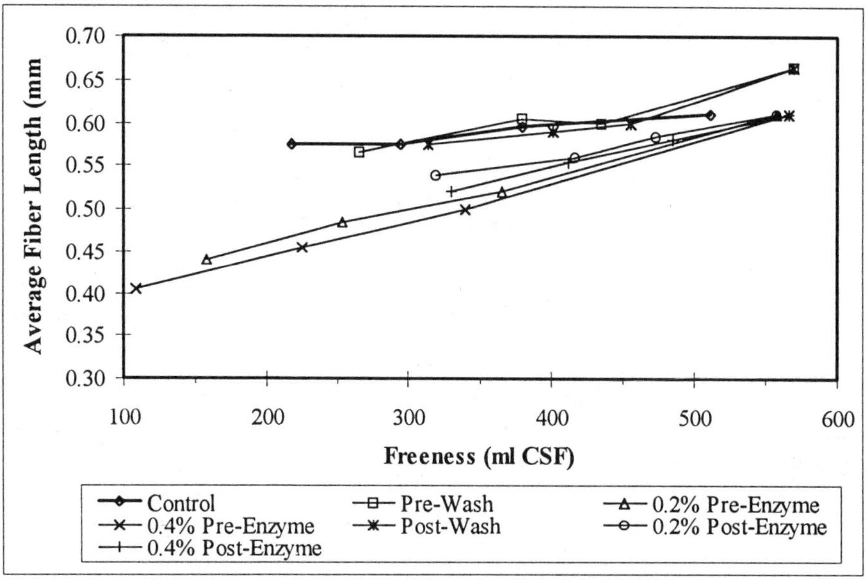

Figure 4. BCL Average Fiber Length Results (Kajaani FS-200)

The changes in the fiber length distribution due to changes in the fines fraction were most evident at the highest level of refining (Figure 5). They were initially noted at the lowest level of refining although they were less pronounced. Treatment with enzymes prior to refining caused a large increase in the 0-0.2 mm fraction as compared to the control. This also held true for the 0.2-0.4 mm fraction. The 0.2% post-refining enzyme treatment slightly increased the 0-0.2 mm fines fraction as compared to the control. The 0.4% post-refining enzyme treatment produced a more significant increase in this fines fraction. Post-refining enzyme treatments did not change the other fines fraction levels as compared to the control. The increase in the 0-0.2 mm fines fraction for post-refining enzyme treatment explains the change in the average fiber length found for those treatments at the highest level of refining (Figure 4).

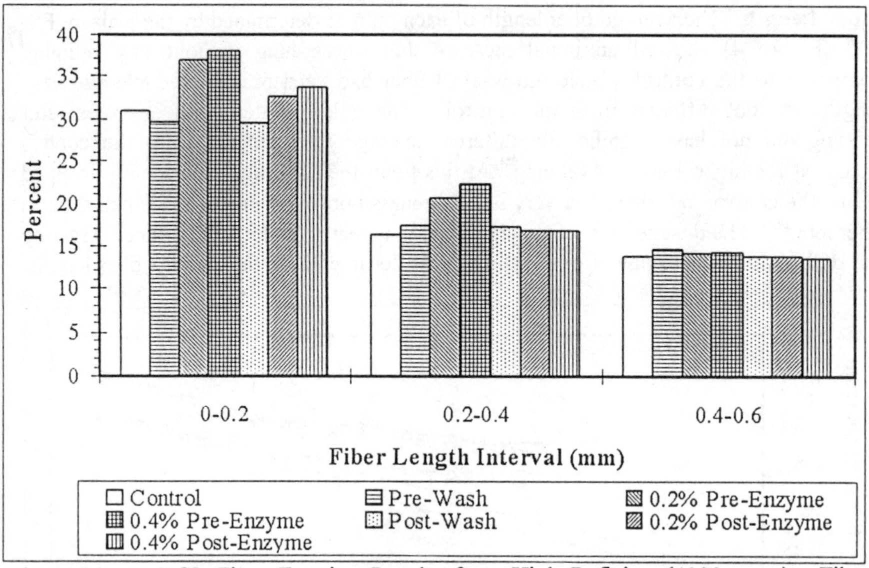

Figure 5. BCL Fines Fraction Results from High Refining (4000 revs.). Fiber fractions above 0.6 mm not shown.

Physical Properties. The overall strength properties of a sheet are often evaluated by comparing the tensile and tear indices (Figure 6). The results from the BCL pulp indicate that strength loss occurred with the enzyme treatments, although it was less significant for the post-refining enzyme treatments. The wash treatments did not significantly affect the strength properties as compared to the control.

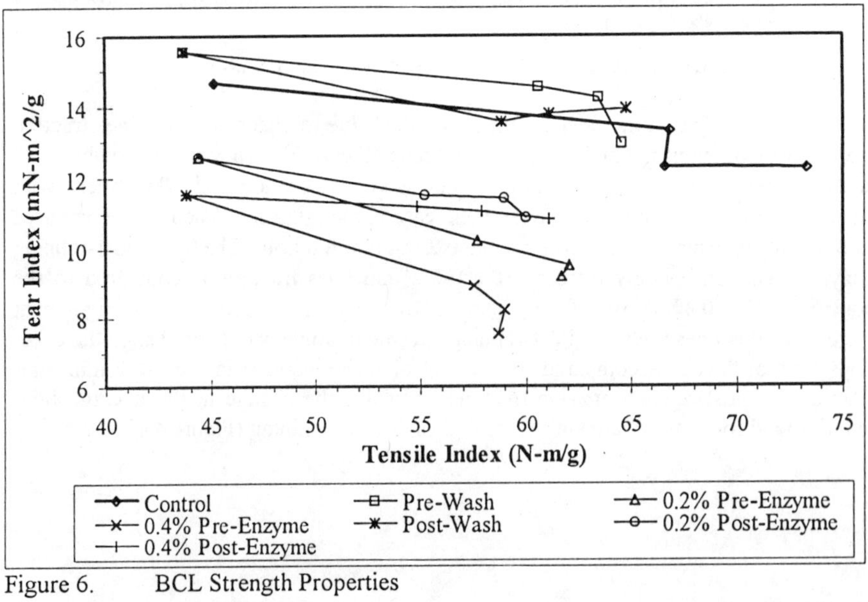

Figure 6. BCL Strength Properties

The general strength loss due to enzyme treatment seen in Figure 6 was influenced by the tear index results. The tear index for the post-refining enzymes treatments appeared to be different from the control and the wash treatments. This was true, except for the 0.2% post-refining enzyme treatments which were not significantly different from the control based on the 95% confidence interval. The treatment of unrefined fibers did significantly decrease the tear as compared to the control, but the tear did not change significantly upon refining for the post-refining enzyme treatments. The wash treatments were also not significantly different from the control, at all points. The tear strength results corresponded well with the results of the fines fraction analysis.

The results from the tensile index data showed that the wash treatments and the 0.2% pre-refining enzyme treatment were not significantly different from the control, based on the 95% confidence interval. This is interesting, as tensile strength is typically adversely affected by decreased average fiber lengths. This assumes that the long fiber fraction decreases when the fines fraction increases. With enzyme treatment the additional fines produced could also come from the stripping of microfibrils from the surface of long fibers.

The apparent densities (Figure 7) of the sheets were not significantly different from the control. The compression strength typically follows a trend similar to that of the density, but in this case there were some differences (Figure 8). There was a slight increase in compression strength above 450 ml CSF for the post-refining wash treatment and the post-refining enzyme treatments. Below this freeness, the post-refining enzyme treatments were not significantly different from the control and the post-refining wash treatment gave a compression strength below the control. The post-refining wash treatment behaved similarly to both of the pre-refining enzyme treatments below 400 ml CSF. The pre-refining wash treatment was not significantly different from the control at any of the refining levels.

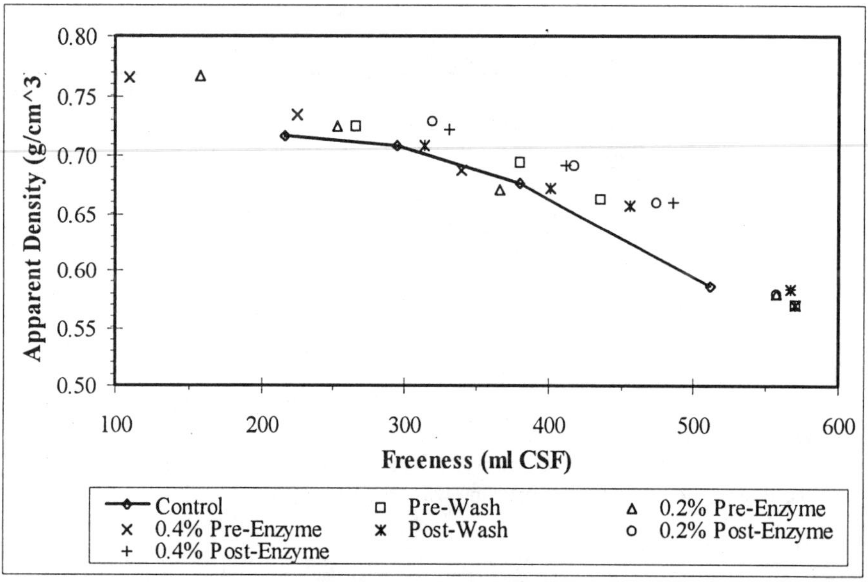

Figure 7. BCL Apparent Density Results

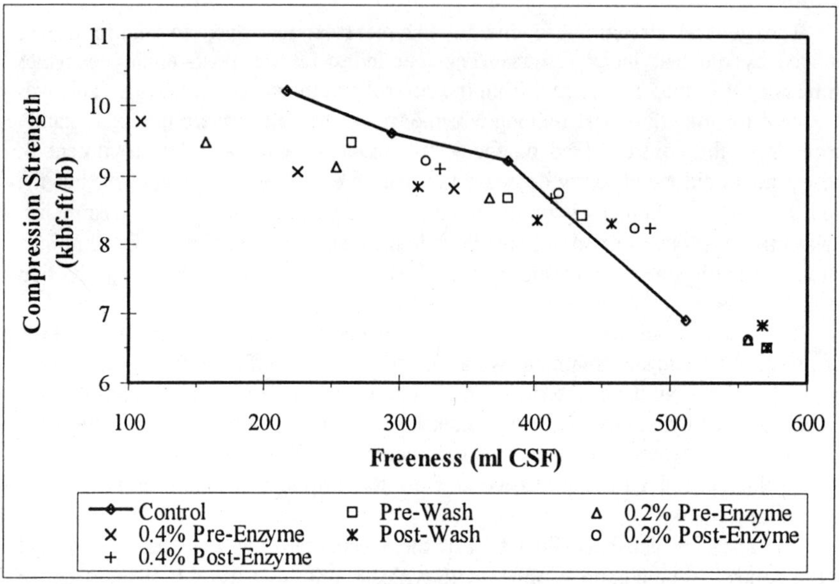

Figure 8. BCL Compression Strength Results

Yield Loss. The yield loss results (Table III) could be of importance when considering the economic value of the treatments. In general the loss due to enzyme treatment was very low, while washing resulted in a much larger yield loss. This did not take into consideration the initial yield loss due to the pretreatment of the pulp (27.9%) shown in Figure 1. The 0.2% enzyme addition level had a virtually insignificant yield loss due to this treatment.

Table III. BCL Percent Yield Loss Results

Revolutions	Control (%)	Pre-Wash (%)	0.2% Pre-Enzyme (%)	0.4% Pre-Enzyme (%)	Post-Wash (%)	0.2% Post-Enzyme (%)	0.4% Post-Enzyme (%)
0	0	3.3	0.17	0.24	3.3	0.17	0.24
1500	0	2.0	0.17	0.26	4.5	0.16	0.24
2500	0	6.4	0.17	0.26	4.8	0.17	0.27
4000	0	4.0	0.17	0.25	5.3	0.16	0.26

The initial ash content of the paper was 12.5% at 525°C and 10.2% at 900°C. After the pretreatment sequence the ash was 0.8% at 525°C and 0.25% at 900°C. Additional washing did not significantly increase the removal of filler material. The enzyme treatments also had no effect on filler removal. The hyperwash treatment was proven to be very efficient in the removal of filler materials. This also shows that the fiber loss due to the pretreatment washing was about 15-16%.

Discussion

The freeness results were promising for the application of Pergalase A40 to BCL after refining. However, doubling the addition level of enzyme did not significantly improve the freeness levels at any level of refining. This implies that the maximum effect was realized at the 0.2% addition level and that lower charges might be effective and more cost efficient. The decrease in freeness found when the fibers were refined after enzyme treatment was a function of the refining action of the enzyme on the fibers (*13*). The fiber length analysis showed that there were only a small number of fines remaining in the starting pulp due to the pretreatment procedures. This accounts for the accessibility of the enzyme to the long fiber fraction of the pulp, which caused a refining effect, or strength loss, or both. The pre-refining enzyme treatment freeness results were probably not being compared at the same refining level. Determination of the PFI mill refining levels of pre-refining enzyme treated pulps that would have been comparable with the control pulp was not within the scope of this study.

The changes in fiber length found due to treating the pulp with Pergalase A40 prior to refining confirmed the ease with which these fibers could be broken down by the refining process. This could be due to a number of factors. One would be the lack of fines in the pulp, due to the pulp preparation process, that would normally protect the long fibers from degradation by the enzymes. Jackson et al. (*3*) showed that hydrolase enzymes attach preferentially to higher surface area materials in the pulp, especially fines and fibrils. Without these materials in the pulp slurry, the long fibers are subject to more severe degradation. It has also been shown by Noé et al. (*13*) that these enzymes produce a refining effect on the fibers. These phenomenon would explain the difficulties found here with treating the pulps with enzymes prior to refining. Recycled fibers break down quite easily in the presence of refining. Pre-refining treatment with enzymes appears to accelerate the rate of refining and has an effect similar to a higher level of refining.

The freeness improvements found by treating the pulps with enzymes after refining also correspond to the findings of Jackson et al. (*3*). The refining process produces more fines material and fibrilates the fibers. These materials act to protect the long fibers from severe degradation by the enzymes. The results of the average fiber length and fines fraction analysis supported this. The wash treatments improved the freeness by removing fines material, which was also supported by the fiber length analysis. The best treatment scheme will depend on the end use requirements, and an economic analysis of the costs involved for each process.

The physical properties of the resulting paper produced some interesting results. The overall strength as determined by the comparison of tear and tensile strength appeared to be adversely affected by all enzyme treatments. This rule held true with the exception of the tensile results of the 0.2% pre-refining enzyme treatments which were not statistically different from the control. This could be due to the refining effect of the enzyme on the surface of the fibers. It is well known that refining improves the strength of recycled fibers to a certain extent. For this reason it is possible that the enzymes could have produced a similar result. The fiber length results, however did not explain the tensile results, as treatment with enzyme after refining did not greatly affect the fiber length distribution except at the highest level of refining. This suggests that the enzymes are changing the surface of the fiber either physically or chemically to decrease the fiber-fiber bond strength.

In a similar study using ONP and OCC as the substrates (*14*), the strength results were different in some respects. The overall strength as judged by comparing tensile and tear strength were greatly influenced by the tear for ONP with the 0.2% post-refining treatment being the closest enzyme treatment to the control. The OCC results appeared to be somewhat better and the 0.2% post-refining enzyme treatment was not significantly different from the control. The OCC tear results were similar to the BCL tear results.

The tensile index results for both ONP and OCC showed that the control was significantly different from most treatments as it was for BCL. These results were not supported by the fiber length distribution results and therefore there must be another explanation for this phenomenon. Most likely it is a surface change that is taking place due to physical modification of the fiber surface or some chemical inhibition of hydrogen bonding to cause the decrease in tensile strength.

The density and compression strength results were of interest when compared to previous results on other grades of recycled fiber (*14*). Both ONP and OCC grades showed direct correlation between apparent density and compression strength, but this was not the case with BCL. The main difference between these grades is lignin content and hardwood content. Either of these could have influenced the effect of the enzymes, but the surface chemistry of the fibers due to wet-end additives in the process of making the original paper could also be a factor. The understanding of this would require further investigations in these areas on a micro scale.

A preliminary consideration of the economic viability of this type of treatment can be made. The most apparent point for economic consideration is the comparison of the cost of the treatment with the cost of fiber lost during the treatment processes. The treatment of pulp with 0.2% Pergalase A40 after refining gave good freeness improvement at all refining levels, while maintaining most strength properties important to board products. The current cost of adding 0.2% Pergalase would be $16/ODT pulp at the current market price of $4/lb. This is quite expensive as compared to the cost of the fiber lost due to the additional washing stage of any of the fiber sources as seen in Table 4. The yield loss costs are based on a 6.7% average yield loss for ONP, 3.6% for OCC, and 4.3% for BCL due to the wash treatment. The net fiber cost is calculated by adding a $20 shipping charge to the dock price and 15% shrinkage due to moisture content and out-throws. It becomes clear from these results that the higher the value of the initial fiber is, the more economically feasible enzyme treatment becomes.

Table IV. Cost of Yield Loss Due to Wash Treatment

	Net Fiber Cost ($/ODT)		Cost for Fiber Loss ($/ODT)	
Paper Grade	**1996**	**1995**	**1996**	**1995**
BCL	112.50	165.43	4.84	7.11
ONP	37.50	67.62	2.51	4.53
OCC	93.39	60.29	3.36	2.17

These calculations did not include the cost of labor, processing, chemicals expended on the lost fiber, effluent treatment cost, or the effect of increased production rate. Due to the improvement in freeness seen in this study, it appears that the increase in production rate alone could cover the cost of the enzyme treatments. This could only be evaluated on a machine and product specific basis. The decrease in yield loss using enzymes as compared to the hyperwasher treatments should also decrease the BOD loading to the effluent treatment plant. To accurately evaluate the economic outlook for the use of enzymes in improving drainage of recycled fiber these other factors should be investigated. This type of evaluation should prove to be beneficial for the application of cellulase enzymes to recycled fiber for drainage improvement.

The results of the preliminary economic considerations and the physical testing showed that lower enzyme dosages should be investigated to further optimize this process. Enzyme charge levels between 0.05% and 0.15% may give substantial freeness improvements without the detrimental physical effects and at a more reasonable price. However, this may vary for the different grades of recycled papers and the freeness at which the enzymes are applied.

Another means of decreasing the cost of the enzyme treatment is recycling of enzymes. Jackson, et al. (*15*) had some success with this, but there are several variables yet to be investigated. The development of enzyme-aided drainage improvement technology for use on a commercial scale will require studies on several fiber types with different compositions. This should include pilot plant and mill trials, since it is difficult to replicate these processes in the laboratory setting. It is especially difficult to determine the effect of recirculation of white water on the use of enzymes in recycled fiber mills.

The results presented in this paper prove that Pergalase A40 can be effectively used as a drainage improvement agent for recycled fibers. There are minimal strength losses and, especially for board grades, product properties are within the proper range to produce high quality products. The economic viability of enzyme-aided drainage improvement will depend on the increased production found for each individual process. The authors expect increased production due to the freeness improvements found here.

Conclusions

1. Treatment of the recycled fiber with 0.2 wt/wt % Pergalase A40 after refining improves freeness significantly with yield losses less than 0.5%.

2. The wash treatments improve freeness to levels similar to the enzyme treatments after refining, but the yield loss is significantly larger.

3. Pretreatment with enzymes while maintaining the same levels of refining is detrimental to the freeness and strength properties of the pulp. The enzyme pretreatment should be coupled with reduced refining to maintain freenesses at the same levels as the control.

4. Treatment with 0.2 wt/wt % Pergalase A40 on the recycled fiber provided the best freeness improvement for the least cost and was least detrimental to the physical properties of the end product. It seems likely that a lower dosage of enzyme would provide a reasonable freeness improvement and be economically more favorable.

5. Enzyme-aided drainage improvement of recycled fiber would be economically feasible, depending on production rate improvements, with process optimization and enzyme recycling.

References

1. Bhardwaj, N. K., P. Bujpai, and P. K. Bujpai. *Appita*, **1995**, 48(5), pp. 378-380.
2. Bhat, G. R., J. A. Heitmann, and T. W. Joyce. *Tappi J.*, **1991**, 74(9), pp. 151-157.
3. Jackson, L. S., J. A. Heitmann, and T. W. Joyce. *Tappi J.*, **1993**, 76(3), pp. 147-154.
4. Pommier, J.-C., J.-L. Fuentes, and G. Goma, 1989. *Tappi J.*, **1989**, 72(6), pp. 187-191.
5. Rutledge-Cropsy, K., S. M. Abubakr, and J. H. Klugness. *1995 Tappi Pulping Conference Proceedings*, TAPPI: Atlanta, GA, pp. 639-643.
6. Stork, G., H. Pereira, T. M. Wood, E. M. Düsterhöft, A. Toft, and J. Puls. *Tappi J.*, **1995**, 78(2), pp. 79-88.
7. Prasad, D. Y., J. A. Heitmann, and T. W. Joyce. *Prog. in Paper Recycling*, **1992**, 1(3), pp. 21-30.
8. Prasad, D. Y., J. A. Heitmann, and T. W. Joyce. *Nordic Pulp and Paper Res. J.*, **1993**, no. 2, pp. 284-286.
9. Prasad, D. Y. *Appita*, **1993**, 46(4), pp. 289-292.
10. Bailey, M. J., P. Biely, and K. Poutanen. *J. of Biotech*, **1992**, 23, pp. 257-270.
11. Miller, G. L. *Analytical Chemistry*, **1959**, 31(3), pp. 426-428.
12. Ghose, T. K. *Pure & Appl. Chem.*, **1987**, 59(2), pp. 257-268.
13. Noé, P., J. Chevalier, F. Mora and J. Comtat. *J. Wood Chem. & Tech.*, **1986**, 6(2), pp. 167-184.
14. Eriksson, L. A., J. A. Heitmann Jr., R. A. Venditti. *1997 Tappi Recycling Conference Proceedings*, TAPPI: Atlanta, GA, **1997**, pp. 423-433.
15. Jackson, L. S., T. W. Joyce, J. A. Heitmann, and F. G. Giesbrecht. *J. of Biotech.*, **1996**, 45, pp. 33-44.

Chapter 5

Enzyme Applications in Conventional Kraft Pulping

C. J. Jacobs-Young[1], Richard A. Venditti[2], and T. W. Joyce[3]

[1]Division of Paper Science and Engineering, University of Washington, Seattle, WA 98195
[2]Department of Wood and Paper Science, North Carolina State University, Raleigh, NC 27695
[3]Department of Paper and Printing Science and Engineering, Western Michigan University, Kalamazoo, MI 49008–5060

A cellulase/hemicellulase mixture was used to increase the diffusion of alkali in sapwood and enhance the efficiency of conventional kraft pulping. Diffusion was enhanced by pretreating sycamore (*Platanus occidentalis*) sapwood with enzymes. Tangential diffusion was increased by 26%, while longitudinal diffusion was increased by approximately 110% in the enzyme treated samples of sycamore sapwood. Sycamore chips were pretreated using cellulase/hemicellulase/pectinase mixtures prior to conventional kraft pulping. After enzyme pretreatment followed by kraft pulping, the kappa number of the control was as much as 10% higher than the treated chips. In addition, enzyme treated pulps required less (approximately 9 kg ClO_2/ ton pulp bleached) than the control. Pulp strength properties for the pretreated chips were comparable to kraft pulps receiving no enzyme pretreatments. Pulp viscosity was lower in pulps treated with cellulase/hemicellulase alone.

Increasing environmental pressures have made it necessary to investigate methods for reducing the amount of energy, sulfur, and chlorine containing compounds utilized during pulping and bleaching. These pressures have also lead to the design of many technologies aimed at lowering the lignin content of the pulp entering the bleach plant. Lowering the kappa number of the pulp by increasing the efficiency of the pulping process has lead to a reduction in chemicals necessary for pulp bleaching and a concomitant reduction in pollutants discharged from the bleach plant.

The objective of this study is to determine the effect of enzyme pretreatment on conventional kraft pulping.

Improving Transport Processes in Wood

Processes which improve liquid transport in wood are of great interest to the pulp and paper industry. Impregnation of cooking chemicals during pulping is a crucial step for producing uniform pulps. Incomplete impregnation results in non-uniform pulps which are characterized by high reject levels, non-uniform delignification, and lower strength properties. The size, nature, and number of intercommunicating structures in wood have a major effect upon movement of water and other materials through wood (*1*). Enzymatic pretreatments can be utilized to alter these structures, thus improving the penetration of chemicals.

Successful implementation of enzymatic pretreatments can provide substantial increases in diffusion, which can possibly be attributed to various factors including the dissolution of pit membranes and the removal of carbohydrates from the lignin-carbohydrate matrix. Numerous studies on the treatment of wood with biological agents have revealed the attack of these substances on pit membranes (*2-4*). In hardwoods, the primary pathways for liquid transport are vessels which are not occluded by tyloses, while fiber tracheids are responsible for conducting fluid in softwoods. In both hardwoods and softwoods, fluid is communicated from one cell to the next through interconnecting pits. The membranes of these pits which offer resistance to flow are composed of primarily lignin, pectin and hemicellulose. For conventional kraft pulping systems, it is imperative that research be conducted that can assist mills in meeting environmental regulations

Enzymes and Kraft Pulping

Delignification in Conventional Kraft Pulping. Conventional kraft pulping is one of several processes used to solubilize and remove lignin from wood used for papermaking. Sodium hydroxide and sodium sulfide are applied to wood and the temperature is raised slowly to a maximum temperature of ≈ 170°C. Kraft delignification occurs in three phases; initial, bulk and residual. The initial phase is characterized by rapid delignification, while a major portion of the lignin is removed during the subsequent bulk phase. The residual lignin removal phase, which occurs at about 90% delignification, is characterized by slow delignification coupled with rapid carbohydrate degradation reactions. The low rate of delignification is believed to be caused by the presence of alkali-stable lignin carbohydrate bonds (*5,6*). In kraft pulping, initial delignification occurs preferentially in the secondary wall. At about 50% delignification, the lignin in the middle lamella and cell corner areas dissolve rapidly, leaving the residual lignin in the secondary wall. This topochemical effect is the result of physical and chemical factors (*7,8*). Experimental evidence suggests that the topochemical variations in delignification can be partially attributed to the difference in the structure of lignin from the secondary wall and middle lamella regions (*9,10*). In addition, important physical factors such as the differential penetration of reagents into various morphological regions, the accessibility of lignin, and the diffusibility of degraded lignin may lead to topochemical effects. The dissolution of SW lignin has been related to the pore size of the cell wall and thus the removal of hemicellulose (*11,12*). The location of residual lignin has a significant bearing on the bleachability and papermaking properties of pulp fibers.

Materials and Methods

The sycamore (Platanus occidentalis) utilized in this study was obtained from an NC State University experimental forest near Raleigh, NC.

Enzyme Pretreatment. Prior to pulping, 650 OD grams of chips were treated with Pergalase A40 (CIBA, Greensboro, NC) which has cellulase and xylanase activity. Xylanase activity was estimated at 407 U/ml using birch glucuronoxylan as substrate according to the method of Poutanen and Puls (*13*). Protein concentration was estimated at 175 mg/ml of enzyme solution according to the method of Lowry et al. (14). Cellulase activity was determined using a filter paper assay described by Mandels (15). Cellulase activity was approximately 70 filter paper units (mmols of reduced sugar) per milliliter. The filter paper activity was determined from liberated reducing sugar as measured by the dinitrosalisylic acid method using glucose as a standard. Pergalase was applied based on the cellulase activity at 5 FPU per oven dry gram of wood. SP 249 (Novo Nordisk Biochem North America, Inc.), an enzyme with pectolytic, cellulolytic and hemicellulolytic activities was also utilized. The supplier reports the activity of pectinase at 1,600 k PU/g determined using citric pectin, arabanase activity at 98 units/g using sugar beet araban, and α-galactosidase activity as 63 units/g using p-Nitro-phenyl-galactoside as the substrate. SP 249 was charged at 1% on OD chip weight. The enzyme preparations were used with a citrate buffer solution (pH ≈ 5) at 50°C. The chips were placed in stainless steel bombs along with the enzyme preparation, citrate buffer (pH 5), and water. Enzyme impregnation was aided by applying 67.73 kPa of vacuum to each bomb. The chips were reacted in a constant temperature device for 24 h at 50°C. The control chips were treated as described above, but without enzyme addition.

Pulping Conditions. Sycamore sapwood chips were pulped in a 7L M&K digester vessel using the following conditions: 15% AA, 30% Sulfidity, 4:1 L/W ratio, and 1000 H Factor. The chips were screened and sorted to remove bark and knots. 650 OD grams of chips were used for all cooks. At the completion of the cook, the spent black liquor was blown into a holding vessel. The chips were not blown under pressure, but removed manually, washed, disintegrated and screened. The screened pulp was centrifuged to approximately 30% consistency and fluffed.

Pulp Testing. Control and enzyme treated pulps were beaten in a PFI mill at 10% consistency with an applied load of 1.8 kg/cm. Samples were collected at predetermined intervals between 0 and 6000 revolutions. Pulp fiber length, coarseness, kink and curl were determined using the Fiber Quality Analyzer (Optest Inc.). Handsheet properties, pulp viscosity, fines analysis and kappa number were determined for each sample using TAPPI standard methods.

Sugar Analysis. Pulp sugars were measured by anion exchange liquid chromatography of their borate complexes, followed by a post column reaction to produce a blue copper complex. The absorbency of the copper complex is measured by a detector set at 560 nm. Quantification was done by comparing sample peak heights to those produced by a standard (*16*).

Results and Discussion

Effect of Enzyme Pretreatment on Diffusion. In a previous study it was determined that a cellulase/hemicellulase (CH) mixture was capable of increasing the diffusion of sodium hydroxide in sycamore sapwood in both the tangential (along the rays) and longitudinal directions (*17*). The diffusion was determined according to a procedure outlined by Talton (*18*) in 1986. Diffusion into the tangential direction was increased from 7.26 x 10^{-6} cm^2/sec to 9.17 x 10^{-6} cm^2/sec. Longitudinal diffusion was 110% higher, with an increase in diffusion from 7.06 x 10^{-6} cm^2/sec to 14.9 x 10^{-6} cm^2/sec. The increase in diffusion was attributed to the degradation of pit membranes which served as a impedance to flow in the sapwood.

Effect of Enzyme Pretreatment on Lignin Removal. Effects of the 24 h enzyme pretreatments on pulp delignification, viscosity and yield are listed in Table I. When compared to the control kraft pulp, both enzyme treatments resulted in higher levels of delignification. The CH mixture resulted in an approximately 1.5 point kappa difference, while maintaining similar unscreened yield values. The level of delignification was increased further by the addition of pectinase to the enzyme mixture. Screened yield values were comparable for all cooks.

Effect of Enzyme Pretreatment on Viscosity and Pulp Strength Properties. Cellulase/hemicellulase/pectinase (CHP) was capable of lowering the kappa number of the pulp, while maintaining similar viscosity values. Pretreatment with CH alone resulted in lower final pulp viscosities than the control and chips treated with CHP. The main differences in the carbohydrate compositions of the pulps were in the xylose and glucose compositions. (Table II) Arabinose and mannose contents were low in each pulp. More xylose was detected in the enzyme pretreated pulps than the pulp receiving no pretreatment. The reprecipitation of xylan during kraft pulping occurs when almost all of the xylan side chains have been cleaved off (*19*). This may indicate that the additional hydrolysis of xylan side chains during pretreatment could have resulted in increased xylan reprecipitation during pulping. As would be expected, hydrolysis of carbohydrates by the cellulase/hemicellulase pretreatments resulted in lower levels of glucose in the final pulps. Filtrate obtained directly after enzyme pretreatment was analyzed using a method outlined by Mandels (*15*). Effluent following enzyme pretreatment had significantly higher levels of reducing sugars than the control. Using the HPLC method, traces of galactose, arabinose and glucose were identified in filtrate following pretreatment. Material also eluted in the tri- and disaccharide region near cellobiose for each sample, but the peaks were too small to quantify.

The physical properties of each pulp were tested before and after refining. In Figure 1, it can be seen that, although the tensile index is not significantly different for the three pulps, the pulp receiving pectinase treatment had a higher tensile index at each

Table I. Pulping Data for Enzyme Pretreated and Conventional Kraft Cooks. Cellulase/Hemicellulase (C/H); Cellulase/Hemicellulase/Pectinase (C/H +P)

	Control (Cook 1, Cook 2)	C/H (Cook 1, Cook 2)	C/H +P (Cook 1, Cook 2)
Pergalase Dosage (Filter Paper Units/g)		5	5
Pectinase Solution Dosage (% on OD basis)			1
Kappa Number	17.6,17.5	16.3, 16	15.6, 15.9
Average Viscosity, mPa·s	50.9	46.7	50.1
Unscreened Yield, %	49.1, 49.8	48.6, 49.9	46.2, 48.2
Screened Yield, %	43.1, 44.9	43.4, 45.3	43.7, 44.9

Table II. Sugar Analysis of Enzyme Pretreated and Conventional Kraft Pulps. Cellulase/Hemicellulase (C/H); Cellulase/Hemicellulase/Pectinase (C/H +P)

	Composition of Carbohydrates			
	% Glucose	% Xylose	% Arabinose	% Mannose
Control	88.6	11.1	<0.1	0.2
C/H	85.6	14.1	<0.1	0.2
C/H +P	85.6	14.2	<0.1	0.2

Table III. Fiber Analysis for Enzyme Pretreated and Conventional Kraft Pulps. Cellulase/Hemicellulase (C/H); Cellulase/Hemicellulase/Pectinase (C/H +P)

Pulp	Avg. Fiber Length, mm	% Fines
Control (15% AA)	0.52± 0.01	13.4± 0.80
Control (16% AA)	0.62± 0.01	9.6± 0.19
C/H	0.67 ± 0.01	8.2± 0.21
C/H +P	0.73 ± 0.00	5.8± 0.35

level of sheet density. At any given breaking length the control pulp was higher in average tear strength than the enzyme pretreated pulps, but was not significantly different (Figure 2).

Enzyme pretreated pulps had higher dry zero-span tensile strength than the untreated pulps (Figure 3). The breaking length of a sheet is influenced by both the fiber length and variations in the level of bonding. Zero-span tensile is decreased by reductions in fiber length, and increased by higher levels of bonding in the sheet (*20*). Higher dry zero-span tensile strength would indicate that the individual fibers resulting from enzyme pretreatment and kraft pulping were stronger than those from kraft pulping alone. After evaluation, it was determined that the untreated pulp had lower average fiber lengths and higher fines contents than the enzyme pretreated pulps (Table III). Both of these factors would lead to lower levels of bonding in the untreated pulps and therefore to lower dry zero-span tensile readings.

The wet zero-span test has been suggested as a method for removing the influence of fiber bonding on zero-span tensile strength (*21*). This is based on the assumption that in wet sheets, all interfiber bonds are effectively broken. The wet zero-span tensile is lower for a wide range of pulps due to the weakening of the individual fibers that make up the sheet. This loss in strength depends on the extent of chemical and mechanical damage to the interfibrillar matrix of fibers during processing. The weaker the supporting lignin-hemicellulose matrix holding the cellulose microfibrils together, the larger is the reduction in zero-span strength upon wetting (*22*). Enzyme pretreatments with cellulase and hemicellulase are capable of degrading the interfibrillar matrix. This can possibly be shown in the manner in which the enzyme pulps behave upon wetting. The enzyme pretreated pulps exhibited approximately an 18% reduction in zero-span strength upon wetting, while the control pulp experienced only a 6% reduction.

The active alkali charge of the conventional kraft process was increased to produce a control pulp with a comparative kappa number to the enzyme pretreated samples (kappa no. 15.2). The pulp was cooked using 16% AA, 1000 H factor and 30% sulfidity. The wet zero-span tensile was 13% lower than the dry zero-span tensile. This pulp exhibited a larger reduction in zero-span strength upon wetting then the control pulp cooked at 15% AA (6%). This may be attributed to the higher loss of hemicelluloses with the increase in active alkali.

The pulps were similar in levels of kink (0.90 kinks/mm), coarseness (16.5 mg/100 m) and curl index (0.06).

Effect of Enzyme Pretreatment on Bleachability. Pulps were bleached using a DED bleaching sequence to determine the effect of the enzyme pretreatment on final bleaching properties. Each pulp was bleached using a kappa factor of 0.18 in the first D stage. The results are listed in Table IV. Results indicate that pulps treated with cellulase/hemicellulase (CH) alone had slightly higher final brightness at each level of chlorine dioxide charged than the control. Figure 4 shows that at 0.5% and 0.9% ClO_2 charge on OD pulp in the final D stage, there was an increase in brightness of up to 1.3% ISO when using pulps treated with CH alone. In order to reach a final brightness of 81% ISO, the CH treated pulp required a ClO_2 charge of 0.88%, while the control required a

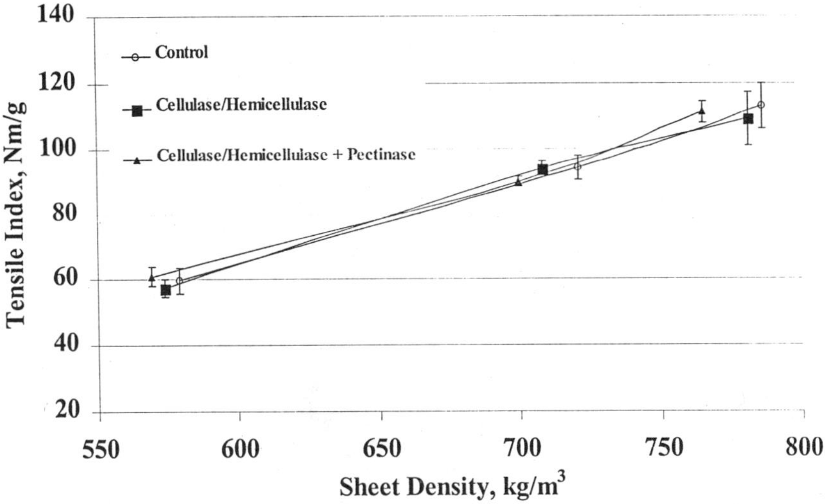

Figure 1. Sheet Density-Tensile Index Analysis of sycamore chips receiving enzyme treatment. The chips were pulped using 1000 Hfactor, 15% AA, and 30% Sulfidity.

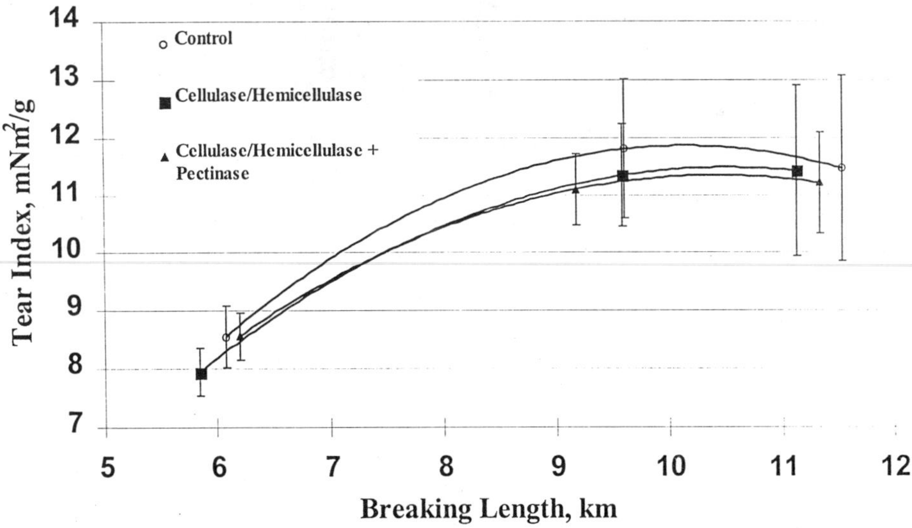

Figure 2. Tear-Tensile analysis of sycamore chips receiving enzyme treatment. The chips were pulped using 1000 Hfactor, 15% AA, and 30% Sulfidity.

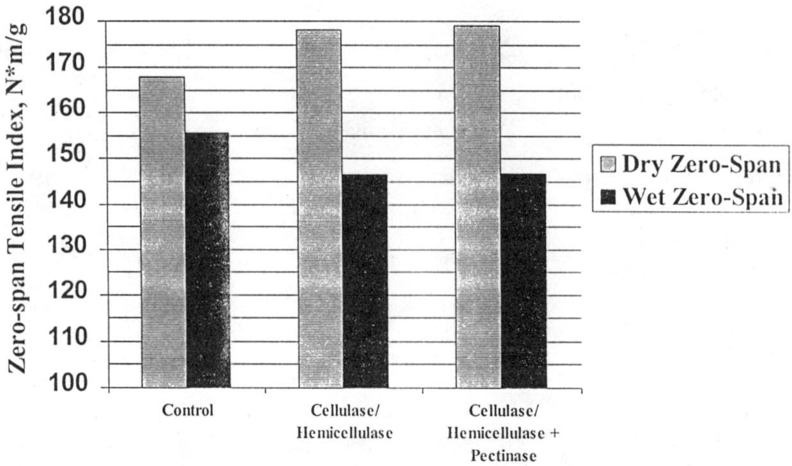

Figure 3. Zero-span tensile of sycamore chips receiving enzyme treatment. The chips were pulped using 1000 Hfactor, 15% AA, and 30% Sulfidity.

Table IV. Bleaching Results. Cellulase/Hemicellulase (C/H); Cellulase/Hemicellulase/Pectinase (C/H +P). DED Bleaching Sequence for Enzyme Pretreated and Conventional Kraft Pulps

Pulp	Brownstock Kappa No.	DE Kappa No.	DE Brightness	Final Brightness		
				0.5 % ClO2	0.9 % ClO2	1.2 % ClO2
Control	17.6	4.68	51.0	74.2	80.1	82.7
C/H	16.2	4.77	52.6	75.5	81.3	83.1
C/H +P	15.8	4.75	52.3	75.0	79.7	83.3

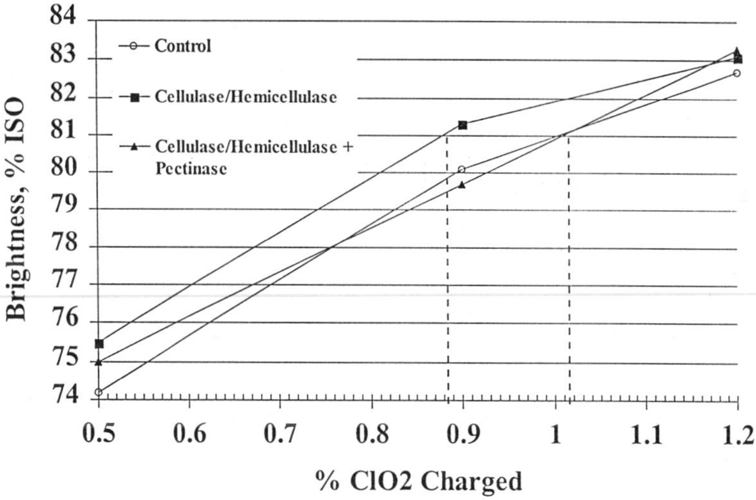

Figure 4. Bleaching of pulps using a DED Sequence.

ClO_2 charge of 1.01%. These differences in ClO_2 charge translate into chemical savings of 8.36 kg ClO_2/ ton pulp bleached. In addition, the lower incoming kappa number of the enzyme pretreated samples allowed reductions of up to .816 kg ClO_2/ton pulp in the first D stage. This indicates that the bleaching of pulps receiving cellulase/hemicellulase treatment was more efficient, resulting in higher brightness pulps.

Conclusions

This study demonstrates that utilizing enzymes to increase the diffusion in sycamore sapwood can enhance kraft pulping while maintaining comparable strength properties. Enzyme pretreatments can increase the delignification in kraft pulping by up to 10%. Utilizing a pectinase preparation in addition to a cellulase/hemicellulase solution was found to be more effective than the cellulase/hemicellulase preparation alone. Pulps treated with a cellulase/hemicellulase solution required less chlorine dioxide than the control and pulps treated with cellulase/hemicellulase in addition to a pectinase solution to reach similar levels of brightness. Reducing the amount of chemicals utilized during bleaching is an integral step in the efforts to minimize the environmental impact of bleach plant effluent.

Acknowledgments

We thank Larry Jervis and William Bryant, NC State University, for supplying the wood used in this work, Kevin Holtman and John Windley, N.C. State University, for assistance with pulping.

Literature Cited

1. Stamm, A. J. *U.S. Dept. Agr. Tech. Bull.* **1946**, No. 929.
2. Nicholas, D. D.; Thomas, R. J. Proc. *Amer. Wood Pres. Assoc.* **1968**, *64*, pp. 70-76.
3. Meyer, R.W. *Wood Science* **1974**, *6*(3), pp. 220-230.
4. Jutte, S.M.; Levy, J. F. *Acta. Bot. Neeri* **1971**, *20*(5), pp. 453-466.
5. Yamasaki, T.; Hosoya, S.; Chen C. L.; Gratzyl, J.S.; Chang, H. In *1981 International Symposium Wood Pulp. Chem.* SPCI, Stockholm, **1981**, *Vol. 2*, p. 34.
6. Gellerstedt, G.; Lindfors, E. *Holzforschung* **1984**, *38*, p. 151.
7. Goring, D. *Appita* **1985**, *38*, p. 31.
8. Proctor, A.; Yean, W.; Goring, D. *Pulp and Paper Magazine Canada* **1967**, *68*, p. T445.
9. Boutelje, J.; Eriksson, I. *Holzforschung*, **1984**, *38*, p. 249.
10. Sorvari, J.; Sjostrom, E.; Klemola, A.; Laine, J. *Wood Sci. Technol.* **1986**, *20*, p.35.
11. Obst, J. *TAPPI J.* **1985**, *68*, p. 53.
12. Kerr, A.; Goring, D. *Can. J. Chem.* **1975**, *53*, pp. 952-959.

13. Poutanen, K.; Puls, J. *Appl. Microbiol. Biotechnol.* **1988**, *28*, pp. 425-432.
14. Lowry, O.; Rosebrough, N.; Parr, A.; Randell, R., J. *Biol. Chem.* **1951**, *193*, pp. 265-275.
15. Mandels, M.; Andreotti, R.; Rosch, C. *Biotech. and Bioeng. Symposium* **1976**, *6*, pp. 21-33.
16. Sinner, M.; Puls, J. *J. of Chromatography* **1978**, *156*, pp. 197-204.
17. Jacobs-Young, C. J.; Joyce, T. W.; Venditti, R. A. Effect of Enzymatic Pretreatment on the Diffusion of Sodium Hydroxide in Wood, *TAPPI J*, forthcoming.
18. Talton, J.H. *The Diffusion of Sodium Hydroxide in Wood at High pH as a Function of Temperature and Degree of Pulping*, MS Thesis, North Carolina State University, **1986**, pp. 86.
19. Viikari, L.; Kantelinen, A.; Sundquist, J.; Linko, M. *1994 Federation of European Microbiological Societies Microbiology Reviews* **1994**, *13*, pp. 335-350.
20. Boucai, E. *Pulp Paper Mag. Canada* **1971**, *72*, pp. 73-76.
21. Cowan, W. F. *CPPA Annual Meeting, Technical Section, CPPA, Montreal*, **1988**, p. A149.
22. Gurnagul, N.; Page, D. H. *TAPPI J.* **1989**, *72*, p. 16.

Chapter 6

Characterization of Fungal Cellulases for Fiber Modification

Martin Schülein, M. S. Kauppinen, L. Lange, S. F. Lassen, L. N. Andersen, S. Klysner, and J. B. Nielsen

Novo Nordisk A/S, Novo Alle, DK 2800 Bagsvaerd, Denmark

Monocomponent recombinant fungal cellulases have recently been introduced in the textile and detergent industries. Via cloning from many fungal sources we have obtained a wide range of cellulases from family 5, 6, 7, 12 and 45 cellulases. Family 45 (EG V) is the most aggressive; family 7 (EG I) the least aggressive endoglucanase. EG V cellulases were found in more than 50 fungal sources, e.g. *Myceliophthora, Thielavia, Macrophomina, Crinipellis* and *Acremonium*. The properties of five of these new cloned endoglucanases were compared with EG V from *Humicola insolens*. They all showed color clarification in detergent and good abrasion on denim. The performance of EG I from *Fusarium, Myceliophthora* and *Humicola* is alike: very little tensile strength loss and only little abrasion; only loose microfibrils are removed. These enzymes can be used in detergents (particularly for soil removal), for stonewashing (backstain control), or in deinking of used paper.

A very important industrial use of cellulolytic enzymes is for treatment of cellulosic textile or fabric. They are used as ingredients in detergent compositions or fabric softener compositions, in textile finishing with the prominent applications within stone-washing of denim jeans and in biopolishing of cellulosic fabric in order to obtain anti pilling effects, softness or special surface finishes. Another important industrial use of cellulolytic enzymes is for treatment of paper pulp, for instance for improving the drainage or for deinking of recycled paper (1).

Humicola insolens cloned monocomponent cellulases (e.g. EG I and EG V) have been introduced both in the textile industry and in the detergent industry.

The present work describes the cloning and characterization of homologous cellulases from other fungal sources and characterization of those on well defined substrates.

Cellulases can be classified into families based on sequence hydrophobe cluster analysis (2). Today, when many 3-D structures of cellulases have been solved, it can be seen that the family system really reflects the motif of the folding and the mechanism or mode of action of the different cellulases.

The *Humicola insolens* EG V is used in the textile industry because of its very efficient removal of fuzz and preventing pilling on the cotton fiber. By using a single endoglucanase for treatment of for instance Tencel blended fabrics only small fibrils are removed rather than whole fibers such as the case when using acid cellulase mixtures from *Trichoderma.* The strength loss of the treated fabric is therefore kept at a minimum when using such cellulase.

The mechanism of *Humicola* EG V has been visualized by cross section and immuno gold labeling of cotton fibers treated with the enzyme. No gold particles could be detected in the interior of the fiber which indicate that the EG V only binds to the surface and hydrolyses the amorphous cellulose on the surface (3).

Cloning of Endoglucanases in Yeast by Expression Method

The endoglucanase V and I from *Myceliophthora thermophila*, EG Va from *Acremonium* sp., EG V from *Macrophomina phaseolina* and EG V from *Thielavia terrestris* were cloned in yeast by monitoring cellulase activity (4). The four fungal sources represent three different orders within the Ascomycetes.

The mRNA was isolated from the fungi, *Acremonium sp., Macrophomina phaseolina, Myceliophthora thermophila* and *Thielavia terrestris*, respectively, after they were grown in a cellulose-containing fermentation medium with agitation to ensure sufficient aeration. Mycelia were harvested after 3-5 days' growth, immediately frozen in liquid nitrogen and stored at -80°C. The double-stranded cDNA was synthesized from 5 μg of poly$(A)^+$ RNA as described in (5) except that 1.5 μg of oligo$(dT)_{18}$-*Not* I primer from Pharmacia was used in the first strand reaction. The adapted cDNAs were digested with *Not* I, size-fractionated by agarose gel electrophoresis, and then ligated into *Bst*XI/*Not* I-cleaved pYES 2.0 vector from Invitrogen, USA. The ligation mixtures were transformed into electrocompetent *E. coli* DH10B cells from Gibco BRL, USA, resulting in libraries consisting of approx. 10^6 individual clones with a vector background of 1%.

Qiagen purified plasmid DNA from cDNA library pools of 10 000 - 15 000 colony forming units was transformed into *S. cerevisiae* strain W3124 (6) by electroporation (7), and the transformants plated on SC-plates (8) containing 2 % glucose. 50-100 plates containing 250-400 yeast colonies were obtained from each cDNA library.

Endoglucanase-positive colonies were identified and isolated on SC-agar (8) plates containing 2 % galactose using an assay with AZCL HE cellulose from (Megazyme International Ltd. Ireland) in the plates and the positive clones were identified by blue halos. The positive yeast colonies were characterized for activity on different substrates like AZCL xyloglucan and AZCL barley beta glucan also from Megazyme.

Total DNA from the yeast colonies was isolated and plasmid DNA was rescued by transformation in *E. coli*. Each cDNA clone was characterized by sequencing the 5'-end. The cDNA clones representing different families of clones were sequenced from both strands and selected for expression in *Aspergillus*. Plasmid DNA was digested with appropriate restriction enzymes, size fractionated on a gel, and fragments corresponding to the endoglucanase genes from *Myceliophthora thermophila, Acremonium sp.*, *Thielavia terrestris* and *Macrophomina phaseolina,* respectively, were isolated and subsequently ligated to pHD414 (an *Aspergillus oryzae* expression vector described in (4)).

Two family 45 endoglucanase genes was found in *Acremonium sp.* However, only one of the genes has been expressed at high levels in *Aspergillus* (see below) and has been further characterized (EG Va).

Cloning of EG V from *Crinipellis scabela*

Cloning of the EG V from the basidiomycetes *Crinipellis scabela* was obtained by construction of a cDNA library as described above, followed by screening by colony hybridization using a random-primed ^{32}P-labeled PCR product from *Crinipellis* as a probe. The PCR primers were based on 2 conserved family 45 sequences. The positive cDNA clones were characterized by sequencing the ends of the cDNA inserts, and by determine the nucleotide sequence of the longest cDNA from both strands. The native EG V cellulase from *Crinipellis* has neither a linker nor a cellulose binding domain. Thus, a gene fusion between the *Crinipellis* cellulase gene and the linker plus CBD from Humicola insolens EG V has been constructed using splicing by overlap extension (9).

Cloning of Fusarium EG I

The cloning of EG I from *Fusarium oxysporium* has been described in (10)

Expression of the Cellulase Genes in A. oryzae Transformants

All the enzymes were produced by cloning the cDNA into a plasmid with the cDNA inserted between the fungal amylase promoter and the AMG terminator from *A. niger* (11), followed by transformation into *Aspergillus oryzae.* The cellulases with CBD were purified using Avicel affinity chromatography and the endoglucanases without CBD were purified using cation exchange chromatography as described for *Humicola* cellulases (12)

Composition of the New Endoglucanases.

In table 1 the composition of the new cloned family 7 and 45 endoglucanases is shown and compared with the published data from *Humicola insolens* EG I and EG V and *Fusarium oxysporium* EG I.

Table I. Domain composition of the cloned fungal endoglucanases

	Number of residues			Molar Extinction coefficient
	Core	Linker	CBD	$M^{-1}*cm^{-1}$
Humicola EG I[a]	415			66,300
Humicola EG V[a]	213	33	38	61,300
Acremonium EG Va	208	36	35	68,020
Crinipellis EG V	204	35	37	54,320
Fusarium EG I [b]	409			58,060
Myceliophthora EG I	436			62,470
Myceliophthora EG V	207	34	38	74,950
Macrophomina EG V	203	34	38	85,340
Thielavia EG V	211	30	37	52,470

[a] (12), [b] (10).

The number of amino acid residues in the 3 domains (the catalytic core, the linker and the cellulose binding domain (CBD)), is based on sequence alignment to homologous cellulases with known function. The molar extinction coefficient was calculated based on the amino acid composition deduced from the DNA sequence.

The family 45 fungal cellulases are characterized by having a catalytic domain of between 202 and 214 amino acid residues folded in the same motif as *Humicola insolens* EG V (13), the new sequences show between 65% and 80% homology with the known *Humicola* sequences. Of the 21 amino acids which directly or indirectly (via water) interact with the cellohexaose in the published structure (13) 20 are conserved in all the 5 new EG V cellulases. The core is followed by a O-glycosylated flexible linker region of 30 to 36 amino acids rich in serines, threonines and prolines in *Acremonium* and *Thielavia*, the other three family 45 cellulases are hybrids with linker and CBD from *Humicola*. At the C-terminal end a fungal type cellulose binding domain (CBD) of 38 amino acid residues with high sequence homology to CBD of *Trichoderma reesei* CBH I (14).

The family 7 cellulase 3-D structures have also been solved, the structure of *Fusarium* EG I with substrate complexes (15, 16) and the 60% sequences homologous *Humicola* EG I structure (17). In contrast to the homologous *Trichoderma* EG I all three cloned genes from *Humicola, Fusarium* and *Myceliophthora* do not code for a cellulose binding domain (CBD).

Catalytic Properties of Endoglucanases

The endoglucanases have catalytic activity against soluble substrates. The EG I cellulases have only 4 subsites in the active site (15, 16, 17) so cellotriose can be used as a substrate for those cellulases (12, 18, 19). Data are shown in Table II.

Table II. Kinetic constant of family 7 cellulases for cellotriose

	k_{cat} s^{-1}	K_M μmole	k_{cat} / K_M
Humicola EG I	3.8	90	0.12
Fusarium EG I	17	400	0.04
Myceliophthora EG I	5.5	379	0.03

Steady state kinetics were determined at 40° C in 0.1 M sodium phosphate buffer pH 7.5 for 60 to 300 sec. using 6 different cellotriose concentrations (25 and 500 μM). The formation of glucose was simultaneously determined by a coupled reaction using glucose oxidase, peroxidases and ABTS, as described (12). The kinetic constants were calculated using Grafit program (20) with standard errors less than 10%. The catalytic activity of *Humicola* EG I at pH 8.0 is published in (12) and is not as high as the catalytic activity at pH 7.5 indicating that the pH optimum for *Humicola* EG I is lower than pH 8.0. Fusarium EG I has also higher catalytic properties at pH 7.5 than at pH 8.5 as published in (19).

The catalytic properties of the 5 new cloned EG V cellulases were determined at pH 7.5 using reduced cellohexaose as substrate and compared with the published data of *Humicola* EG V, see table III.

Table III Kinetic constants for family 45 cellulases using reduced cellohexaose (cellohexaitol) as substrate at pH 7.5.

	K_{cat} s^{-1}	K_M μmole	k_{cat} / K_M s^{-1} $\mu mole^{-1}$
Humicola EG V	18[a]	140[a]	0.13[a]
Acremonium EG Va	13	49	0.27
Crinipellis EG V	3	700	0.004
Macrophomina EG V	15	52	0.3
Myceliophthora EG V	17	100	0.17
Thielavia EG V	3	130	0.02

[a] Data from (12).

The results were obtained using a coupled assay with cellobiose dehydrogenase and the colored electron acceptor 2,6-dichloroindophenol as described by Schou et al 1993 (18). Experiments were performed in 0.1 M sodium phosphate buffer pH 7.5 with 82 μM 2,6-dichloroindophenol. The substrate was cellohexaitol in concentrations between 0.2x K_M to 5 x K_M. Incubation 200 to 600 sec at 40°C. The kinetic constants were calculated using Grafit program (20) with standard errors less than 10%.

The new EG V from *Myceliophthora* and *Acremonium* have identical catalytic properties as *Humicola*, whereas *Thielavia* and *Crinipellis* has lower k_{cat} indicating that pH 7.5 is not optimal.

The activities with CMC as substrate were determined under steady state conditions. Using varying concentrations of substrate the apparent kinetic constants could be calculated. The assay was done at two different pH values both under acid conditions pH 5.5 using 0.1 M sodium acetate and at neutral conditions (pH 7.5) using 0.1 M sodium 3-[N-Morpholino]propanesulfonic acid (MOPS) buffer. The incubation was for 20 min at 40°. By using a very sensitive detection of reducing sugars with the reagent *p*-hydroxy benzoic acid hydrazide (PHBAH) (12, 21) we assume steady state conditions due to the small total turnover of substrate. For each CMC concentration (from 0.1 g l^{-1} to 15 g l^{-1}) the assay was done in duplicate with double blind background, table IV.

Table IV. Apparent kinetic constants on CMC at pH 5.5 and 7.5

	pH 5.5		pH 7.5	
	$k_{cat\ (app.)}$ s^{-1}	$K_{M\ (app.)}$ g l^{-1}	$k_{cat\ (app.)}$ s^{-1}	$K_{M\ (app.)}$ g l^{-1}
Humicola EG I	89	1.5	86	3.0
Humicola EG V	38	0.7	55	0.9
Acremonium EG Va	11	0.3	26	1.5
Crinipellis EG V	39	1.8	27	0.9
Fusarium EG I	85	5.6	61	1.5
Macrophomina EG V	31	0.7	70	1.4
Myceliophthora EG I	75	5.6	58	13
Myceliophthora EG V	35	1.3	22	0.5
Thielavia EG V	20	1.1	19	1.4

After determination of reducing sugars, the apparent kinetic constants were calculated using the Grafit program (20). The concentrations of substrate were between 0.2 * K_M to max. 15 g/l with standard errors less than 10%.

The result indicate that the family 7 cellulases, without CBD, have higher catalytic activity on CMC compared with family 45. However, the family 45, with the CBD has the lowest K_M

Thielavia has much higher catalytic activity on CMC compared with the data from reduced cellohexaose at pH 7.5.

Finally the apparent kinetic constants were determined using phosphoric acid swollen cellulose (PASC) as substrate, see table V.

The results were obtained after incubation for 20 min at 40°C at either pH 4.5 in 0.1 M sodium acetate buffer or at pH 8.5 in 0.1 M barbiturate buffer. After determination of reducing sugars using the PHBAH reagent (21), the apparent kinetic constants were calculated using the Grafit program (20). The concentrations of substrate were between 0.08 g/l to 8 g/l with standard errors less than 10%. The data indicate that family 45 endoglucanases with CBD have the highest catalytic rate on insoluble PASC whereas the family 7 endoglucanases without CBD have the highest rate on soluble CMC although the determination are not done at the same pH. The presence of the CBD does not change the catalytic rate, this has been shown by removal of the CBD (12). However, on V, which is not active at high pH levels has a higher rate on CMC at pH 7.5 than on PASC at pH 8.5. The *Thielavia* and *Crinipellis* EG V are candidates for application at low pH, whereas *Humicola, Acremonium, Myceliophthora* are candidates for neutral and alkaline industrial applications.

Table V. Apparent kinetic constants on phosphoric acid swollen cellulose (PASC).

	pH 4.5		pH 8.5	
	$k_{cat\,(app.)}$ s^{-1}	$K_{M\,(app.)}$ $g\,l^{-1}$	$k_{cat\,(app.)}$ s^{-1}	$K_{M\,(app.)}$ $g\,l^{-1}$
Humicola EG I	43	2.7	16*	2.5*
Humicola EG V	46	0.3	57*	0.5*
Acremonium EG Va	36	0.2	21	0.5
Crinipellis EG V	37	0.1	26	0.9
Fusarium EG I	>40	>8	16	2.5
Macrophomina EG V	40	0.3	44	0.5
Myceliophthora EG I	>30	>8	15	3.0
Myceliophthora EG V	33	0.2	39	0.5
Thielavia EG V	17	0.1	8	0.2

*The *Humicola* data at pH 8.5 is from (12).

The native structure of family 45 cellulase *Humicola insolens* EG V was determined in 1993 (22) and later 3 other structures were determined, two of which contained substrate bound in the active site (13). The family 45 endoglucanases have an active site which is a long (not deep) groove (40 Å) giving space for 7 glucose units. The groove is covered by a flexible loop during catalysis. The data obtained here indicate that all six family 45 endoglucanases, which all must have this fold, have the optimal activity with PASC as substrates. Using substrates of lower molecular mass such as reduced cellohexaose, the catalytic rate is reduced (all the 7 subsites not in contact with substrate). Using a high molecular mass, soluble substituted substrate like CMC, which is cellulose with a degree of polymerization around 200 (like the PASC), the catalytic rate is also reduced a little.

This could be due to either steric hindrance from the substitutions or to the electrostatic repulsion from the charges of the carboxyl groups of CMC.

The family 7 endoglucanase structures (14,15,16) have a different folding motif. Here the active site groove is much deeper with no flexible loop to cover the groove and a much shorter groove with space for only 4 glucose units. These endoglucanases have optimal activity on CMC, around twice as high as on PASC indicating that the solubility of the substrate is more important than the substitution of the glucose units.

Conclusions

All family 45 endoglucanases can be used for color clarification, fuzz removal in detergent applications and for increased abrasion in denim finishing due to their efficient removal of fuzz and indigo from the surface of cotton. The pH level during application will decide the selection of enzyme. The use of well defined substrates for characterization of the endoglucanases is necessary for selection of the right candidate for different industrial applications. The family 7 endoglucanases accomplish very little color clarification and abrasion but they are very efficient in detergent as cleaning cellulases by stain removal and for preventing back staining in Denim wash. They have also shown efficient performance in de-inking of office waste paper. The use of the well defined substrate for characterization of an endoglucanases can be useful for the choice of the right candidate for industrial application.

Literature Cited

1. Jeffries, T. W.; Klungness, J. H.; Sykes, M. S.; Rutledgecropsey, K. R. *TAPPI J.,* **1994,** *77*, pp 173-179.
2. Henrissat B.; Bairoch A. *Biochemical J.* **1993,** *293*, pp 781-788.
3. Boisset, C.; Chanzy, H.; Schülein, M.; Henrissat, B. *Cellulose,* **1997**,*4*, pp 7-20.
4. Dalboge, H.; Heldthansen, H. P.; *Molecular General Gen.* **1994**, *243*, pp 253-260.
5. Kofod, L. V.; Kauppinen, S.; Christgau, S.; Andersen, L. N.; Heldt-Hansen, H. P.; Dörreich, K.; Dalbøge, H. J.; *Biol. Chem..* **1994**, *269*, pp 29182-29189.
6. van den Hazel H. B.; Kielland-Brandt, M. C.; Winther, J. R. *Eur. J. Biochem.* **1992**, *207*, pp 277-283.
7. Becker, D. M.; Guarente, L. *Methods in Enzymol.* **1991,** *194*, pp 182-187.
8. Sherman, F. *Methods in Enzymol.* **1991,***194*, pp 3-21.
9. Horton, R. M.; Hunt, H. D.; Ho, S. N.; Pullen, J. K.; Pease, l. R. *Gene* **1989,** *77*, pp 61-68.
10. Sheppard, P.O.; Grant, F. J.; Oort, P. J.; Sprecher, C. A.; Foster, D. C.; Hagen, F. S.; Upshall, A.; Mcknight, G. L.; Ohara, P. J. *Gene*, **1994,** *150*, pp 163-167.
11. Christensen, T.; Wöldike, H.; Boel, E.; Mortensen, S. B.; Hjortshøj, K.; Thim, L.; Hansen, M.T.; *Biotechnology*, **1988,** *6*, pp 1419-1422.
12. Schülein, M. *J. Biotechnology,* **1997,** in press.
13. Davies, G. J.; Tolley, S. P.; Henrissat, B.; Hjort, C.; Schülein, M. *Biochemistry* **1995,** *34*, pp 16210-16220.
14. Kraulis, P.; Clore, G. M.; Nilges, M.; Jones, T. A.; Pettersson, G.; Knowles, J.; Gronenborn, A. M. *Biochemistry*, **1989,** *28*, pp 7241-7257.
15. Sulzenbacher, G.; Driguez, H.; Henrissat, B.; Schulein, M.; Davies, G. J. *Biochemistry,* **1996,** *35*, pp 15280-15287.
16. Sulzenbacher, G.; Schülein, M.; Davies, G. J. *Biochemistry*, **1997,** *36*, pp 5902-5911.

17. Davies, G. J.; Tolley, S. P.; Divne, C.; Alwyn, J.; Wöldike, H. F.; Schülein, M. Paper submitted.
18. Schou, C., Rasmussen, G., Kaltoft, M. B., Henrissat, B. and Schülein, M. European Journal of Biochemistry **1993,** *217*, 947-953
19. Mackenzie, L. F., Davies, G., Schülein, M. and Withers, S. G. Biochemistry **1997,** 36, 5893-5901
20. Leatherbarrow, R. J. Grafit version 3.0 Erithacus Software Ltd. Staines, U.K. **1992**.
21. Lever, M. Anal. Biochem. **1972,** *47*, 273-279
22. Davies, G. J., Dodson, G. G., Hubbard, R. E., Tolley, S. P., Dauter, Z., Wilson, K. S., Hjort, C., Mikkelsen, J. M. and Schülein, M. Nature **1993,** *365*, 362-364

Chapter 7

The Synergistic Effects of Endoglucanase and Xylanase in Modifying Douglas Fir Kraft Pulp

Catherine D. Edgar, Shawn D. Mansfield, Georg M. Gübitz, and John N. Saddler[1]

Chair of Forest Products Biotechnology, Department of Wood Science, University of British Columbia, 270-2357 Main Mall, Vancouver, British Columbia V6T 1Z4, Canada

Kraft pulp from Douglas-fir (Pseudotsuga menziesii) was treated with an enzyme preparation containing both cellulase and xylanase activities, as well as individual endoglucanase and xylanase components to assess their potential to enhance both pulp characteristics and paper properties. The use of individual components greatly reduced both yield and strength losses when compared to treatments with the crude cellulase preparation. Xylanase treatments enhanced fiber flexibility, resulting in denser paper sheets with a lower light scattering coefficient. Although treatments with the endoglucanase decreased individual fiber integrity, enhanced fiber collapsibility was obtained. When the purified xylanase and endoglucanase preparations were used in combination, the action of the xylanase increased fiber accessibility to the endoglucanase. The combined action of the two enzymes resulted in enhanced bonding, indicated by as much as 10% increase in burst index over the level obtained by xylanase alone.

Enzyme applications in the pulp and paper industry have been steadily increasing over the last twenty years. Although most areas are still evolving, several applications are being aggressively pursued, including the reduction of pitch deposits (*1*), drainage enhancements (*2*), deinking of secondary fibers (*3*), enhancing pulp bleaching (*4*), and modifying fiber characteristics (*5, 6*). Enhancing bleaching with the use of xylanases has been the most successful endeavor to date, with several mills routinely using xylanase prebleaching (*7*).

The continued use of xylanases in alternative bleaching strategies has inspired considerable research to try to both elucidate the enzymatic bleaching mechanism (*8-10*) and provide a better understanding of fiber morphology (*11, 12*). This increased knowledge about the action of xylanase enzymes can also be used to enhance enzymatic fiber modification.

[1]Corresponding author.

Various theories have been proposed to explain the enzyme mechanism involved in fiber modification, with the predominant suggestion being that the fibrils and fiber bundles are attacked on the surface, peeling off subsequent layers and eventually leading to disintegration of the fibers (*13, 14*). This enzymatic cleavage can lead to increased freeness (*2, 6, 15-17*) and enhanced fiber flexibility, resulting in denser paper sheets (*6*). However, enzyme treatments can also result in a decrease in yield and fiber strength. One probable reason for the observed strength loss is that commercial cellulase systems, originally intended for use in cellulose hydrolysis, have been used to try to obtain selective fiber modification. One possible solution to this problem would be to use individual cellulase components to obtain beneficial fiber changes while minimizing yield and strength losses in the pulp.

Previous work by other researchers found that pure endoglucanases alone did not give rise to high levels of hydrolysis (*5*) and indicated that the action of endoglucanases was necessary to achieve improved pulp freeness (*17*). Related work also indicated that xylanases may play an important role in fiber modification (*18*) and that xylanase enzymes selectively attack xylan, leaving the cellulose relatively intact, preserving the strength of the fibers (*19*).

In the work reported in this paper, the effects of a cellulase/hemicellulase preparation were compared to the action of pure endoglucanase and xylanase enzymes when added to Douglas-fir kraft pulp. In this way, the individual components were assessed for their ability to enhance the fiber characteristics of the pulp while limiting both yield and strength losses. The effects of the xylanase and endoglucanase preparations were assessed both separately and in combination to distinguish their individual actions and determine any synergistic effects.

Materials and Methods

Pulp. Chemical pulp (kraft) derived from Douglas-fir (Pseudotsuga menziesii) was obtained from the Croften mill (Fletcher Challenge), British Columbia, Canada. The sugar composition of the pulp was determined by acid hydrolysis and subsequent HPLC analysis, yielding 72.6% glucose, 5.95% xylose, 0.41% arabinose, 0.56% galactose, and 6.66% mannose. The total lignin content was 5.4%.

Enzymes. Novozyme SP342, SP476, and SP613, derived from Humicola insolens, were obtained from Novo Nordisk, Denmark. Ecozyme was obtained from Thomas Swan and Co., England. Xylanase I was obtained from Primalco, Finland. The enzyme activities of these preparations were measured on carboxymethylcellulose (2% CMC, Sigma), xylan (1% birchwood xylan, Sigma), and filter paper (No. 1 Whatman) using methods described previously (*20*). The assays were carried out at pH 7.0. Total protein in solution was determined using the bicinchoninic acid protein assay (*21*).

Enzyme treatment of pulp. The pulp slurries (3% consistency in 50 mM phosphate buffer, pH 7.0) were treated for 1 hour at 50 °C under continuous agitation, with a range of enzyme loadings based on activity (CMC or Xylanase IU) per gram oven dry fiber.

All enzyme reactions were stopped by boiling for 10 minutes. Control pulps were similarly treated, without the addition of enzymes. The reaction filtrates were acid hydrolyzed and analyzed for sugars released by anion-exchange chromatography on a CarboPac PA-1 column using a Dionex DX-500 HPLC system (Dionex, Sunnyvale, California, USA), using fucose as an internal standard.

Pulp Testing. Pulp freeness was measured at 20 °C according to TAPPI T 227 om-94 (CSF tester, R. Mitchell Co.). Handsheets were prepared according to TAPPI T 205 om-88, with the white water recycled to ensure that the fines were included in the resultant handsheet. Light scattering coefficient was measured at 681 nm (Carl Zeiss, Elrepho). For the determination of burst index, tear index, zero-span breaking length, and density tests were conducted according to TAPPI standard methods.

Results and Discussion

Although several endoglucanase and xylanase preparations were initially evaluated, subsequent analysis of the individual activities and specificities of the various enzymes (Table I) indicated the high purity of the SP613 (endoglucanase) and Ecozyme (xylanase) preparations. Two other preparations (SP476 and Xylanase I) were also evaluated, as they were used to compare the effects of different enzymes with similar activity profiles. The crude cellulase enzyme (CX) and the endoglucanase (EG) were loaded at 15 IU CMC/g of pulp. At this dosage, the crude cellulase preparation (CX) had 800 IU xylanase activity. Therefore, the xylanase enzyme (X) was loaded at 800 IU xylanase/g of pulp, to ensure that the activities of the individual components were comparable to those of the crude cellulase preparation. Similarly, xylanase I was loaded at 800 IU xylanase/g of pulp.

Table I. Protein content and activities of various enzyme preparations

Commercial Name	*Description*	*Protein (mg/mL)*	*Endoglucanase (IU/mL)*	*Xylanase (IU/mL)*	*Filter Paper (IU/mL)*
SP342	Cellulase/ xylanase (CX)	44.1	30	1610	2
SP613	Endoglucanase (EG)	24.2	20	2.1	0
Ecozyme	Xylanase (X)	5.7	0	6010	0
SP476	Endoglucanase (EG4)	22	17	9.8	0
Xylanase I	Xylanase (XI)	27	6.0	2690	0

The crude cellulase preparation greatly reduced burst index, while the endoglucanase and xylanase components caused no significant decrease, as compared to the control (Figure 1A). It is probable that the synergistic action of the endoglucanases and cellobiohydrolases within the crude cellulase preparation (CX) contributed to the observed strength reduction. Other workers have also shown that the combined action of cellobiohydrolases and endoglucanases was more damaging to pulp fibers than was the action of the endoglucanase alone (*17, 22*). The use of the individual enzymes, especially the endoglucanase, successfully curtailed pulp hydrolysis as only 0.37% and 1.9% yield loss was observed after respective endoglucanase and xylanase treatments (Table II). Both the crude cellulase preparation and the xylanase released similar amounts of xylose (Table II), solubilizing 27% and 29% respectively of the original xylose present in the pulp. However, when equal loadings of endoglucanase were added in either the form of the crude cellulase preparation or the purified endoglucanase, the crude cellulase preparation released 30 mg glucose/g pulp (4%), while the endoglucanase only released 1.4 mg glucose/g pulp (0.2%). This emphasized the key role that the synergistic action of the cellobiohydrolases plays in cellulose hydrolysis when it acts in combination with the endoglucanase.

Table II. Sugars released by enzymatic treatments of Douglas-fir kraft pulp

Enzyme Preparation	*Amount of sugar solubilized (mg/g pulp)*					*Yield loss (%)*
	Glucose	*Xylose*	*Arabinose*	*Galactose*	*Mannose*	
Cell/Xyl (15 CX)	29.5	15.5	1.29	0.77	3.35	5.0
Endo (15 EG)	1.37	1.04	0.12	0.45	0.72	0.37
Xylanase (800 X)	0.26	17.4	1.42	0.17	0	1.9
15 EG + 800 X	2.64	14.9	1.28	2.99	2.6	2.4
Xylanase I (800 XI)	2.18	12.5	1.06	0.26	0	1.6

Even at this low level of hydrolysis(0.2%), the endoglucanase was still able to modify the fiber characteristics, as indicated by reductions in both tear (32%) and wet zero-span (22%) strength (Figure 1B, C). This would seem to confirm the concerns about cellulase contamination in xylanase preparations reducing the strength of enzyme bleached pulps (*22*). However, some decrease in fiber strength may be the trade-off required to obtain gains in other pulp characteristics. For example, the endoglucanase caused a 22% decrease in individual fiber strength, yet it did not decrease bonding strength, as indicated by burst index (Figure 1A). The key to enzymatic fiber modification is achieving a compromise between the degradation of individual fibers and

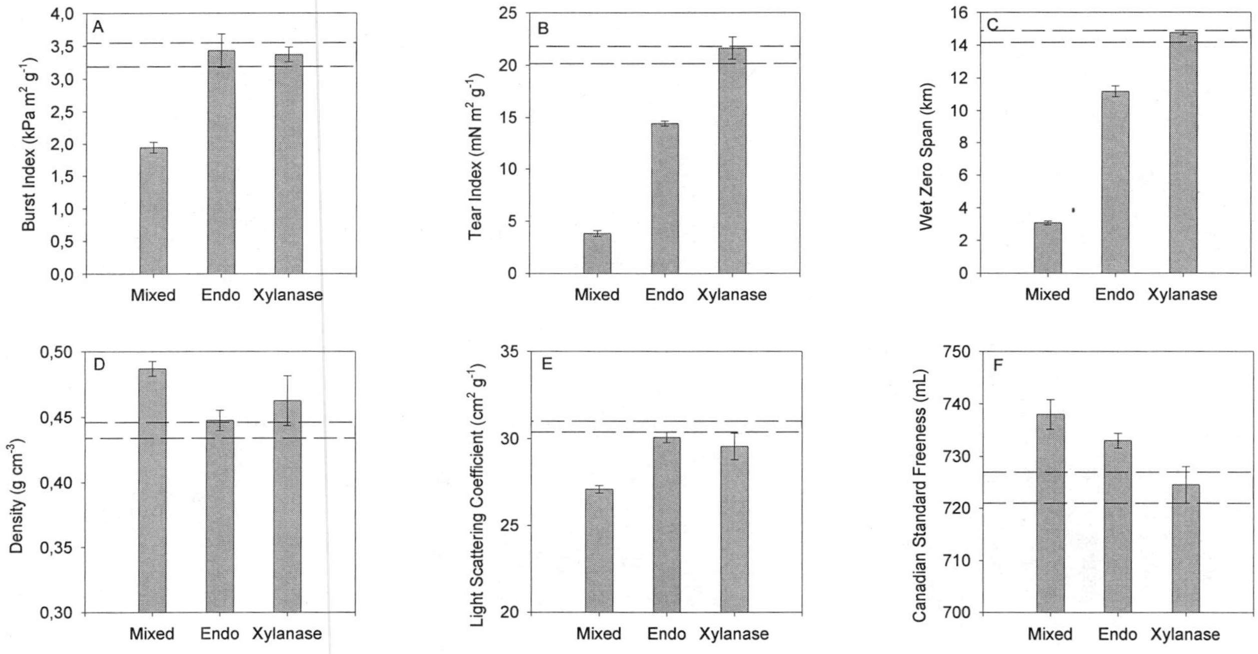

Figure 1. The effects of adding a crude cellulase preparation ("mixed"), an endoglucanase ("endo"), and a xylanase on the (A) Burst Index, (B) Tear Index, (C) Wet Zero-span, (D) Density, (E) Light Scattering Coefficient, and (F) Freeness of a Douglas-fir kraft pulp. (Error bars represent standard deviation of duplicate experiments, while the dashed lines represent the standard deviation of six control treatments).

the enhanced fiber flexibility/collapsibility gained through these treatments, resulting in enhanced interfiber bonding. This compromise has a sensitive balance point which is affected both by the type of enzyme and the enzyme loading that is used. Although the endoglucanase used in this investigation caused a decrease in fiber strength, bonding strength was not affected. It appeared that the enzymatically modified fibers lead to enhanced bonding which was able to compensate for the loss in fiber strength. In contrast, the crude cellulase preparation caused a substantial reduction in fiber strength as well as paper strength and it was probable that the large loss in fiber strength tipped the balance too far. As a result, any improvements in fiber bonding were not able to compensate for the observed loss in fiber strength.

It has been shown that the degree of fiber bonding within a sheet can be inferred from measurements of sheet density and light scattering coefficient, as more densely packed fibers are expected to have a greater degree of bonding (*23*). As light is scattered at air-fiber interfaces within a sheet, a decrease in light scattering coefficient indicates an increase in fiber-fiber bonding (*24*). Typically, fine fibers are more flexible and collapsible, forming denser sheets upon papermaking (*23*). Work by Mansfield et al. (*6*) has shown that the coarseness of Douglas-fir fibers can be reduced by enzymatic treatments, leading to denser paper sheets. The increases in density and decreases in light scattering coefficient (Figure 1D, E) indicated that the enzymatic treatments could enhance fiber flexibility, making the fibers more collapsible upon papermaking. Although the endoglucanase treatment was expected to provide increased bonding, compensating for the loss in fiber strength, this increase was not evident as there was no significant change in the density or light scattering coefficient. The most marked changes in both density and light scattering coefficient occurred with the crude cellulase preparation. However, the fiber strength was also greatly reduced after this treatment, compromising any beneficial effects arising from increased interfiber bonding.

Xylanase treatments seemed to have more potential than endoglucanase for increasing sheet density and decreasing light scattering coefficient (Figure 1D, E). Our previous work also found increases in sheet density following xylanase treatment (*18*). The xylanases were solubilizing xylan (Table III), likely from the surface of the pulp, due to the low levels of fiber degradation that were observed. Previous studies have shown that both xylan and lignin are precipitated on the surfaces of the fibers in the final stages of the kraft cook (*25, 26*). It has also been suggested that there are chemical links between the redeposited lignin and xylan in the form of lignin-carbohydrate complexes (*10, 27, 28*). Hence, the enzymatic removal of xylan may be accompanied by the release of lignin, whether as a result of direct linkages with the xylan, or through physical leaching resulting from increased fiber porosity after xylanase treatment (*9, 11, 29, 30*). It has also been shown that xylanase treatments result in an increased solubilization of lignaceous material, as detected by measuring the UV absorbency of the reaction filtrates at 280 nm (*18*). Furthermore, our previous work indicated that xylanase treatments appeared to increase accessibility to the pulp as it not only increased the overall liberation of lignin, but also the proportion of higher-molecular-mass compounds, as detected by size-exclusion chromatography (*18, 31*).

It appears that the removal of the stiff lignaceous material during xylanase treatments helps increase fiber flexibility, accounting for increased density and decreased light scattering coefficient (Figure 1D, E). This is in contrast to the results of Kibblewhite et al. (*32*) who found that the fibers were stiffer and had less bonding potential after xylanase treatment. However, Kibblewhite used another xylanase which may have acted differently, as it has been shown that different xylanases, even those produced by the same organism, can display quite different behaviors on pulp fibers (*33-35*). It has also been reported that different xylanase treatments result in different effects on paper strength, with some treatments showing increases (*36*), decreases (*37*), and no change in strength (*8, 38*).

Table III. Sugars released after enzymatic treatment of Douglas-fir kraft pulp with various doses of xylanase (X) alone and in combination with endoglucanase (EG)

Enzyme Preparation	*Amount of sugar solubilized (mg/g pulp)*					*Yield loss (%)*
	Glucose	*Xylose*	*Arabinose*	*Galactose*	*Mannose*	
100 X	0.11	10.7	0.82	0.10	0	1.2
400 X	0.18	14.7	1.07	0.10	0	1.6
800 X	0.26	17.4	1.42	0.17	0	1.9
100 X+15 EG	1.87	9.25	0.73	1.50	2.60	1.6
400 X+15 EG	1.95	11.6	0.94	1.46	2.96	1.9
800 X+15 EG	2.64	14.9	1.28	2.99	2.60	2.4

In addition to changes in paper properties, enzyme treatments have also been shown to increase pulp freeness (*2, 6, 15-17*). In our work, the crude cellulase preparation increased freeness by 14 CSF points, while the endoglucanase brought about a 9 CSF point increase. Although other workers (*17*) had found that the action of endoglucanases was synergistically enhanced by the addition of xylanases, when our xylanase and endoglucanase were used in combination the freeness values were lower than those achieved by the endoglucanase alone. This was similar to the work reported by Jackson et al., (*16*) who also found that a smaller freeness gain was obtained when a cellulase and hemicellulase preparation were used in combination. This result is consistent with the mechanism that the xylanases are liberating some lignin containing material, as removal of this particulate matter could contribute to the observed decrease in pulp freeness.

Having established that both the individual endoglucanases and xylanases resulted in some limited fiber modifications, we next assessed the possible synergistic effects of endoglucanase-xylanase interaction, as found in the crude cellulase preparation, without the possible interference of the exoglucanase component (Figure 2). Although the xylanase treatments alone did not significantly affect the burst index (Figure 2A), when it was supplemented with the endoglucanase, the burst index increased 7 to 10% above the xylanase level obtained with xylanase alone. As the greatest increase in burst

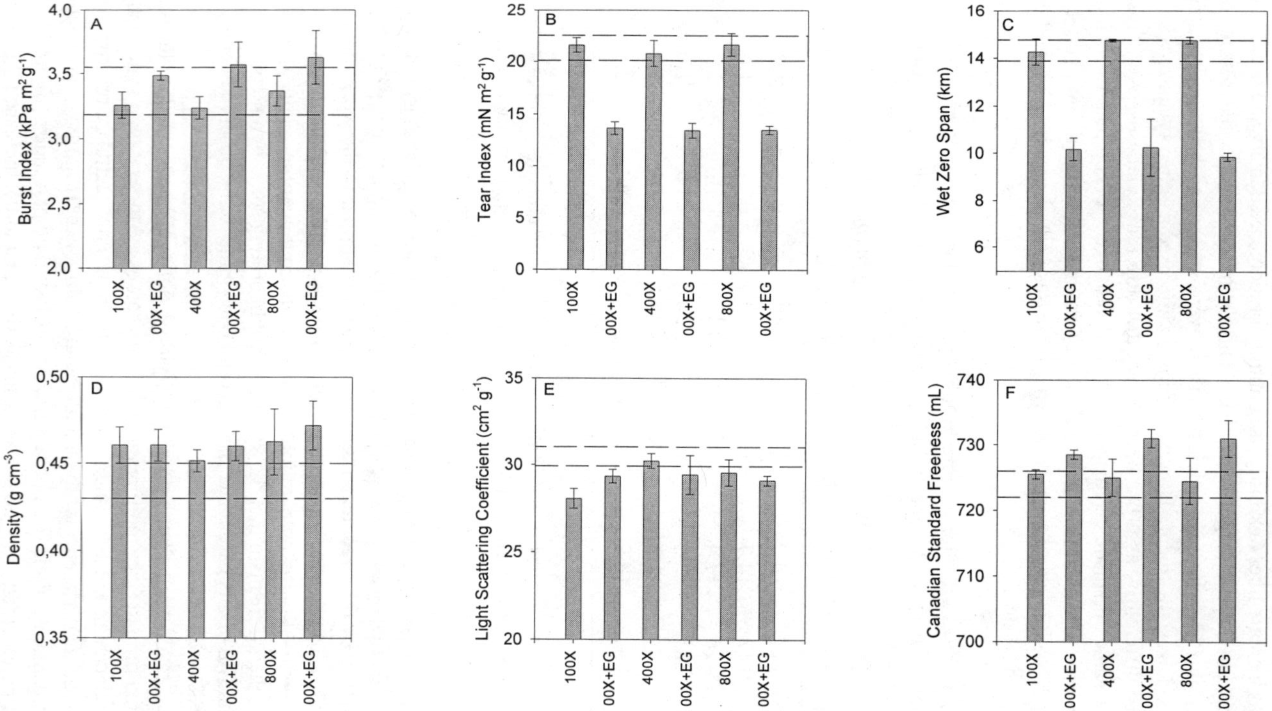

Figure 2. The effects of different xylanase (X) loadings alone and in combination with 15 IU/g endoglucanase (EG) treatments on the (A) Burst Index, (B) Tear Index, (C) Wet Zero-span, (D) Density, (E) Light Scattering Coefficient, and (F) Freeness of a Douglas-fir kraft pulp. (Error bars represent standard deviation of duplicate experiments, while the dashed lines represent the standard deviation of six control treatments).

strength was obtained with the 800 IU dose, this loading was selected for the subsequent work where the endoglucanase doses were varied while keeping the xylanase loading constant (Figure 3). Although there were no significant changes in burst index between the 10, 15, and 20 IU endoglucanase loadings, larger changes are evident when the endoglucanase was used in combination with the xylanase (Figure 3A). The combination of xylanase with the 15 IU endoglucanase dose gave the largest increase in burst index. It appears that these enzymes acted together synergistically to enhance interfiber bonding. It is possible that the xylanase increases fiber flexibility by removing lignin-carbohydrate complexes, while the endoglucanase altered fiber integrity by cleaving glucose bonds, resulting in more collapsible fibers.

It was apparent that the nature of the endoglucanase used is very important as under the same conditions and loading (15 IU/g pulp) another endoglucanase (SP476) resulted in a 56% reduction in fiber strength. Similarly, the purity of the xylanase can influence the results. Although Xylanase I (XI, 800 IU/g pulp) was contaminated with only a small amount of endoglucanase activity (Table I) treatment with this enzyme resulted in a 42% reduction in wet zero-span strength. It seemed likely that the endoglucanase activity contaminating this xylanase preparation was quite active as treatments with Xylanase I increased freeness by 8 CSF points. It was also apparent that even with only 1.8 IU/g endoglucanase activity, Xylanase I liberated more glucose than did the 15 IU/g endoglucanase activity in the pure EG preparation (Table II).

While the pure xylanase (X) did not affect the tear and wet zero-span values, the addition of the endoglucanase significantly decreased these properties (Figure 2B, C). This effect was more pronounced at both higher endoglucanase and higher xylanase doses. These results suggest that the action of the xylanase helps increase the accessibility of the substrate to the endoglucanase. This has been suggested previously by Suurnäkki et al. (*30*) who showed that xylanase treatments increased the median pore width of kraft pulp fibers. We found that increasing the xylanase dosage at a constant endoglucanase concentration lead to the liberation of more glucose (Table III), which is consistent with this proposed mechanism.

Although the action of the xylanase enzyme was the more important mechanism for increasing fiber flexibility, the combined effect of the xylanase and the endoglucanase seems to be required to significantly increase pulp density. The addition of the xylanase increased the paper density by 4 to 5% above the level achieved with the action of the endoglucanase alone (Figure 3D). The combined enzymes also gave a slight increase (1 to 2%) over the level attained with xylanase alone.

The synergistic interaction of the enzymes was also evident by the amount of galactose, mannose, and glucose liberated, with the combined treatment (xylanase and endoglucanase) solubilizing more than the additive amounts of the enzymes separately (Table III). The light scattering coefficient also showed a similar trend to that observed for density, with the combined enzyme treatments lowering light scattering coefficient more than did treatment with xylanase alone. It appears that the combined action of the xylanase and endoglucanase activities leads to increased fiber flexibility and enhanced fiber bonding within the sheet.

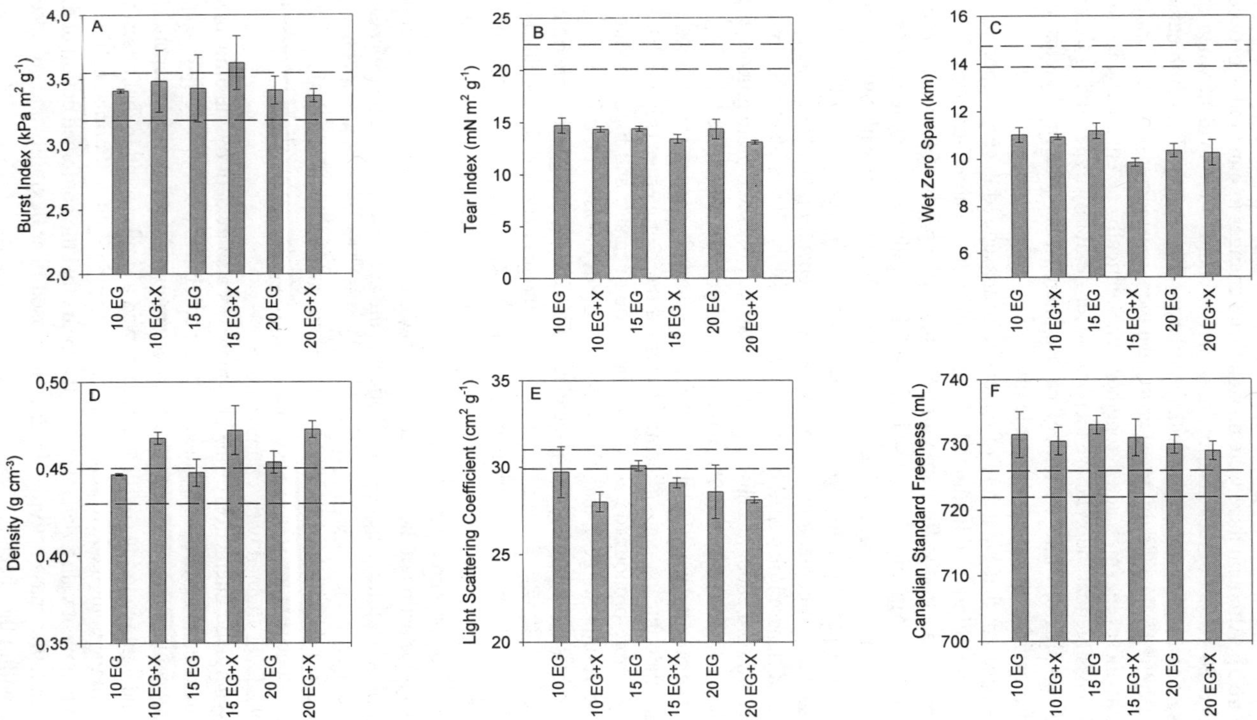

Figure 3. The effects of different endoglucanase (EG) loadings alone and in combination with 800 IU/g xylanase (X) treatments on the (A) Burst Index, (B) Tear Index, (C) Wet Zero-span, (D) Density, (E) Light Scattering Coefficient, and (F) Freeness of a Douglas-fir kraft pulp. (Error bars represent standard deviation of duplicate experiments, while the dashed lines represent the standard deviation of six control treatments).

Conclusions

It is apparent that the endoglucanase activity has the greater effect on burst index, tear index, wet zero-span, and freeness, while the action of the xylanase enzyme is essential for enhancing density and increasing fiber-fiber bonding, as indicated by light scattering coefficient. Even greater changes are obtained when these enzymes are used in combination, with their synergistic interaction causing significant changes in all of these properties. It is likely that the xylanase enzyme acts by releasing lignin-carbohydrate complexes, thus increasing both the flexibility of the fibers as well as increasing the access of the pulp to the endoglucanase. The endoglucanase acts by decreasing the integrity of the fibers and making them more collapsible under the physical pressure applied during papermaking. In the case of burst index, density, and light scattering coefficient, the level of fiber bonding seems to be increased by the action of the combined enzymes. Although the tear index and wet zero-span values decreased, these losses were not detrimental to paper quality as the enhanced bonding compensated for these changes. However, some endoglucanase and xylanase preparations were shown to be excessively destructive in their action on the pulp fibers. It can be concluded that the action of both xylanases and endoglucanases are required to obtain beneficial fiber modifications, however, the nature of the specific xylanase and endoglucanase used, as well as the loading of these enzymes, has a major influence on the resultant changes.

Acknowledgments

We would like to thank the Pulp and Paper Research Institute of Canada (Vancouver) for the use of their handsheet making and testing facilities and MacMillan Bloedel for financial support.

Literature Cited

1. Farrell, R. L.; Blanchette, R. A.; Brush, T. S.; Gysin, B.; Hadar, Y.; Perollaz, Y.; Wendler, Y.; Zimmerman, W. 5th International Conference on Biotechnology in the Pulp and Paper Industry, Kyoto Japan, May 27-30, **1992**, pp 27-32.
2. Pommier, J.-C.; Fuentes, J.-L.; Goma, G. *TAPPI J.,* **1989**, *72(6)*, pp 187-191.
3. Prasad, D.; Heitmann, J. A.; Joyce, T. W. *Nordic Pulp Paper Res. J.*, **1993**, *8(2)*, pp 284-286.
4. Viikari, L.; Sundqvist, J.; Kettunen, *J. Paperi Ja Puu,* **1991**, *73(5)*, pp 384-389.
5. Pere, J.; Siika-aho, M.; Buchert, J.; Viikari, L. *TAPPI J.*, **1995**, *78(6)*, pp 71-78.
6. Mansfield, S. D.; Wong, K. K. Y.; de Jong, E.; Saddler, J. N. *TAPPI J.,* **1996**, *79(8)*, pp 125-132.
7. Tolan, J. S.; Olson, D.; Dines, R. E. *Pulp Paper Can.*, **1995**, *96(12)*, pp T403-T406.
8. Paice, M. G.; Gurnagul, N.; Page, D. H.; Jurasek, L. *Enzyme Microb. Technol.* **1992**, *14*, pp 272-276.

9. Kantelinen, A.; Hortling, B.; Sundquist, J.; Linko, M.; Viikari, L. *Holzforschung,* **1993**, *47*, pp 318-324.
10. de Jong, E.; Wong, K. K. Y.; Saddler, J. N. Holzforschung 1997, 51, pp 19-26.
11. Buchert, J.; Carlsson, G.; Viikari, L.; Ström, G. *Holzforschung*, **1996**, *50*, pp 69-74.
12. Laine, J.; Buchert, J.; Viikari, L.; Stenius, P. *Holzforschung*, **1996**, *50*, pp 208-214.
13. Lee, S. B.; Kim, I. H.; Ryu, D. D. Y.; Taguchi, H. *Biotechnol. Bioeng.*, **1983**, *23*, pp 33-51.
14. Chanzy, H.; Henrissat, B. *FEBS Lett.,* **1985**, *184(2)*, pp 285-288.
15. Bhat, G. R.; Heitmann, J. A.; Joyce, T. W. *TAPPI J.*, **1991**, *74(9)*, pp 151-157.
16. Jackson, L. S.; Heitmann, J. A.; Joyce, T. W. *TAPPI J.,* **1993**, *76(3)*, pp 147-154.
17. Stork, G.; Pereira, H.; Wood, T. M.; Düsterhöft, E. M.; Toft, A.; Puls, J. *TAPPI J.,* **1995**, *78(2)*, pp 79-88.
18. Mansfield, S. D.; Wong, K. K. Y.; de Jong, E.; Saddler, J. N. *Appl. Microbiol. Biotechnol.*, **1996**, *46*, pp 319-326.
19. Saake, B.; Clark, T.; Puls, J. *Holzforschung,* **1995**, *49*, pp 60-68.
20. Wood, T. M.; Bhat, K. M. *Methods Enzymol.*, **1988**, *160*, pp 87-112.
21. Stoscheck, C. M. *Methods Enzymol.*, **1990**, *182*, pp 50-68.
22. Tolan, J. S. *J. Pulp Paper Sci.*, **1995**, *21(4)*, pp J132-J137.
23. Seth, R. S. In *Materials Research Society Symposium Proceedings*. Caulfield, D. F. et al. Ed.; Materials Research Society: Philadelphia, 1990, Vol. 197, pp 143-161.
24. Scallan, A. M.; Borch, J. In *The Fundametal Properties of Paper Related to its Uses*. Bolam, F., Ed.; Transactions of the Symposium held at Cambridge, 1973, Vol. 1, pp 152-163.
25. Yllner, S.; Enström, B. *Svensk Papperstidning*, **1956**, *59(6)*, pp 229-232.
26. Yllner, S.; Östberg, K.; Stockman, L. *Svensk Papperstidning*, **1957**, *60(21)*, pp 795-802.
27. Iversen, T.; Wännström, S. *Holzforschung,* **1986**, *40*, pp 19-22.
28. Karlssson, O.; Westermark, U. *J. Pulp Paper Sci.*, **1996**, *22(10)*, pp J397-J401.
29. Yu, X.; Minor, J. L.; Atalla, R. H.; Labbauf, M. M.; Farrell, R. L. *1994 Intl. Pulp Bleaching Conf.,* **1994**, pp 63-67.
30. Suurnäkki, A.; Li, T.-Q.; Buchert, J.; Tenkanen, M.; Viikari, L.; Vuorinen, T.; Ödberg, L. *Holzforschung,* **1997**, *51*, pp 27-33.
31. Yokota, S.; Wong, K. K. Y.; Saddler, J. N.; Reid, I. D. *Pulp Paper Can.*, **1995**, *96(4)*, pp T131-T133.
32. Kibblewhite, R. P.; Clark, T. A. *Appita*, **1996**, *49(6)*, pp 390-396.
33. Wong, K. K. Y.; Nelson, S. L.; Saddler, J. N. *J. Biotechnol.*, **1996**, *48(1-2)*, pp 137-145.
34. Elegir, G.; Sykes, M.; Jeffries, T. W. *Enzyme Microb. Technol.*, **1995**, *17*, pp 954-959.
35. Gübitz, G. M.; Schnitzhofer, W.; Balakrishnan, H.; Steiner, W. *J. Biotechnol.* **1996**, *50*, pp 181-188.

36. Clark, T.; Steward, D.; Bruce, M. E.; McDonald, A. G.; Singh, A. P.; Senior, D. *J. Appita,* **1991**, *44*, pp 389-404.
37. Roberts, J. C.; McCarthy, A. J.; Flynn, N. J.; Broda, P. *Enzyme Microb. Technol.* **1990**, *12*, pp 210-213.
38. Viikari, L.; Kantelinen, A.; Poutanen, K.; Ranua, M. In *Biotechnology in Pulp and Paper Manufacture*. Kirk, T. K.; Chang, H.-M., Eds.; Butterworth-Heinemann: Boston, 1990, pp 145-151.

Chapter 8

Laccase Catalyzed Bonding of Wood Fibers

Claus Felby[1,3], Per Ole Olesen[1], and Tomas Tage Hansen[2]

[1]Plant Fiber Laboratory, Royal Veterinary and Agricultural University, Taastrup, Denmark
[2]Novo Nordisk A/S, Novo Alle, 2880 Bagsvaerd, Denmark

The phenoloxidases laccase and peroxidase can be used for bonding of wood fibers. Binderless fiberboards are made by an oxidative pretreatment of fibers with laccase. Strength enhancement of paper handsheets is accomplished by mixing fibers with laccase and a phenolic polysaccharide. The bonding mechanism is linked to the formation of stable radicals in the fiber lignin matrix. The extent of laccase catalyzed oxidation of fibers, and thereby the performance of the fiberboards, is dependant on the amount of phenolic extractives in the wood species used and the lignin colloids generated by the defibration process. For implementation of the techniques in present-day production of fiberboards and paper and packaging materials, new methods for enzyme treatment under semi-dry conditions need to be developed and low-cost sources of phenolic polysaccharides must be available.

The increasing demand for wood-based products with low emission of VOC's has generated a need for new, environmentally friendly adhesives. In conventional production of wood-based materials, synthetic adhesives such as urea- and phenol formaldehyde, are used in combination with hot pressing in order to obtain boards with good mechanical properties. Besides the environmental and health aspects, synthetic adhesives contribute significantly to the manufacturing costs of boards and panels. There is thus a need for new adhesive systems. An attractive possibility which may provide both environmental and economic advantages is to use the catalytic effect of oxidative enzymes for bonding of wood fibers. Recently these enzymes have become commercially available in bulk quantities making industrial applications economically feasible. This paper describes the use of a phenol oxidizing enzyme for bonding of fibers in boards and for strength enhancement of paper.

[3]Current address: Plant Fiber Laboratory, Agrovej 10, 2630, Taastrup, Denmark.

Enzymatic Bonding

Bonding of wood fibers and particles by adhesives is accomplished by forming an adhesive matrix in which the particles or fibers are crosslinked by e.g. mechanical entanglement or covalent bonding. However, the bonding is caused not only by the added adhesive but also by the auto-adhesive properties of the wood components. Under properly selected conditions of heat, pressure and moisture, wood fibers or particles will bond through auto adhesion (*1*). Important factors related to auto-adhesion are a flow of lignin and hemicellulose which increases the bonding area (*2*), a breakdown of hemicelluloses to crosslinkers such as furfural (*3*) and condensation reactions of lignin (*4*). These factors are all contribute to the adhesion in wood-based products prepared from wood particles or wood fibers which are hot pressed. However, the binding achieved by auto-adhesion is insufficient to eliminate the use of synthetic adhesives.

Oxidative enzymes such as the phenoloxidases laccase and peroxidase may be used for polymerization or crosslinking of wood components in order to bond these. Furthermore the auto-adhesive properties of wood fibers may be enhanced, thereby eliminating the need for synthetic adhesives.

The enzymes peroxidase (EC 1.11.1.7) and laccase (EC 1.10.3.2), will catalyze a one-electron oxidation of phenolic hydroxyl groups while reducing H_2O_2 and O_2, respectively, yielding phenoxy radicals and water. Both enzymes have low specificity with respect to the reduced substrate and have been found to catalyze oxidation of a number of organic compounds including *o*- and *p*-diphenols, polyphenols, lignin, aminophenols, polyamines, aryl diamines and certain inorganic ions (*5*). The enzyme catalyzed reactions of lignin compounds can be written as follows:

$$\text{Phe-OH} + H_2O_2 \xrightarrow{\textit{Peroxidase}} 2\text{Phe-O} + 2H_2O \tag{1}$$

$$4\text{Phe-OH} + O_2 \xrightarrow{\textit{Laccase}} 4\text{Phe-O} + 2H_2O \tag{2}$$

The enzymes occur in both plants and insects. Their physiological role in higher plants is associated with bio-synthesis and bio-degradation of lignin through radical reactions.

The concept of using lignin-oxidizing enzymes for bonding applications is based on the reactivity of radicals in the plant cell wall (*6*). *In vivo,* phenoloxidases participate in polymerization of lignin through crosslinking of radicals, and it might thus be possible to utilize the same type of reaction for bonding in lignocellulosic materials *in vitro*.

The concept of enzyme catalyzed bonding of wood has been applied as a one-component system for activation of lignin on wood fibers by phenoloxidases, or as a two-component system where wood particles or fibers are mixed with a co-substrate e.g. technical lignin followed by phenoloxidase catalyzed polymerization (curing), which is similar to a conventional process using adhesives *(7)*.

The possibility of curing a lignosulphonate-based adhesive by the use of horseradish peroxidase was initially investigated in (*8*), but no polymerization of the lignosulphonate was observed.

A two-component adhesive system using laccase treated lignosulphonates for bonding of solid wood was described in a patent application (*9*), and the system was further used for wood laminates and particleboards (*10*). The adhesive formulation employed for bonding of particleboards contained one part of laccase in solution and two parts of technical lignin (spent sulphite liquor), with a total dry matter content of approximately 50%.

An adhesive system based on brown rotted lignin and laccase or peroxidase was described for bonding of wood laminates (*11*).

Peroxidase and laccase in combination with DHP (dehydrogenative polymer), synthetic "lignin" made from vanillic acid, was applied to improve the plybond strength of paper boards (*12*). The DHP was precipitated on the surface of fibers made by thermomechanical pulping, and paperboards with improved strength properties were prepared from the treated fibers. The application of a similar technique to precipitate DHP made by polymerization of vanillic acid, catechol and different tannins onto a TMP pulp was investigated in (*13*); for comparison the TMP pulp was either untreated or was pretreated with laccase for 72 hours. A positive effect on paperboard plybond strength was found for the TMP with precipitation of DHP as well as for the TMP pulp subjected to a laccase treatment only (one-component system).

The idea of a one-component system for bonding of wood fibers via phenoloxidase catalyzed activation of the middle lamella was first put forward in (*14*), and a one component system for bonding of wood fibers has been described in the patent application (*15*). Laccase treated wood fibers with no adhesive added were used for making MDF-boards. The fibers were submerged in an aqueous laccase solution for 2-7 days. Following the laccase treatment the fibers were either used directly for making wet-process fiberboards or they were dried for making fiberboards by a dry-process. Control boards were made in the same way, but replacing the enzyme solution with tap water. Boards made from the enzyme treated fibers had superior physico-mechanical properties compared to the control.

Bonding of wood fibers by a laccase treatment was investigated in further details (*16*). Wet- and dry-process fiberboards were made from laccase oxidized beech wood TMP fibers. The fibers were suspended in water at room temperature, laccase was added and the enzyme treatment was performed for 1 hour. For dry-process boards, the fibers were dried for 18 hours, air-laid fiber mats were formed, and 3 mm fiberboards were pressed at 200°C. Wet-process boards pressed at 180°C were made directly from the enzyme treated fibers. The enzyme treated fiberboards had significantly higher modulus of rupture and modulus of elasticity than the control as well as better dimensional stability. A control series of boards using heat inactivated enzyme showed no sign of any effect attributable to the protein and carbohydrate content of the enzyme solution on the mechanical properties and the dimensional stability. The bonding effect is thus caused by the catalytic effect of the enzyme only. Note that the laccase treatment is only 1 hr compared to 2-7 days reported in the above mentioned patent application (*15*).

In (*16*) it was also found that the laccase action could be correlated to the amount of stable radicals in the fiber lignin matrix, and a theory was put forward that the bonding mechanism is associated with crosslinking of laccase generated radicals situated in the lignin matrix.

The bonding mechanism of enzyme treated DHP and thermomechanical pulp in paperboards was described in *(17)* as crosslinking of fibers due to an increased bonding area, and an enzyme generated depolymerization/loosening of the three-dimensional lignin structure, thus improving the accessibility and reactivity of the lignin. A similar theory, which include degradation of lignin followed by precipitation of lignin fragments on the fiber surfaces as a part of the bonding mechanism in laccase bonded fiberboards, was put forward in *(18)*. A loosening or degradation of the lignin structure does not appear likely, as no depolymerization of lignin is known to take place solely by the action of laccase.

However, precipitation of colloidal lignin and phenolic extractives on the fiber surfaces was shown to enhance the surface compatibility and thereby the potential bonding area of laccase treated fibers (*19*).

In this paper we describe the concept of laccase catalyzed bonding using either a one-component system for bonding of fiberboards or a two-component system for bonding of paper and packaging materials.

Fiberboards bonded by laccase will be made at lower pressing times than previously applied. Strength enhanced paper handsheets will be made by mixing a mechanical pulp with a phenolic polysaccharide and laccase.

The extent of the laccase catalyzed oxidation will be measured by electron spin resonance-spectrometry (ESR), and the fiberboard and handsheet properties will be evaluated by mechanical testing.

Experimental

Materials.

Fibers for Fiberboards. Beech (*Fagus sylvatica*) wood fibers produced by a thermomechanical (TMP) pulping process were obtained from the MDF-board plant of Junckers Industries, Koege, Denmark. The fibers used for the experiments were freshly pulped with a dry matter content of approximately 40 %.

Fibers for Paper Handsheets. Radiata pine *(Pinus radiata)* fibers produced by thermomechanical pulping were supplied by The Forest Research Institute, Rotorua, New Zealand. The fibers were dried for shipment and stored at 4 C.

Enzyme. A *Trametes villosa* laccase, SP504 (EC 1.10.3.2), was supplied by Novo Nordisk A/S, Bagsvaerd, Denmark. The enzyme had a molecular weight of 55 kDa. Activity was measured in laccase units (LACU). 1 LACU was defined as the amount of enzyme which under standard conditions (30°C; pH 5.5 in 0.05 M potassium buffer) oxidize 1 μmole syringaldazine per minute. The enzyme product had a carbohydrate content of 21 g/l and a protein content of 2.0 g/l. Specific activity was 100 LACU/mg protein.

Ferulated Arabinoxylan. Ferulated arabinoxylan extracted from corn bran was obtained under the name α-Supergel from GB Gels Ltd, Swansea, Wales, UK.

Methods.

Enzyme Treatment for Fiberboards. Wood fibers were treated in the form of an aqueous suspension containing 5% w/w of fibers. Temperature and pH of the suspension were 20 C and 4.5, respectively. No pH adjustment was made. To ensure a sufficient supply of oxygen for the enzyme reaction, the water must be saturated with atmospheric air at the start of each experiment. Laccase was added at a dosage of 3 LACU/g fiber dry substance, and the suspension was stirred for a few minutes in order to obtain a homogeneous pulp suspension. Residence time for the enzyme treatment was 1 hour. After enzyme treatment, the water was drained from the fibers, and the fibers were dried at 40°C for approximately 18 hours. A series of control experiments was carried out in an identical manner except that the fibers were suspended in demineralized water only.

The extent of the enzyme catalyzed oxidation was evaluated by the radical activity in the fibers as measured by ESR-spectrometry. For ESR-analysis of the radical activity following the enzyme treatment, the fibers were drained and mildly dried by forced air at 30°C for 20 min, whereupon samples were transferred to 710-SQ quartz tubes from Wilmad, Buena, NJ and frozen in liquid nitrogen in order to stop further reactions of residual laccase. An ECS 106 X-band ESR spectrometer (Bruker, Karlsruhe, Germany) equipped with an X-band ER 4103TM cavity was used for the measurements. The radical activity on frozen fibers were quantified by the amplitude of the first derivative ESR signal corrected for the sample density. No determination of the exact number of unpaired spins were made.

Strength Enhancement of Paper Handsheets. Handsheets of radiata pine TMP pulp (160 g/m^2) were made in a sheet mould. The sheets were dewatered by pressing in a sheet press for 5 minutes at a pressure of 400 kPa. After pressing, the wet handsheets were placed on a net and immersed in different solutions of laccase and ferulated arabinoxylan. The pH and temperature of the solutions were 4.5 and 20°C. Adjustment of pH was done by sulphuric acid and sodium hydroxide. For all treatments the sheets were immersed for 90 seconds, and the temperature of the solutions was 20°C. The applied treatments were as follows:

- -Control. Immersed in 200 ml water only.
- -Laccase. Immersed in 200 ml solution of laccase (0.1 LACU/ml).
- -Ferulated arabinoxylan. Immersed in 200 ml 0.6% solution of ferulated arabinoxylan.
- -Ferulated arabinoxylan + laccase. Immersed in 200 ml freshly made solution of ferulated arabinoxylan (0.6% w/w) and laccase (0.1 LACU/ml).

After immersion, the sheets were left at room temperature for 5 minutes and then pressed in the sheet press for 5 minutes at a pressure of 400 kPa. After pressing, the sheets were dried in a sheet dryer. The drying lasted 5 minutes.

Fiberboards. Air laid mats 30 x 30 cm were made from the fibers, and the initial absolute moisture content before hot pressing was 12%. Target density for the boards was 850 kg/m^3, and 6 mm boards were made by hot-pressing to stops at 200°C for 2.5 and 5 min.

Testing of Fiberboards and Handsheets. The fiberboards were tested according to ASTM D 1037 for modulus of rupture (MOR), modulus of elasticity (MOE), tensile strength perpendicular to the board surface (IB), as well as weight gain and thickness swell following a 24 hour cold water soak.
The handsheets were tested for thickness and dry and wet tensile index according to the SCAN standards SCAN-P7 and SCAN-P16. Wet tensile index was measured following a 24 hour cold water soak.

Results and Discussion

Laccase treatment of beech fibers in an aqueous suspension changes the color of the fiber suspension from brown to dark-brown/red within a few minutes after the enzyme addition. The color change is the only visual effect of the laccase treatment, and it correlates with the formation of radicals in the fiber lignin matrix, see Table I and Figure 1. The laccase generated radicals are very stable, and no decrease in the amount of radicals can be seen after more than two weeks. The relatively high amount of radicals present in untreated fibers, are probably caused by UV-radiation or mechanical stress during the fibration process.

The mechanical and dimensional properties of the fiberboards bonded by laccase catalyzed oxidation are shown in Table I. Compared to the control boards the mechanical strength and the dimensional stability of laccase treated boards increase significantly. The bonding effect can be seen for both 2.5 minutes and 5 minutes pressing time. By increasing the pressing time from 2.5 to 5 minutes the mechanical properties of both laccase treated and control boards are increased. This effect may be caused by a higher flow and entanglement of the hemicellulose and lignin components in the fibers, but apparently a longer pressing time has no effect on the relative strength increase caused by a laccase treatment, see Table I.

As the glass transition temperature of lignin is lowered by water, an important parameter for the bonding effect is the initial moisture content before the hot pressing of fibers to boards. If the moisture content of the fibers is lower than 10% (data not shown), very little or no effect of the laccase treatment can be seen. This effect indicates that lignin is the active component in the bonding mechanism for laccase treated fibers. Another important factor for the bonding effect is the interaction between the fiber lignin matrix and colloidal lignin or phenolic extractives.

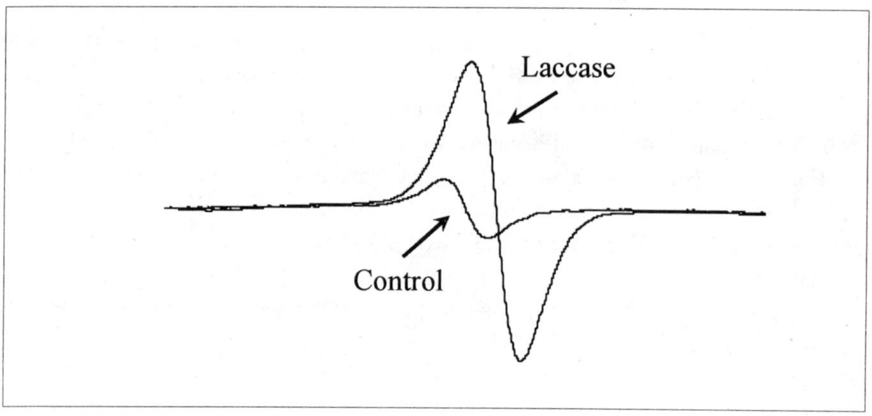

Figure 1. ESR-spectra of radical activity on laccase treated and untreated control beech wood fibers. No hyperfine structure can be recorded. Enzyme treatment is identical to the fibers used for fiberboards.

Table I. Radical activity (arbitrary units) of laccase treated beech wood fibers and mechanical properties of fiberboards. Modulus of rupture (MOR), modulus of elasticity (MOE) and tensile strength perpendicular to the surface (IB) are shown for 6 mm dry process fiberboards made from laccase treated and control (untreated) wood fibers. Thickness swell (T.S) and water absorption (W.A) are measured following a 24 hour cold water soak. (Figures in each column marked by different letters are different at a significance level of 0.05).

Treatment	Density (kg/m^3)	Radical activity	MOR (MPa)	MOE (GPa)	IB (MPa)	T.S (%)	W.A (%)
Control 2.5 min pressing	845	75a	32.1a	7.1a	0.29a	107a	147a
Control 5 min pressing	862	81a	37.8a	8.3a	0.41b	96a	104b
Laccase 2.5 min pressing	837	506b	53.8b	13.6b	0.81c	44b	68c
Laccase 5 min pressing	841	535b	58.3c	15.1c	0.96c	40b	59c

Removal of low molecular weight lignin compounds by dialysis prior to the laccase treatment has been found to decrease the amount of laccase generated radicals and the tensile strength perpendicular to the surface of laccase treated fiberboards, whereas no effect on the tensile strength perpendicular to the surface is seen for control boards (*20*). Moreover it has been found that native phenolic extractives and colloidal lignin, similar to the interaction between e.g. veratryl alcohol and lignin (*21; 22*), function as radical mediators (*23*). This means that when beech wood fibers in an aqueous suspension are oxidized by laccase, a cyclic radical interchange mechanism between fiber lignin and lignin substances in solution extends the oxidation to domains not accessible to the large enzyme molecule (*20;23*).

The role of low molecular weight lignin compounds on the bonding effect as found in the work cited above shows the importance of the chemical composition and physical structure of the fibers, i.e. the amount of phenolic extractives in the fibers and lignin colloids generated by the defibration process. Further investigations into the bonding mechanism are likely to improve the performance of fiberboards made from laccase treated fibers.

The results on strength enhancement of paper handsheets made from radiata pine fibers by mixing laccase and ferulated arabinoxylan are shown in Table II. The handsheets made with ferulated arabinoxylan and laccase show a significant increase in tensile strength, and approximately 25 % of the strength is retained following a 24 hour cold water soak, whereas the handsheets made from the other treatments disintegrated.

Table II. Tensile strength of handsheets made from unbleached TMP softwood fibers. The handsheets are either untreated, made with laccase only or immersed in a 0.6% solution of ferulated arabinoxylan with or without laccase present. The wet tensile index is measured following a 24 hour cold water soak. (Figures in each column marked by different letters are different at a significance level of 0.001).

Treatment	Dry tensile index (Nm/g)	Wet tensile index (Nm/g)	Thickness (μm)
Control	8.1a	Disintegrated	694a
Arabinoxylan	7.3a	Disintegrated	639a
Laccase	5.3a	Disintegrated	682a
Arabinoxylan + Laccase	47.0b	11.8	350b

The amount of ferulated arabinoxylan adsorbed by the handsheets is approximately 0.8% (w/w), and compared to the bonding of particleboards by laccase and lignosulphonate co-substrate (*9*) only about 10 % of the co-substrate ferulated arabinoxylan is needed for bonding of the fibers.

The bonding mechanism of laccase and ferulated arabinoxylan is believed to be both a formation of a lignin-ferulate-polysaccharide complex between the ferulic acid substituents on arabinoxylan and lignin on the fibers, and an embedding of the fibers in a matrix of polymerized ferulated arabinoxylan. These mechanisms are also supported by the high wet strength of handsheets bonded by laccase and ferulated arabinoxylan. Furthermore, the reduction in thickness upon treatment with laccase and ferulated arabinoxylan increases the density of the handsheets, causing a higher level of fiber to fiber contact and intramolecular interactions.

Compared to bonding of fibers through a laccase catalyzed oxidation of fibers, the use of a co-substrate such as ferulated arabinoxylan only involves temperatures and pressures compatible with paper manufacturing. Polysaccharides with phenolic substituents may be well suited as co-substrate for enzyme catalyzed bonding, as the polysaccharide chain may serve as an effective gapfiller crosslinking the fibers. Ferulated arabinoxylan is almost colorless and will not result in any significant darkening or discoloration as it is seen with technical lignins. This makes it of interest for strength enhancement of paper and packaging materials. A major limitation for further use of the technique is that ferulated arabinoxylan is presently not available in bulk quantities. However, other sources of phenolic polysaccharides e.g. modified technical starches may be an alternative.

The presented results and the techniques described in this paper show that phenoloxidases can be used in various ways for bonding of wood fibers.
However, for bonding of fiberboards on a laboratory scale the enzyme treatment is done in an aqueous suspension of wood fibers at 2.5-5 % consistency. For industrial implementation of e.g. laccase catalyzed bonding in present-day production of dry-process fiberboards, such high water contents would require additional drying which will raise the production costs, making the concept economically unattractive. For industrial applications under semi-dry conditions new methods for enzyme applications have to be developed.

Conclusions

Laccase can be used for bonding of wood fibers by an oxidative pretreatment of the fibers or by mixing wood fibers with laccase and a phenolic polysaccharide.

The extent of laccase catalyzed oxidation of fibers and thereby the performance of the fiberboards, is dependant on the amount of phenolic extractives in the wood species used and the lignin colloids generated by the defibration process.

For implementation of the described techniques in present day production of fiberboards and paper and packaging materials new methods for enzyme treatment at semi-dry conditions needs to be developed and low-cost sources of phenolic polysaccharides must be available.

Acknowledgements

This work was made possible by a grant from The Danish Environmental Protection Agency.

Literature Cited

1. Avella, T.; Lhoneux, B. *Journal of Applied Polymer Science: Applied Polymer Symposium*. 1984, *40*, pp 203-207.
2. Back, E. L. *Holzforschung*, 1987, *41*, pp 247-258.
3. Stofko, J. I. In *Proceedings of 1980 Symposium: Wood Adhesives -Research, Application and Needs*, Editors W.E. Johns; R.H. Gillespie. USDA Forest Service FPL and Washington State University. Madison, 1981, pp 44-55.
4. Nimz, H. H. In: *Wood Adhesives, Chemistry and Technology*. Editor A. Pizzi. Marcel Dekker New York, Basel, 1983, pp. 247-288.

5. Yaropolov, A. I.;Skorobogatko, O. V; Vartanov, S. S; Varfolomeyev, S. D. *Applied Biochemistry and Biotechnology,* 1995, *49*, pp 257-280.
6. Glässer, W. G.; Sandermann, W. *Holzforschung*, 1970, *24*, pp 73-76.
7. Johns, W. E. (1989) In *Wood Adhesives: Chemistry and Technology vol. 2*, Editor A. Pizzi. Marcel Dekker New York, Basel, 1989, pp. 75-98.
8. Nimz, H. H.; Gurang, I.; Mogharab, I. *Liebigs Ann Chem*, 1976, pp 1421-1434.
9. Haars, A.; Hüttermann, A. *German patent application*, DE 30,37,992,C2, 1983.
10. Haars, A.; Kharazipour, A.; Zanker, H.; Hüttermann, A. In: *Adhesives from Renewable Resources*. Editors R.W. Hemingway; A.H. Conner. ACS Symp. ser. 385. American Chemical Society, Washington DC,1989, pp 126-134.
11. Jin, L.; Nicholas, D. D.; Schultz, T.P. *Holzforschung*, 1991, *45*, pp 467-468.
12. Yamaguchi, H.; Maeda, Y.; Sakata, I. *Mokuzai Gakkaishi*, 1992, *38*, pp 931-937.
13. Yamaguchi, H.; Nagamori, N.; Sakata, I. *Mokuzai Gakkaishi*, 1993, *37*, pp 220-226.
14. Körner S. *Dissertation*, Technischen Universität Dresden. 1990.
15. Kharazipour, A.; Hüttermann, A.; Kühne, G.; Rong, M. *European patent application* 0,565,109,A1, 1993.
16. Felby, C.; Pedersen, L. S.; Nielsen, B. R. *Holzforschung,* 1997, *51,* pp 281-286.
17. Yamaguchi, H.; Maeda, Y.; Sakata, I. *Mokuzai Gakkaishi*, 1994, *40*, pp 185-190.
18. Kharazipour, A.; Hüttermann, A.; Luedemann,HD. *Journal of Adhesion Science and Technology*, 1997, *11*, pp 419-427.
19. Thygesen, L. G.; Felby, C.; Sanadi, A. R.; Hassingboe, J.; Olesen, P.O. *Holzforschung,*1997, *submitted.*
20. Felby, C; Hassingboe, J. In: *Toward the new generation of bio-based composite products*; Editors H, Kajita; K, Tsunoda, Kyoto, 1996. pp 283-291.
21. Harvey, P. J.;Gilardi, G. F.; Palmer, J. M. In: *Enzyme systems for lignocellulose degradation*; Editor M.P, Coughlan, Elsevier Applied Science, London and New York, 1989, pp 111-119.
22. Shoemaker, H. E.; Piontek, K. *Pure and Applied Chemistry*. 1996, *68,* pp 2089-2096.
23. Felby, C.; Nielsen, B. R.; Olesen, P. O.; Skibsted, L. H.; *Applied Microbiology and Biotechnology*, 1997, *43*, in press.

Chapter 9

Enzymatic Activation of Wood Fibers for Wood Composite Production

A. Kharazipour[1], K. Schindel[2], and A. Hüttermann[2]

[1]Pfleiderer Industrie D 59759 Arnsberg Germany
[2]Forstbotanisches Institut der Universität Göttingen, D 37077 Göttingen Germany

The incubation of wood fibers with a phenol oxidizing enzyme results in an oxidative activation of the lignin crust on the fiber surface. When such fibers are pressed together, boards are obtained which meet the required German Standards for medium density fiber boards (MDF). The wood composites made by this process are bound together in a way which comes closer to the situation in the naturally grown wood than any other process used in the present production of wood composites. This process will for the first time yield wood composites which are produced solely from naturally grown products without any addition of resins. Since they will produce no emission whatsoever a high acceptance of the public can be expected.

The production of wood composites like fiber or particle boards always follows the same basic rationale: solid wood is fragmented into pieces of usually small sizes like strands, chips or fibers, supplemented with a binder and pressed to form a wood-like structure again. By this general record, the anisotrophy of wood is reduced and wood of small dimensions or used wood can be converted to a useful product. In this process basically, only one new component is added which was not present in the original wood: the binder, which usually is either urea-formaldehyde, phenol-formaldehyde or isocyanate. Several strategies have been developed to utilize the binding material which glues natural wood together, for the binding of the wood composites, too. The first attempts in this direction were published by (*1*) who demonstrated that water resistant bonds can be obtained by pressing at high forces (100 kg cm^{-2}) and temperatures (200 °C) for rather long times. During those conditions, the binding is obtained by the reaction of pyrolytic degradation products of cell wall constituents. The authors calculated the stable bonds (0.0572 g/100 g of wood fragments) and the thickness of the glueline would be rather thin, 0.015 nm compared to 0.34 nm of the resin-bonded particle boards. More recent reviews on this approach have been published by (*2*) and (*3*). With procedures that are described above wood composites with satisfying mechanical properties can be achieved. The problem which usually remains to be solved is the high swelling of these board in the presence of water. Therefore those composites have not yet been produced commercially.

The use of enzymes for the bonding of wood particles was first suggested by (*4*) who used peroxidases for the cross linking of lignin for binding processes. Hüttermann and co-workers (*5-9*) used phenoloxidase and a mixture of water insoluble lignin with small amounts of resin polyphenyl-methane-di-isocyanate (PMDI) to produce particle boards. Jin et al., (*10*) used brown-rotted lignin as binder together with either laccase or peroxidase. Kühne and co-workers, finally, developed a method where living fungi were directly integrated into the processing of wood chips. They incubated wood chips with either white- or brown-rot fungi in solid state fermentation prior to refining. As a result of such pretreatment two positive effects were observed: First, the energy demand for refining the wood chips to fibers could significantly be reduced, as it was found by Bar-Lev and co-workers earlier (*11, 12*). Second, the amount of petrochemical resins required to produce fiber boards, could be decreased (*13-16*).

During the processes for the production of wood fibers, the lignin of the middle lamella, which is the natural glue between two woody cells, is plastified at temperatures above its glass-transition-point in order to get the cells separated. After cooling down to room temperature, the lignin solidifies again and forms a glassy crust on the surface of the wood fiber. The crust forms a barrier which reduces the binding strength of any added resin (*13*), resulting into the need for rather high amounts of binders for the production of fiber boards. Any process which would use the former middle lamella as a binding agent has to reactivate the glassy crust of middle lamella lignin and then use it as a binder via in-situ polymerization.

In this communication we will present evidence that by treatment with ligninolytic enzymes from white-rot fungi it is indeed possible to reactivate the lignin on the surface of the wood fibers and that the lignin can be used as the sole binder for wood-composites. Preliminary results of these studies have been published before (*17-20*).

Materials and Methods

Materials. Commercially produced wood fibers were obtained from Bestwood, Ribnitz-Damgarten. The fibers, consisting of about 80 % softwood (spruce and pine) and 20 % hardwood (beech), were produced in a defibrator at 8 - 10 bar pressure at a temperature of 180°C. Laccase, SP504 (EC 1.10.3.2) produced from *Coriolus versicolor* and peroxidase (Mn-dependent peroxidase, EC 1.11.1.7) were supplied by NOVO-NORDISK, Bagsvaerdt, Denmark. The latter enzyme was genetically cloned from a Basidiomycete and produced by recombinant techniques.

Treatment with Laccase. Wood fibers were sprayed with 100 %, 150 %, or 200 % (on dry weight) of enzyme buffer solution (Mc Ilvaine buffer, pH 5.0) with always the same amount of enzyme present in the solution. The wet fibers were incubated at room-temperature for 12 hrs. and dried at 50 °C to a final water content of 20-25 %. Afterwards a fleece was formed from the defelted fibers. The fleece was pressed at a temperature of 190°C, pressure of 25 bar to 5 mm thick fiber boards. The boards were tested for their mechanical properties by Pfleiderer Industrie, Werk Arnsberg, according to DIN 52350, 52351, and 52352.

Treatment with Peroxidase. 2.500 kg of wood fibers were supplemented with 35 l of a buffered enzyme solution (universal buffer, pH 7.0) containing the indicated amounts of enzyme units and incubated in a stainless-steel vessel at room temperature. H_2O_2 was added at time intervals of 10 min. For the determination of the release of lignin-like material, aliquots were taken from the supernatant and the OD_{280} was measured in a HP-spectrophotometer. After the incubation, the fibers were dewatered, defelted and formed to a fleece.

Analysis of the supernatant. For the study of the kinetics of the liberation of substances from the fibers, 5 g of fibers were incubated in 500 ml of buffered enzyme solution and the absorption in supernatant was measured at 280 nm. For the isolation of the substances, the incubation was done for 24 hrs. and after filtration of the fibers the solution was concentrated in a rotary evaporator. The solution, about 5 ml, was centrifuged for 5 min at 2,000 rpm and the supernatant processed for the determination of the molecular weight distribution (*21*). 2 ml of the supernatant were extracted with 2 ml of 50 mM N-Methyl-tridecylammoniumchloride in ethylacetate by shaking for 30 min on a horizontal shaker. An aliquot of 1 ml of the organic phase was shaken several times with 1 ml of 1 % NaCl in water for purification. 0.5 ml of the washed organic phase were transferred into HPLC-vial, dried in a vacuum at -24°C and stored at -24°C. After addition of 0.55 ml of tetrahydrofurane, the sample was analyzed with a HPLC (Waters) equipped with an diode array spectrometer on a TSK-gel column, by gel permeation chromatography (GPC).

Scanning Electron Microscopy. For scanning electron microscopy, fibers were prepared using an adaptation of the method of Cole (*22*). The commercially available or with enzyme solution incubated fibers were dehydrated in alcohol, the alcohol was gradually exchanged with acetone and the acetone finally replaced by liquid carbon dioxide in a pressure chamber (Critical Point drying Apparatus E 3 000, Polaron Equipment Ltd.). After complete saturation with carbon dioxide was achieved, the temperature in the pressure chamber was raised above the critical point of carbon dioxide, about 45 °C, and the liquid was evaporated. The specimen were coated with gold in a sputter and inspected in a Philips 510 SEM.

Transmission Electron Microscopy and EDAX measurements of lignin. Samples of medium density fiber boards (MDF), 1x1x5 mm size, manufactured with and without laccase were cut with a razor blade. For the specific labelling of lignin *in situ* the method from Eriksson et al. (*23*) was used. The incubation mixture contained 20 mg MDF fibers and 0.15 g mercuric acetate dissolved in 6 ml methanol and 0.25 ml acetic acid. The mercurization was performed at 95°C for 8 hrs. in sealed flasks. Excessive mercury was washed out by refluxing twice in methanol for a total time of 8 hrs. Methanol was replaced in a graded series by acetone, followed by Spurr′s medium (*24*). The samples were cured for 16 hrs. at 70°C. Transverse sections (1 µm) of the embedded fibers were cut with a diamond knife mounted on an ultramicrotome. They were collected on formvar-covered copper grids and coated with carbon before analysis.

Mercury detection and quantification was carried out with a Philips EM 420 electron microscope connected to an energy-dispersive system EDAX 9100. The apparatus was operated under the following conditions: The accelerating voltage was 120 kV, emission current 20 µA, tilting angle of the specimen holder towards the detector 5°, condenser aperture shut, objective aperture open and counting time 30 live seconds. The Hg-L-α signal (9,98 keV) of the X-ray spectra was processed with the SEMI program of the EDAX 9100 software after manual fitting of the background.

Thickness corrections of the sections were made measuring the Cl-K-α emission (3.32 keV) of spots in the embedding medium in close vicinity to the analysed fibers. In each section five measurements were made per morphological region (S2-cell-wall, middle lamella of free fibers, middle lamella of joined fibers, and middle lamella in corners between joined fibers). Four sections were investigated of laccase and control MDF-fibers, respectively. The mean and standard deviation of 20 measurements per morphological region was calculated (n=20).

Results and Discussion

Morphology of the Surface of Wood Fibers. When commercially processed wood fibers are inspected in a scanning electron microscope, a rather conspicuous structure can be seen, a distinctive crust which covers the surface of the fiber (Fig. 1). When such fibers are reacted with mercury for the quantitative localization of lignin, embedded, cut in the microtome and looked at in a transmission electron microscope, lumps of mercurized material are visible on the surface of the fibers (Fig. 2). Measurements of the relative lignin content using EDAX revealed that these materials had a twofold higher lignin concentration than the secondary wall (Fig. 3). This twofold increase in lignin content of the middle lamella compared to the S2-layer of the woody cell wall is in accordance with the values published in the literature (*25*), where also two times more lignin was found in the middle lamella than in the S2-layer of the cell wall.

These data indicate that during the refining process the middle lamella lignin of the wood fibers is plastcized and sheared off, resulting into the separation of the individual cells. After cooling, the former middle lamella forms a distinct crust on the surface of the fibers.

Reaction between Laccase and the Lignin on the Fiber Surface. Fig. 4 shows that a treatment of the fibers with laccase leads to a gradual increase in the A_{280} of the supernatant. Depending on the enzyme activity of the solution, after two (with lower activities) or five days a maximum is reached, after which no further increase in A_{280} is achieved. When compared to N-Methyl-trioctylamine, the material obtained from solutions which were incubated at a pH-value which is close to the optimum of the

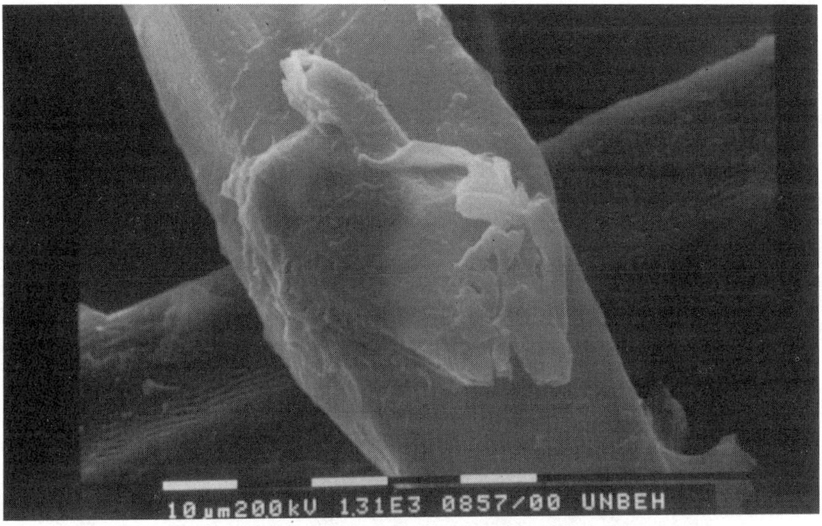

Fig. 1. SEM picture of the surface of a commercially available wood fibre.

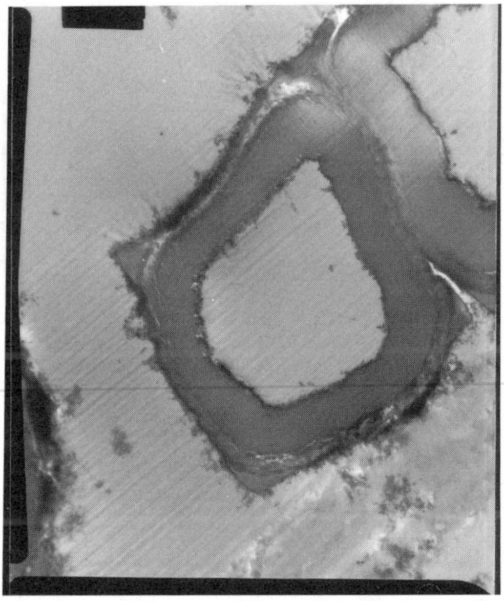

Fig. 2. TEM picture of a cross-section of a mercurized wood fibre. The preparation was not treated with usual contrast chemicals. The differences in the absorption are due to the relative mercury contents of the different layers of the cell wall (scale bar 1 5/8 µm).

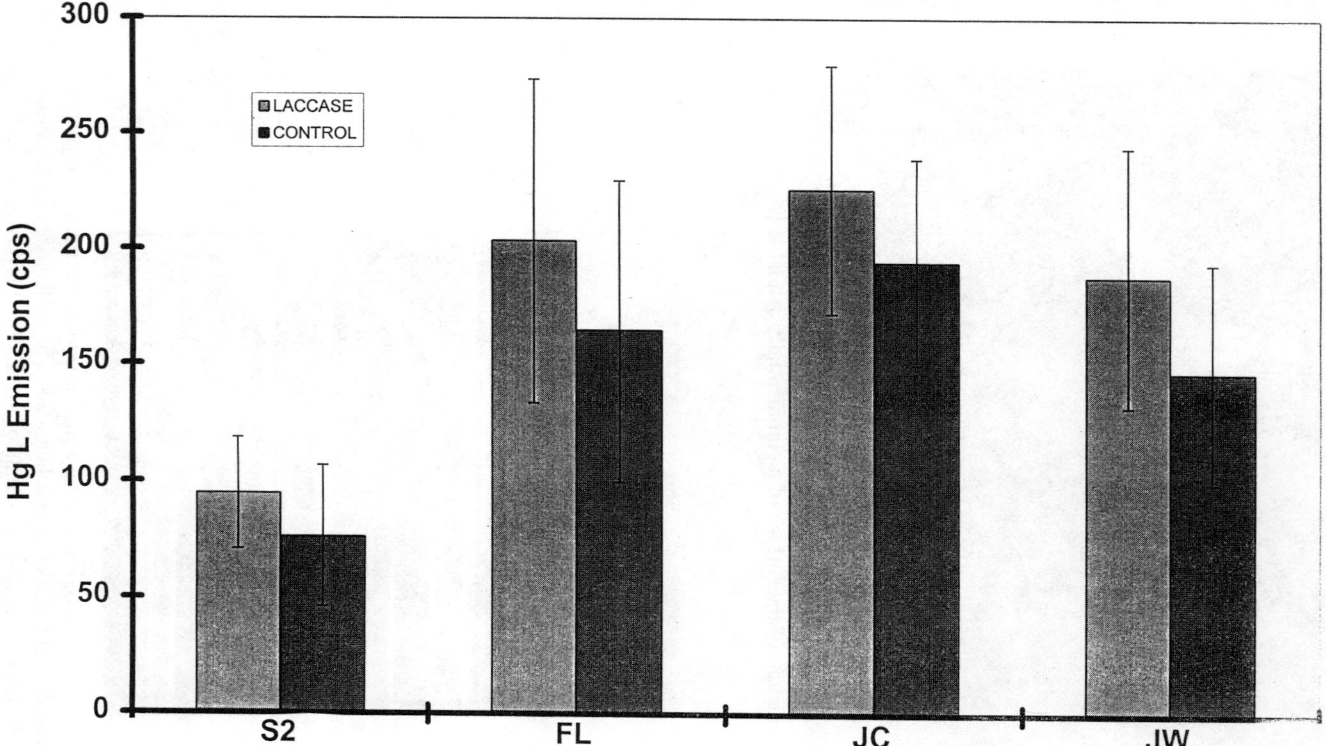

Fig. 3. Lignin content of cross-sections of wood fibres and enzymatically produced fibre boards measured via mercurization of the samples and EDAX-analysis of the Hg-signals. The symbols indicate the regions where the measurements were done. Key: S2, S2 layer of the cell wall; FL, middle lamella on the surface of the unreacted fibres; JC, region in the corner where three different wood fibres are bound together; and JW, region where the two lignin layers of wood fibres are bound together.

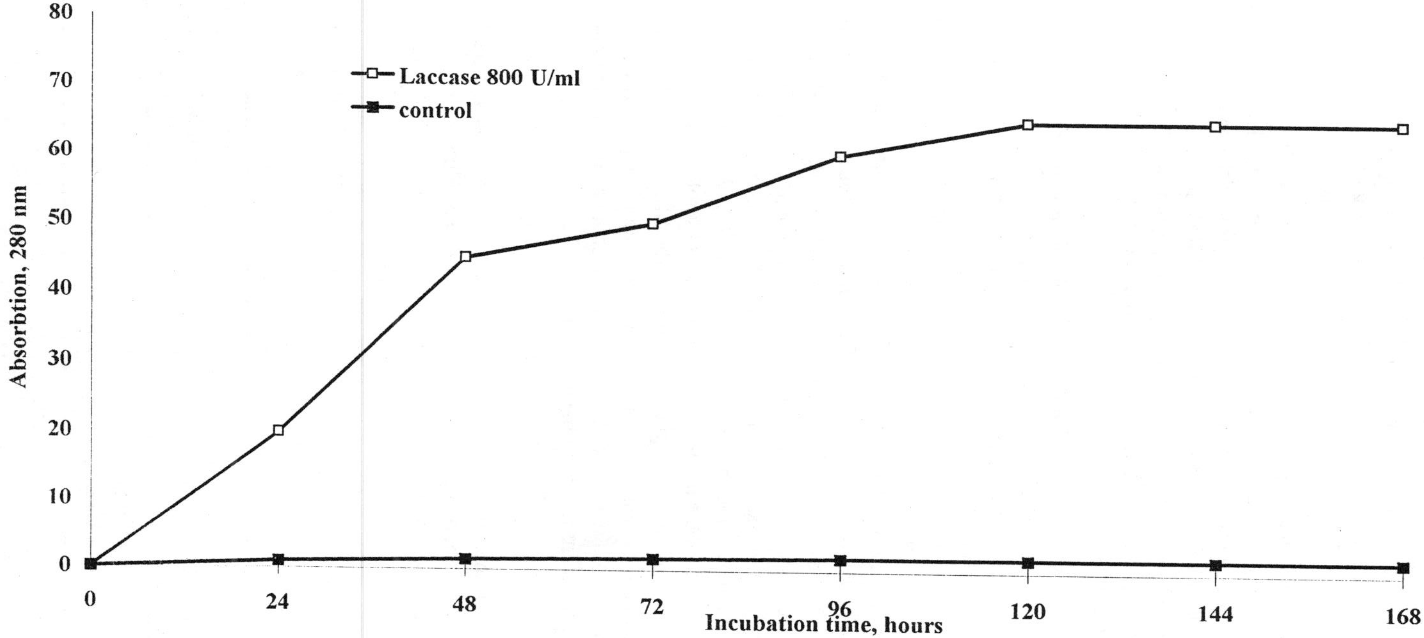

Fig. 4. Kinetics of the release of lignin from the fibre surface with regard to enzyme concentration and time of encubation.

enzyme, had a MW of about 3.600 Da (Fig. 5). ^{13}C NMR analysis (data published in *20*) revealed that the material had the typical spectrum of lignin, although highly oxidized, a fact supported by the high content of carboxylic groups (about 13%).

The combined data indicate that the lignin from the fiber surface is susceptible to react with the enzyme. On the basis of our present knowledge on the reaction of laccase with lignin (*26-32*), the following interpretation of the above presented data can be offered: During the incubation with laccase, different reactions take place with the former middle lamella lignin: (i) oxidation, (ii) partial polymerisation with perhaps concomitant degradation and possible repolymerization, (iii) cleavage of lignin-carbohydrate bonds.

Fig. 6 shows that after a relatively short incubation with laccase, the lignin crust of the fiber surface was loosened and presumably rendered susceptible to further reactions. A prolonged treatment with the enzyme even removed completely the crust from the fiber surface (Fig.7).

Utilization of the Lignin from the Fiber Surface as a Binder for Wood Composites. The data presented above indicate that the lignin from the fiber surface can indeed be activated by the laccase reaction. From our experience with the laccase-lignin system used as a binder for particle boards (*5*) it was tempting to test whether this reaction might be suitable to serve as the basis for the formation of strong physico-chemical and perhaps even covalent bonds between fibers in wood composites. We therefore attempted to find conditions suitable for the production of fiber boards by sole incubation of the fibers with laccase. It was found that a simple incubation of the fibers with laccase was enough to activate the lignin present there and to use it as sole binder for fiber boards. Table I shows the results of the basic experiment and the enhancement of the binding strength as a result of the treatment with the enzyme.

Table I. Effect of incubation of fibers with enzyme solution on the mechanical properties of the board. The fibers were sprayed with the enzyme solution, incubated for 12 hrs., and pressed for three minutes at 190 °C and 10 MPa to 5 mm thick boards of a density of 800 kg m^{-3}.

	Internal Bond Strength ($N\ mm^{-2}$)	**24 h swelling (%)**
control: incubation with denatured enzyme	**0.1**	**the fiber board was not stable in water and disintegrated**
incubation with active enzyme (800 U ml^{-1})	**0.52**	**23**

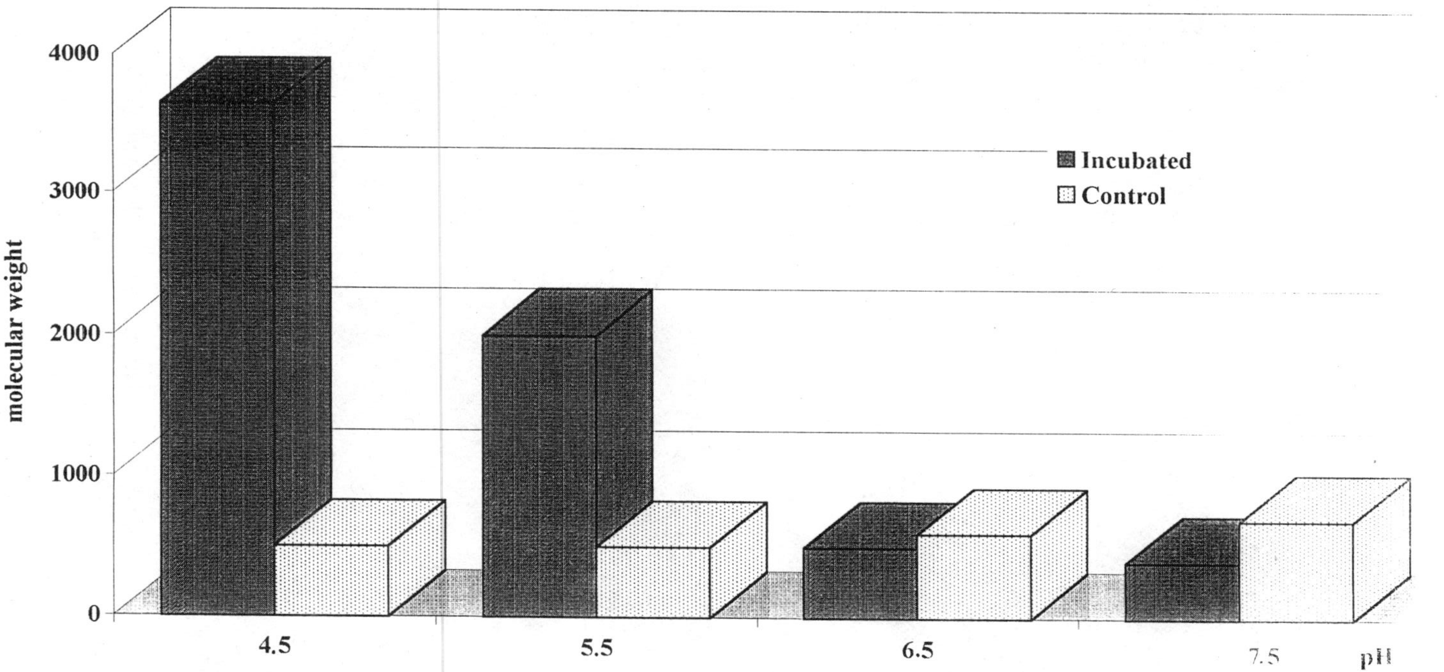

Fig. 5. Mean molecular weight of the lignin released into the supernatant after incubation for 72 hrs. with 800 U/mL of laccase at different pH-values in the incubation medium. The controls were obtained by analysing the supernatant of incubations without the enzyme.

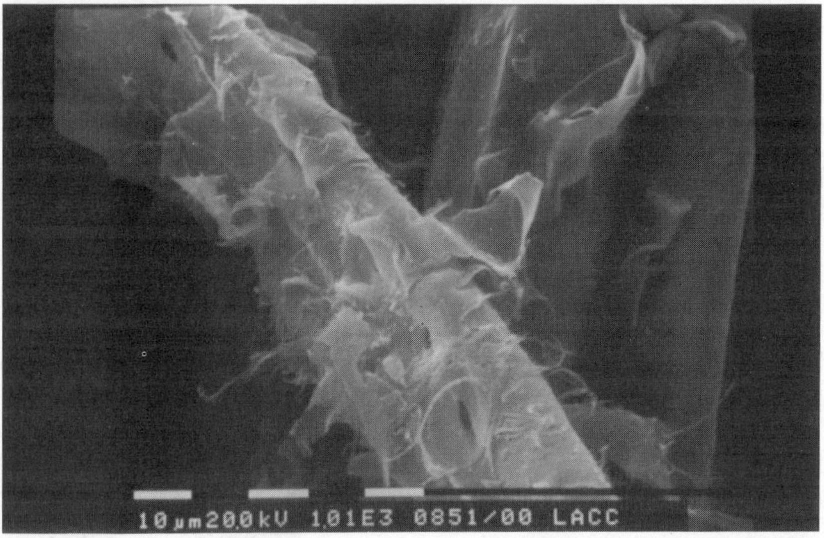

Fig. 6. SEM picture of the surface of a commercially available wood fibre after 2 hrs. of incubation.

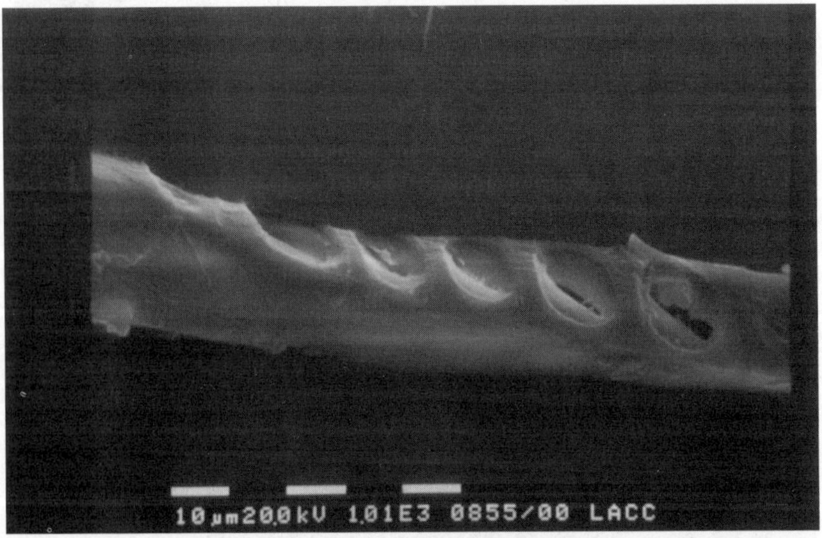

Fig. 7. SEM picture of the surface of a commercially available wood fibre after 12 hrs. of incubation.

When the fibers were incubated at different pH-values in the incubation buffer, a clear dependency of the technical values of the boards on the pH was observed, with the highest tensile strength at the pH-optimum of the enzyme (Fig. 8). From this it can be concluded that the binding of the fibers definitely is the result of the activity of laccase. The technical values obtained for the best enzymatically bound MDF-boards meet the requirements of the German Standards (DIN).

Peroxidase as an activating Agent of the Lignin. When the fibers were incubated at different pH-values in the incubation buffer, a clear pH-dependency of the technical values of the boards was observed with the highest tensile strength at the pH-optimum of the peroxidase enzyme (Fig. 9). When the process was employed at different H_2O_2-concentrations, an optimum result was obtained with a concentration of 15 mM (Fig. 10). At higher concentrations the enzyme is probably denatured by the peroxide.

Whereas the results obtained for the treatment with laccase were somewhat hoped for in view of the known polymerising activity of this enzyme, the data obtained with the peroxidase treatment are surprising because of the data in the literature on lignin as a substrate for Mn-dependent peroxidase (c.f. 32). To our knowledge, no studies on possible polymerisation activities of Mn-dependent lignin-peroxidase have been published so far. On the opposite, several authors have described only activities of this enzyme which lead to depolymerization of lignin (*34, 35*) and chlorolignin (*36*). Therefore it was surprising to find out that with this enzyme the same process can be catalyzed as was described for laccase. From our data it can be concluded that at least under certain conditions, Mn-dependent peroxidase also may catalyze the formation of stable bonds which are strong enough to produce fiber-boards which can meet the German Standards required for fiber boards.

Micromorphology of the boards and lignin distribution. When enzymatically bound boards were mercurized, embedded, cut with an ultra-microtome and looked at in a Transmission Electron Microscope, a very close connection between the different fibers was to be seen (Fig. 11), which is similar to the contact between fibers in a native wood tissue (e.g. 36. p.19, Fig. 16). Analysis of the lignin distribution in the EDAX-system revealed that indeed the contact zone between the fibers contained two times more lignin than the S2-layer of the walls of fibers (Fig. 3). Thus it is evident that indeed the lignin of the former middle lamella of the wood fibers is the active binder in this process.

A comment to the EM-pictures. We know that the TEM pictures are apparently not of the best quality, this is due to the fact that these are pictures of the mercurized samples, without any usual contrast treatments. The reason for including them is to show that the different layers of the cell walls can be clearly distinguished and thus measured precisely. The SEM-pictures are presented as colour prints. They need not to be printed in colour, they still reproduce very well in black and white. To take colour photographs from SEM is an old trick to get the best reproductions, direct b/w would not give all the details.

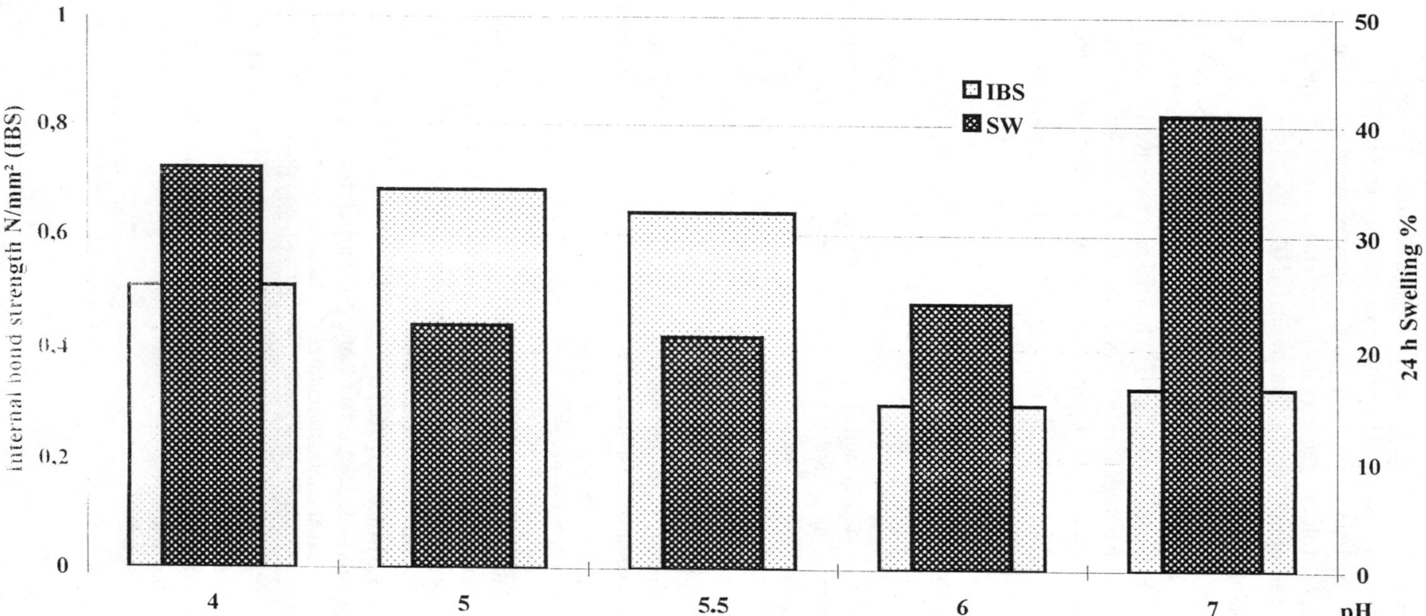

Figure 8. Technical properties of 5 mm MDF-boards, pressed at 190 °C (800 kg/ m^3) with regard to the pH of the incubation solution with the enzyme laccase.

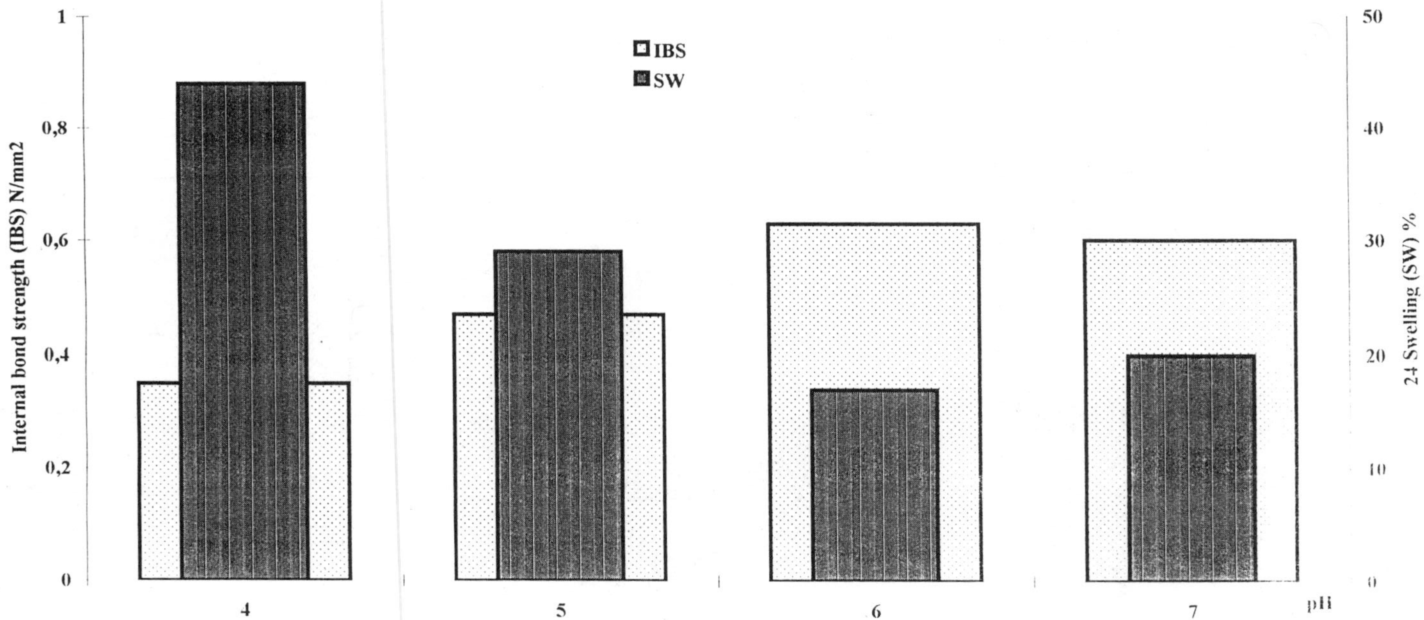

Fig. 9: Technical properties of 5 mm MDF-boards with regard to the pH of the incubation solution with the enzyme peroxidase.

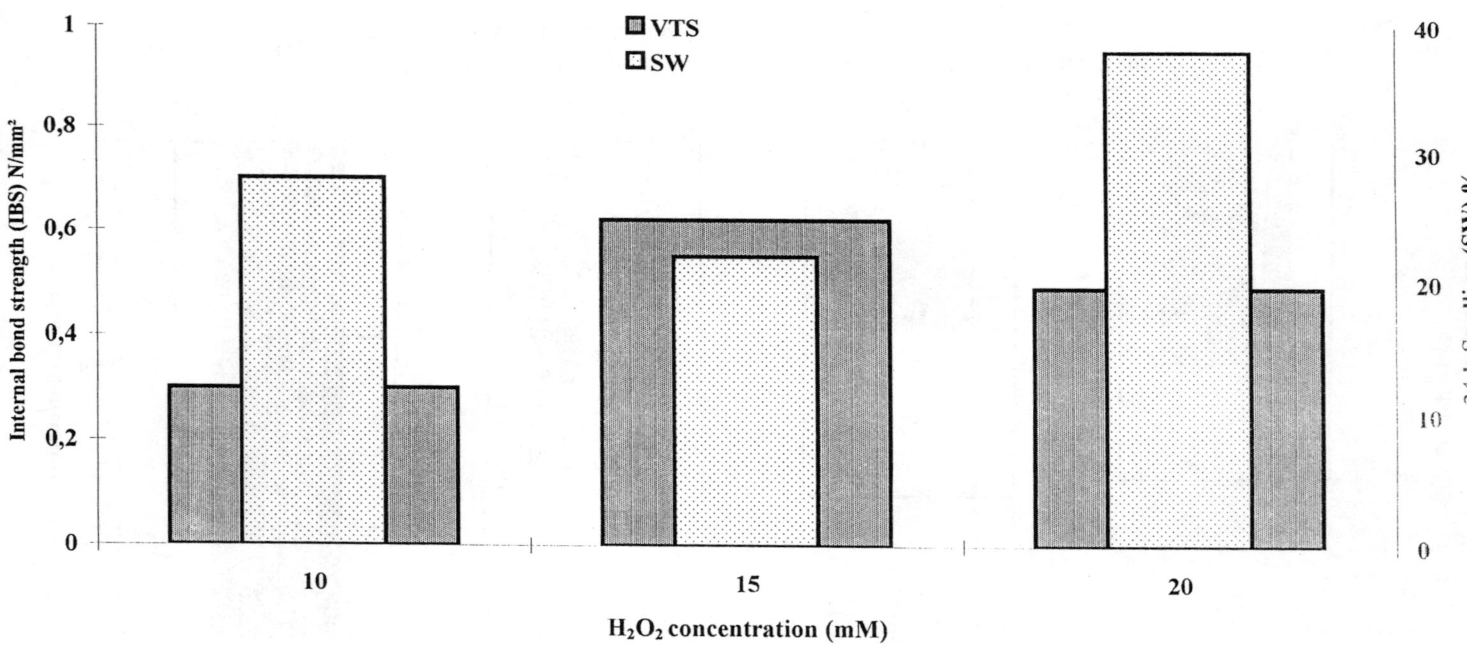

Fig. 10: Technical properties of 5 mm MDF-boards with regard to the concentration of the H_2O_2.

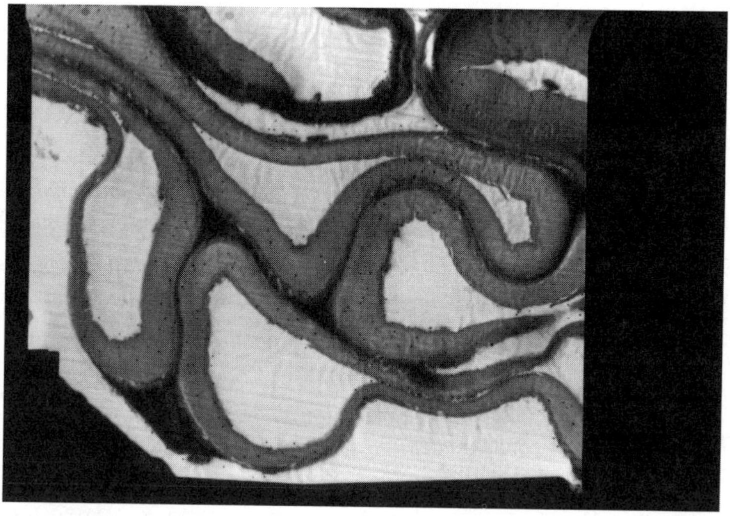

Fig. 11. TEM picture of a cross-section of a mercurized fiber board obtained by pressing of fibers incubated in laccase-solution.

Literature Cited

1. Klauditz W.; Stegmann, G. *Holz Roh -Werkstoff* **1955,** *13,* pp. 434 - 440.
2. Zavarin, E. In: *The chemistry of solid wood,* Rowell, R Ed. Advances in Chemistry Series 207, American Chemical Society, Washington, DC, 1984 pp. 372-400.
3. Ellis, S.; Paszner, L. *Holzforschung* **1994** *48 Suppl.*, pp. 82-90.
4. Nimz, H.; Razvi, A.; Marquharab I.; Clad, W. *German Patent* **1972** 222 1353
5. Haars, A.; Hüttermann, A. *Arch. Microbiol.* **1980** *125,* pp. 233 - 237.
6. Haars, A.; Hüttermann, A. *Arch. Microbiol.* **1983** *134,* pp. 309-313.
7. Haars, A.; Hüttermann, A. U. S. Patent **1984** 4432 921.
8. Haars, A.; Kharazipour, A.; Zanker, H.; Hüttermann, A. In *Adhesives from renewable resources*. R. W. Hemingway A. H. Conner S. J. Braham Eds. ACS-16 Symposium series No 385, American Chemical Society, Washington, D.C. 1989 pp. 126-134.
9. Hüttermann, A.; Haars, A.; Trojanowski, J.; Milstein, O.; Kharazipour, A In *New Polymeric Materials from Lignin* W. G. Glasser and S. Sarkanen Eds. ACS-Symposium Series No. 397, American Chemical Society Washington, D.C. 1989 pp 361-371.

10. Jin, L.; Sellers, T.; Schultz, T. P.; Nicholas, D. D. *Holzforschung* **1990** *44,* pp. 207-210.

11. Eriksson, K.-E. *Wood Sci. Technol.* **1990** *24,* pp. 70-102.

12. Akhtar M.; et al. Adv. In Biochem. Eng./Biotechnol.

13. Wagenführ, A. *Praxisrelevante Untersuchungen zur Nutzung biotechnologischer Wirkprinzipien bei der Holzwerkstoffherstellung.* Dissertation, TU Dresden 1988.

14. Körner, S. *Verfahren zur stofflichen Modifikation des Rohholzes für die Holzwerkstoffherstellung.* Dissertation TU Dresden.1990.

15. Körner, I.; Kühne, G.; Pecina, H. *HK-International, Holz- Möbelind.* **1993** Heft 4.

16. Körner, I.,*Untersuchungen zur Inokulation und Fermentation von unsterilen Hackschnitzeln mit Braunfäulepilzen für die Herstellung von MDF-Platten.* Dissertation TU Dresden 1993.

17. Kharazipour, A.; Hüttermann, A.; Kühne, G.; Rong, M. *German Patent* **1993** P 43 05 411.

18. Hüttermann, A.; Kharazipour, A.: In: A. Maijanen and A. Hase (eds.): *New catalysts for clean environment.* VTT Symp. 163, pp. 143 - 148, Espoo: VTT, 1996.

19. Kharazipour, A.; Hüttermann, A. In *Biotechnology in the pulp and paper industry* Srebotnik, E., Messner, K. Eds., Facultas Universitätsverlag, Wien, 1996 pp. 617 - 620.

20. Kharazipour, A.; Hüttermann, A.; Lüdemann, H.-D. *Adhesion Sci.* **1997** *11,* pp. 419 - 427.

21. Majcherczyk, A.; Hüttermann, A. *J. Chromatogr.* **1997** *764,* 183 - 191.

22. Cole, G. T. 1973. *Cytobios* **1975** *12,* pp. 115-121.

23. Eriksson, I.; Lidbrandt, O.;Westermark, U. *Wood Sci. Technol.* 1988, *22*, pp. 251-257.

24. Spurr, A. R. *J. Ultrastr. Res.* **1969** *26*, pp. 31-43.

25. Fergus, B. J.; Procter, A. R.; Scott, J. A. N.; Goring, D. A. I. *Wood Sci. Technol.* **1969** *3,* pp. 117 - 138.

26. Hüttermann, A.; Gebauer, M.; Volger, C.; Rösger, C., *Holzforschung* **1977**,*31*, pp. 83-89.

27. Haars, A.; Hüttermann, A. *Naturwissensch.* **1980** *67,* pp. 39-40

28. Eriksson, K.-E.; Kirk, T. K. In: C. L.Cooney, A. E. Humphrey (Hg).*Comprehensive Biotechnology* vol 3, pp. 271 - 294, Toronto (1985).

29. Eriksson, K.-E.; Blanchette, R. A.; Ander, P. *Microbial and Enzymatic Degradation of Wood and Wood Components*. Springer, Berlin, 1990.

30. Higuchi, T. *Wood Sci. Technol.* **1990** *24,* pp. 23-63.

31. Duarte, J. C. (ed.) *FEMS Microbiol. Rev.* **1994** *13*, pp. 121-390.

32. Hatakka, A. FEMS *Microbiol. Rev.* **1994** *13,* pp. 125-135.

33. Wariishi, H.; Valli, K.; Gold, M. H. *Biochem. Biophys. Res. Comm.* **1991** *176,* pp. 269-275.

34. Perez, J.; Jeffries, T.W. *Appl. Environm. Microbiol.* **1992** *58,* pp. 2402-2409.

35. Lackner, R.; Srebotnik, E., Messner, K. *Biochem. Biophys. Res. Comm.* **1991** *178,* pp. 1092-1098.

36. Higuchi, T. *Biochemistry and Molecular Biology of Wood* 362 p. Springer, Berlin 1997.

Chapter 10

Visualization of the Action of Ligninolytic Enzymes on High-Yield Pulp Fibers

K. Ruel[1], V. Burlat[1], J. Comtat[1], S. Moukha[2], J. C. Sigoillot[2], M. Asther[2], and J. P. Joseleau[1]

[1]Centre de Recherches sur les Macromolécules Vegetales (CNRS), BP 53, 38041 Grenoble cédex 9, France
[2]INRA, LBCF, CP 925, 13288 Marseille, France

High-yield pulps from wheat straw and from poplar wood were treated with manganese-peroxidase (MnP) or with laccase (Lac), before and after a second refining stage. The micromorphology of the degradation performed by the enzymes on the fibers was examined using electron microscopy. The preferential site of attack and the diffusability of the enzymes in the fiber walls were evaluated. It appears that MnP and Lac produce typical defibrillation patterns. In addition, Lac shows a particular ability to degrade lignin from the middle lamella. A correlation between the observed patterns, the type of enzyme and the nature of lignins present in the different parts of the fiber wall, as demonstrated with immunocytochemical probes, is discussed.

The purpose of the pulping process is the separation of wood fibers from each other. Pulping can be achieved either by chemical treatments intended to degrade and remove the material that glues together the fibers, or by mechanical treatments which physically tear apart the fibers from the wood tissue, or also by a combination of chemical and physical action. Obviously, the nature of the treatments influences the resulting pulp properties, and also the pulp yield. A characteristic difference between mechanical and chemical pulps is the higher lignin content of the former. This gives mechanical fibers their stiffness and opacity, but enhances their tendency to yellow and their poor flexibility. In search of an alternative to chemical treatments to improve mechanical pulp properties, biological treatment with ligninolytic fungi has been envisaged (*1-3*). Several reports described the application of lignin-degrading fungi at different stages of refining, and demonstrated the effectiveness of biomechanical pulping with white-rot fungi in saving energy and in increasing pulp strength (*3-7*). Fungal treatment with white-rot fungi by stationary solid state cultivation before soda pulping on graminaceous material resulted in a higher yield of pulps with decrease in kappa number and less fine formation(*6*).

Although the residual lignin in unbleached chemical pulps is substantially modified, mostly by alkaline condensation occuring during the kraft pulping process, some fungi can nevertheless delignify it (*8,9*). It was suggested that manganese peroxidase (MnP) plays an important role in the bleaching of hardwood kraft pulp by white-rot fungi (*10*). Other enzymes known to take part in lignin oxidative breakdown are also potential candidates for improving pulp properties (*10,11*). Thus, several attempts to apply ligninolytic enzymes to pulps have been made, most of them on chemical pulps with the aim of developing a chlorine-free biobleaching process. Lignin peroxidase (*12*), manganese peroxidase (*13,14*) and laccase (*15*) all increased pulp brightness, and demonstrated therefore their ability to degrade or modify lignin. In the enzyme action there is contact between the pulp and the enzymes (*16*) which adsorb on the fibers during solid state fermentation (*17*).

In this work, we studied the action of MnP and laccase (Lac) treatments on the structure of fibers of high-yield pulps. Micromorphological changes in the cell walls of mechanical and chemimechanical pulps (CMP) were examined by transmission electron microscopy (TEM). The relationship between the enzymes and lignin distribution was investigated using immunocytochemical methods.

Materials and Methods

Pulps and Enzymes. Wheat straw CMP (yield: 85%) and refined CMP were from the SAICA mill (Zaragosa, Spain). Poplar wood CMP and refined CMP pulps were from SICEM-SAGA spA (Canossa, Italy). MnP was produced from *Phanerochaete chrysosporium* I-1512 (CNCM, Institut Pasteur, Paris, France) using a new bioreactor design combining an airlift column and biofilm immobilization as previously described (18). After 5 days of incubation, the culture supernatant was harvested and concentrated by ultrafiltration using a 10 000 Da membrane (Millipore S.A., Molsheim, France). Concentrate activity was about 15 000 U/L MnP and 3 000 U/L LiP.

Lac was produced from *Pycnoporus cinnabarinus* MIC 11 (600 U/L Lac) in the presence of ferulic acid as an inducer (*19*). The cultures in the bioreactor were harvested after 8 days, corresponding to maximal enzyme activity. The biomass was removed by filtration and the filtrate was concentrated by ultrafiltration followed by diafiltration.

Treatment of Pulps with Enzymes. For MnP treatment the pulps (40g dry weight) were resuspended at 2% weight consistency (w/v) in lactate buffer (0.1 M, pH 4.5) and the final pH adjusted to 4.5 when needed. Four volumes of enzymatic solution (300 U each) were added at 1 hour intervals. The suspension was mechanically agitated and hydrogen peroxide (0.01 M) was introduced in the suspension at a rate of 5 mL/hour. After 4 hours, the pulp was filtered in a Buchner funnel and rinsed twice with distilled water. For controls, samples were treated with lactate buffer plus hydrogen peroxide and MnP plus hydrogen peroxide in the absence of Mn^{2+}, respectively.

For Lac treatment the pulps (40g dry weight) were suspended in lactate buffer (0.1 M, pH 4.5) at 2% consistency. Lac (150 U) was added in the presence of 2,2'-azinobis (3-ethyl-benzothiazoline-6-sulfonate) (ABTS) (1 mM). The mixture was kept under gentle stirring at 40 °C for 16 hours. At the end, the pulp was filtered in a Buchner funnel and rinsed with distilled water. For control, samples were treated with ABTS and laccase alone, respectively.

Fixation of Samples for Electron Microscopy. Small samples of the pulps were fixed in a freshly prepared mixture of 0.2% glutaraldehyde, 2% paraformaldehyde in 0.1 M phosphate buffer pH 7.0. Samples dehydrated through a series of increasing ethanol concentrations up to 70% (v/v) were embedded in London Resin white (LRW) (hard mixture) and polymerized 24 h at 50 °C.

Chemical Staining. Polysaccharides were stained on thin sections by the periodic acid-thiocarbohydrazide-silver proteinate (PATAg-method of Thiery), as modified by Ruel *et al.* (*20*).

Immunocytochemical Labelings. Labelings were performed on thin sections floating in plastic rings. The protocol was as described in Joseleau *et al.* (*21*) with a few modifications: BSA was replaced by 5% non-fat dried milk both in TBS_{500} (0.5 M NaCl in 0.01 M tris-phosphate buffer, pH 7.4) and TB (0.01 M tris-phosphate buffer, pH 7.4). The secondary marker was protein A-gold (pA_{10} or pA_5) (Amersham) diluted in tris buffer containing fish gelatin. When pA_5 was used, the technique of silver enhancement was applied (Amersham kit). After rinsing in TB, the sections were fixed in 2.5% glutaraldehyde, rinsed in water and transferred onto copper carbon coated grids. Post staining was in 2.5% aqueous uranyl acetate. Observations were performed with a Philips CM 200 cryo transmission electron microscope operating at 80 kV.

Labeling of manganese-peroxidase and laccase used an immunocytochemical method. Two polyclonal antisera raised in rabbits against the MnP complex and Lac from *Ceriporiopsis subvermispora* were used. Due to the strong homology among fungal enzymes, the antisera could be applied for labeling MnP from *P. chrysosporium*, and Lac from *P. cinnabarinus*. Immunocytochemical controls were the following: omission of the primary antibody, substitution of the primary antibody with the non-immune serum, and incubation with primary antibody that had first been saturated with the corresponding antigen. All the control tests were negative.

Results and Discussion

Pulps from wheat straw and poplar wood were used. For straw the pulping treatments were first a coarse primary defibration in the presence of 1M NaOH which gave chemimechanical pulp (CMP), then a secondary refining with soda on the preceding material, yielding refined pulp (R-CMP). From poplar wood a chemimechanical pulp was prepared by 1M NaOH impregnation of wood chips and primary defibration (CMP), followed by a second refining with soda and hydrogen peroxide to give R-CMP.

Changes in the Micromorphological Structure of Fibers Induced by Pulping Processes. Ultrastructural description of pulps is difficult due to their heterogeneity. However general consistent trends can be observed and predominating morphological characteristics may be related to a type of pulping process, to a type of tissue and in some cases to the specificity of enzyme treatment. Thus, CMP are characterized by bundles of associated cells still showing a tissue organization. This is illustrated with CMP wheat straw pulp (Fig. 1A) showing aggregates of fibers, sclerenchyma, vessels and pluristratified parenchyma. The action of soda impregnation is visible in the slightly swollen secondary walls of fibers and on cell corners and middle lamellae exhibiting the beginnings of tissue breakdown. The effect of the second refining can be seen on R-CM pulp. The number of isolated fibers is higher, and more fibers are present in the aggregates. The alkaline cooking favors swelling of the S_2 layer and dissociation between S_1 and compound middle lamella (Fig. 1B). CMP of poplar consisted mainly of aggregates of fibers and vessels. The more fragile ray cells have been disrupted (Fig. 1C) and produced fines. The effect of secondary refining of this pulp was to produce R-CMP in which the alterations observed in CMP were accentuated. In this pulp more dissociation between fibers and vessels is visible, and the disruptions have been enhanced with,in some parts of the secondary wall, zones where cellulose microfibrils have been dissociated (Fig. 1D). This resulted in a great heterogeneity in which almost unmodified fiber and vessel walls were found as well as highly defibrillated parts of secondary walls. In addition, numerous fines originating from parenchyma and peeled off middle lamellae can be seen.

Straw Pulps Treated with Manganese Peroxidase. The compactness and structural heterogeneity of CMP pulps rendered difficult the identification of the effects due to the enzyme action. However, observation of a great number of images allowed us to ascribe the effects due to the enzymes. The visualization of MnP by immunolabeling (*22*) with an antiserum directed against MnP isoenzymes from *C. subvermispora* (*23*) and which cross-reacted with MnP from *P. chrysosporium* showed the interaction between the enzyme and the pulp. The labeling was most abundant on the fibers located at the periphery of the tissues (Fig. 2A). It demonstrates that in such a compact and highly aggregated pulp, the diffusion of the enzyme is low. In fact the enzyme was only seen interacting with fibers in desorganized areas, where a slight defibrillation had occured (Fig. 2B).This may be a crucial point in the use of enzyme pretreatment of mechanical pulps. Morphologically, the effects due to MnP appeared more clearly on the R-CMP. Here again the heterogeneity of the pulp prevented the enzyme from performing evenly . However, MnP-treated fibers showed definite signs of delamination and defibrillation due to the enzyme action (Fig. 2C). Such modifications were not observed in controls where Mn^{2+} or enzyme were omitted in the treatment. This confirms previous observations of the typical defibrillating effect of MnP (*22*). The labeling with the anti-MnP antiserum was restricted to these modified areas. In zones where an advanced and intense defibrillation had occured, it is interesting to note that the labeling of MnP was low when not totally absent (Fig. 2D). This suggests that when lignin has been extensively removed the enzyme either does not bind to the area or has been removed

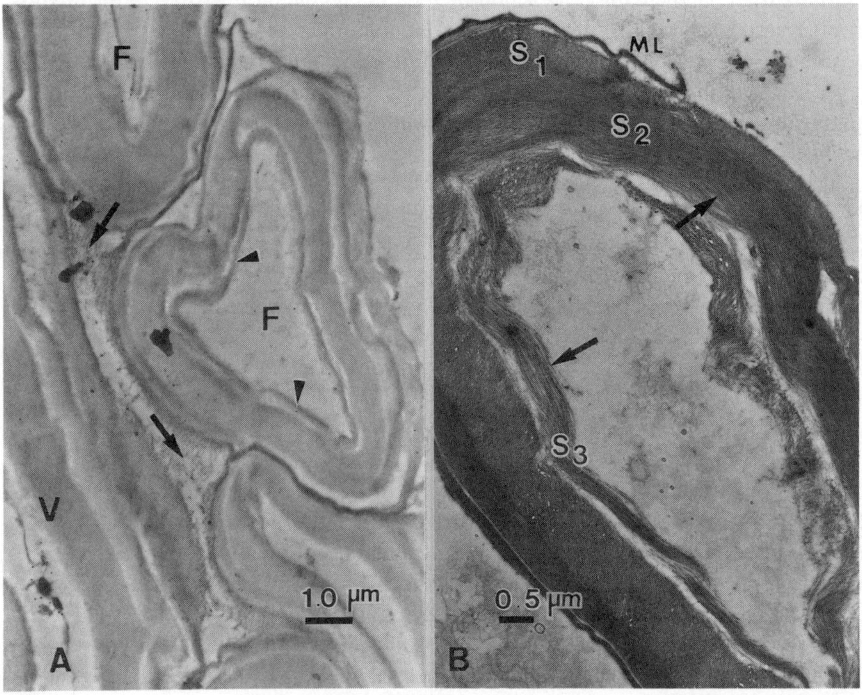

Figure 1. A and B: Micromorphology of wheat straw pulp; A, CMP. Three fibers (F) are still associated a vessel (V). Cell corners have been delignified (arrow) and S_3 detached (arrowhead) by the first soda treatment. Uranyl acetate staining; B, R-CMP. Structure of an isolated fiber: the compound middle lamella is detached and a defibrillation is clearly visible in S_3 and S_2 (arrows). Uranyl acetate.

C and D: Micromorphology of poplar pulp. C, CMP. Aggregates of fibers (F) are separated from a highly loosened ray cell (RC), the isotropic layer appears in black (arrow). PATAg staining; D, R-CMP. Fibers are more dissociated and some of them have been defibrillated (arrows). Uranyl acetate.

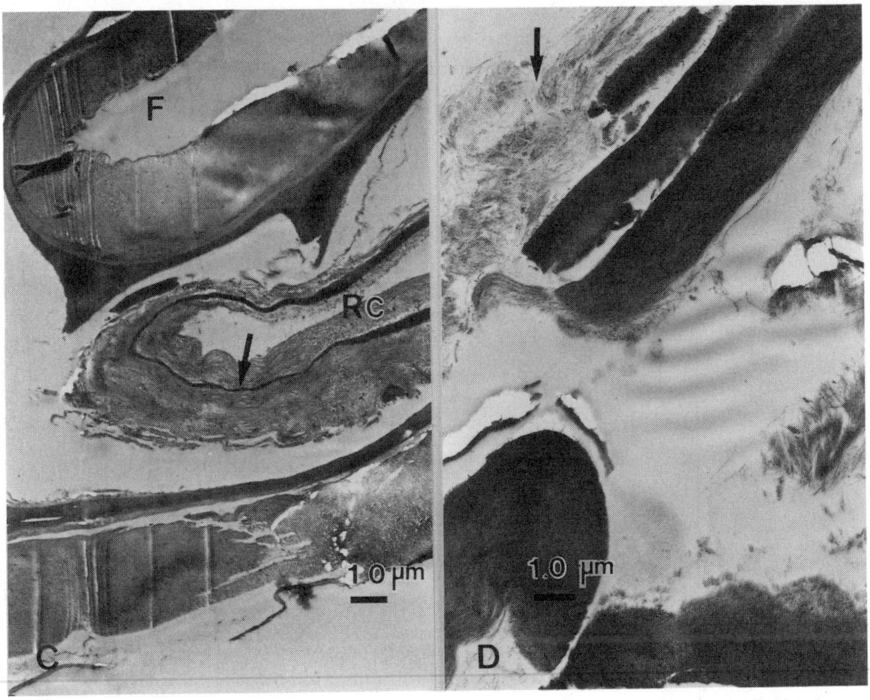

Figure 1. *Continued.*

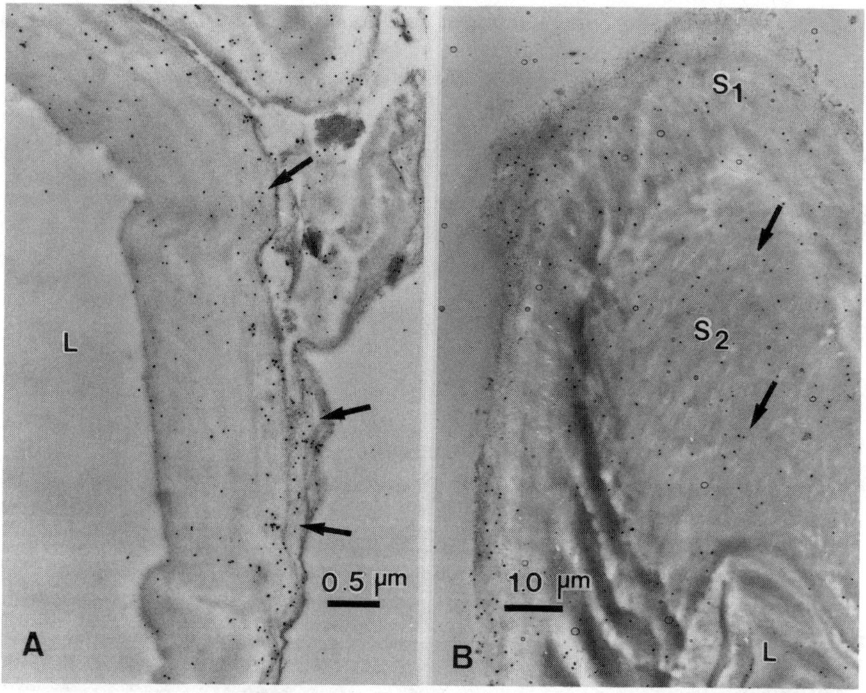

Figure 2. Wheat straw pulp treated with MnP: immunogold labeling of the enzyme with anti-MnP. A, CMP, fiber seating outside of a fiber aggregate: MnP is concentrated in the outer layer (arrows); B, CMP; part of a fiber in which MnP has penetrated inside the defibrillated areas (arrows); C, R-CMP; part of a fiber showing a strong delamination of the S_2 layer; D, R-CMP. Highly defibrillated fiber: the separation of cellulose microfibrils is clearly visible around the lumen and MnP is no longer visible.

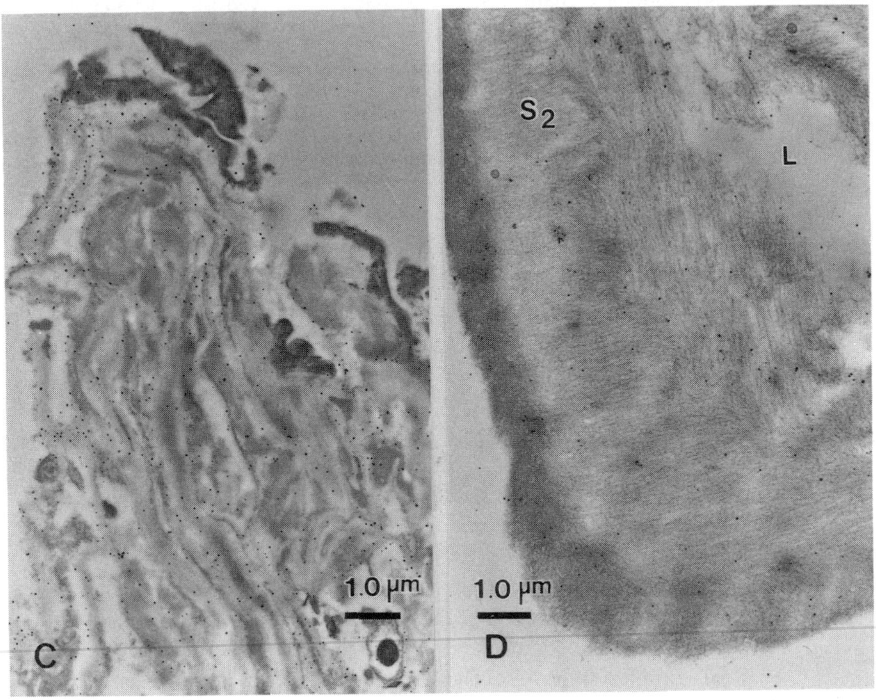

Figure 2. *Continued.*

during washing of the pulp. Extensive removal of lignin is supported by the great fragility of the microfibrils of these areas under the electron beam. This is a well established observation in TEM that cellulose microfibrils which are no longer protected by lignin and hemicelluloses show a shortened life time when working at 80 kV (*24*).

Poplar Pulps Treated with Manganese Peroxidase. CMP from poplar treated with MnP showed slight defibrillation with the action of the enzyme localized at the internal face of fibers. The defibrillated areas correspond to the most accessible zones. These ultrastructural alterations occurred within a narrow band which can be depicted as a fringe in which cellulose microfibrils became conspicuous and acquired random orientations (Fig. 3A). The modifications of fiber microstructure due to MnP on R- CMP were of the same type, but greater, probably because of the better accessibility.

Immunocytochemical visualization of the enzyme showed that, here again, the action of MnP was largely dependent on the morphology of the fibers in the pulp. On compact fibers, the labeling showed that the penetration of the enzyme was almost nil. On broken fibers showing clear traces of fracture, the enzyme penetration was limited to the surface of the fracture. Accumulation of MnP, however, could be observed wherever the treatment had resulted in an opening of the fiber ultrastructure, particularly in zones undergoing significant defibrillation (Fig. 3B). The ability of MnP to penetrate was considerably enhanced by the second refining treatment.

Straw and Poplar Pulps Treated with Laccase. Straw pulps (CMP and R-CMP) were treated with Lac from *P. cinnabarinus* in the presence of mediator 2,2'-azinobis (3-ethyl-benzothiazoline-6-sulfonate) (ABTS) (*14*). As for MnP, fibers treated with Lac-ABTS showed ultrastructural modifications, the most distinct effect being delamination accompanied with a clear separation of cellulose microfibrils (Fig. 4A). These modifications were not observed on the reference treatment with ABTS and laccase alone, respectively. The involvement of Lac in defibrillation is suggested by the association of the enzyme with the microfibrils in zones where defibrillation was enhanced (Fig. 4A). As was observed with MnP, the immunolabeling of Lac revealed only a weak labeling in the areas where defibrillation is the most pronounced. This again suggested that the enzyme is not retained when delignification has been extensive.

An interesting result of the action of Lac-ABTS is the degradation of cell corners and adjoining middle lamellae. Lignin degradation in these parts of the fiber walls gives rise to spherical particles (Fig. 4B). These electron-dense particles resemble those observed in steam exploded wood (*25,26*) and described as resulting from the tendency of lignin molecules to undergo "coalescence" once they have been released from their native state in the walls (*27*). Another explanation could be the formation of repolymerized compounds by the phenoxy radicals generated by Lac from the phenolic moieties of lignin. This is compatible with the enzyme's mechanism of action and has been shown to occur *in vitro* (*28*).

The effect of the Lac-ABTS system on high-yield pulps from poplar appeared to consist mainly of enhanced defibrillation. This again is supported by the presence of the enzyme in cell walls in which prior chemimechanical fragmentation facilitated a large

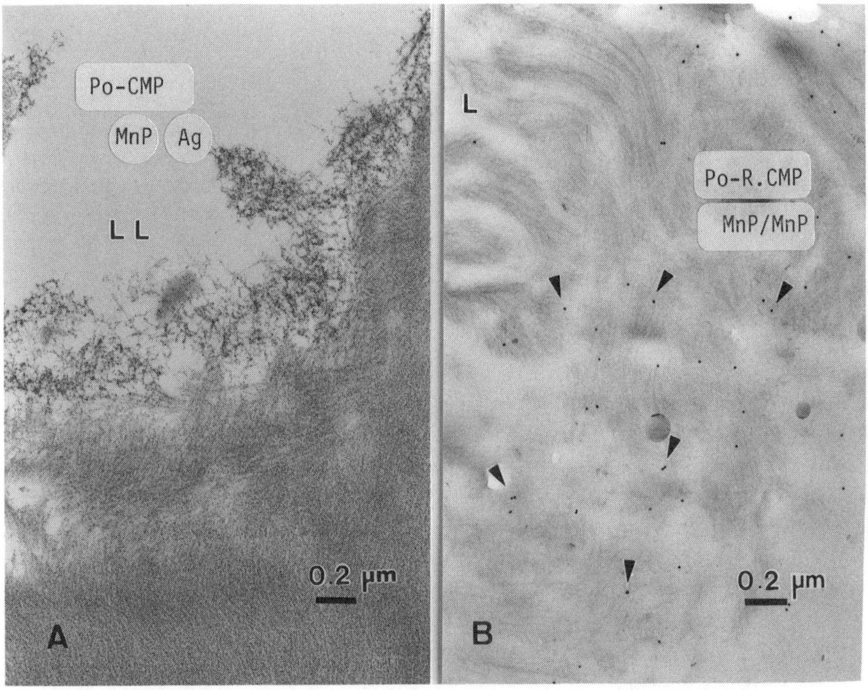

Figure 3. Poplar pulp treated with MnP. A, CMP. Inner part of a fiber showing cellulose microfibrils unmasked and randomly orientated. PATAg staining; B, immunogold labeling of MnP. The enzyme is associated with the fiber undergoing defibrillation.

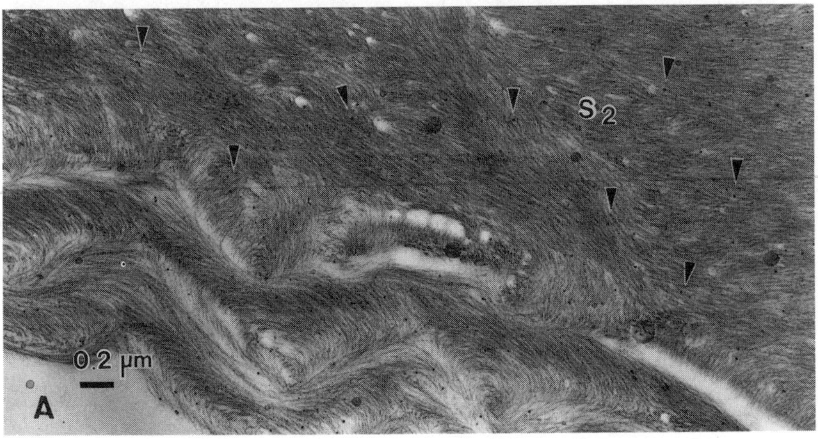

Figure 4. Effects of laccase on pulps. Immunogold labeling with anti-Lac. A, wheat straw CMP. Loosening of the inner part of S_2 depicting the defibrillation effect of Lac. The enzyme is present in the less opened areas (arrowheads); B, wheat straw R-CMP. Loosening of the whole fiber. Cell corners (Cc) and middle-lamella (ML) have been degraded leaving lignin spherical particles (arrows); C, poplar CMP. The dense gold labeling shows the penetration of Lac into the loosened areas. *Continued on next page.*

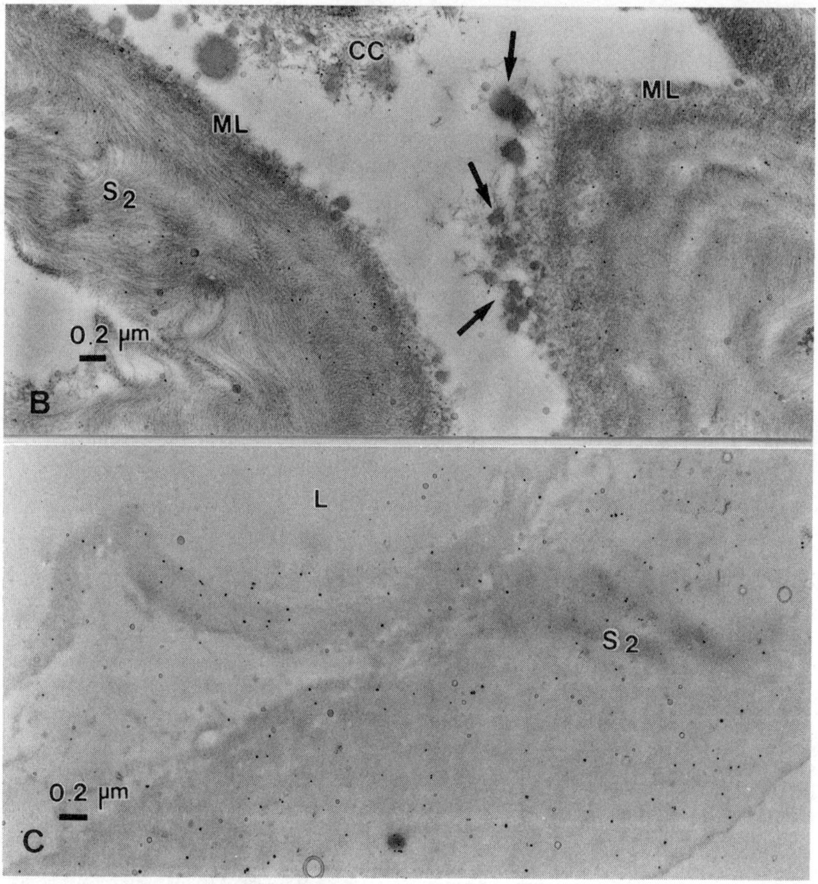

Figure 4. *Continued.*

penetration (Fig. 4C). Another feature which seemed to be associated with the action of Lac on poplar pulps was attack on middle lamellae. However, in contrast to what was observed with wheat straw pulps, no formation of spherical particles occurred. This suggests that the lignin moieties in the middle lamellae of hardwoods and grasses may be structurally different, and not undergo Lac-mediated oxidation with the formation of the same degradation intermediates.

Altogether, the action of Lac on high-yield poplar pulps was not as effective as that observed on wheat straw pulps. This difference in susceptibility to Lac oxidation between hardwood and grass lignin is consistent with hardwood lignins being extensively etherified at their phenolic groups (*29*). This should make them less susceptible to Lac oxidative cleavage by one-electron abstraction from phenolic substructures.

Relationship between the Enzyme Effects and the Nature of Lignin. Several studies have indicated that syringyl and guaiacyl moieties in lignin have different susceptibilities to oxidation by MnP and Lac systems (*28,30*). The nature of lignin substructures, whether condensed or noncondensed, also influences lignin depolymerization by the oxidative enzymes (*28,30-32*). It was therefore interesting to investigate the action of MnP and Lac on fibers in relation to lignin structures. This was done using antibodies directed against synthetic lignins (*32)*. Thus, labeling of guaiacyl-syringyl (GS) lignin with immunogold probes directed against dehydrogenation polymers (DHPs) respectively obtained by end-wise (ZT) or by bulk (ZL) polymerizations allowed us to distinguish the noncondensed lignins corresponding to the former from the more condensed ones corresponding to the latter (*33-34)*. Figures 5A and 5C show the dense labeling of untreated CMP from wheat straw with both probes. This indicated that GS lignin with noncondensed and condensed interunit linkages are common in the high-yield pulps (*35*). The noncondensed GS lignin appears more abundant in the S_2 than in the S_1 layer and is virtually absent in middle lamellae (Fig. 5A). This shows that different types of lignins are present in the different morphological zones of fibers. Labeling of MnP-treated straw pulps with the same probes resulted in a less intense gold particle distribution both for noncondensed and condensed GS lignins (Fig. 5B and 5D). In fibers whose structure had been opened by the enzyme, there was still a significant amount of noncondensed lignin present, whereas the labeling was much lower in zones in which defibrillation had occured (Fig. 5B). Labeling of the more condensed type of GS lignin showed that the diminution of number of gold particles due to MnP treatment of the pulp was also greater in the more defibrillated areas of fibers. This shows that MnP had an action both on condensed and noncondensed GS lignins. The stronger decrease in the labeling intensity provided by the anti-GS_{ZL} probe on MnP-treated pulp suggests that the enzyme removed more easily the condensed lignin units. This does not agree with the higher susceptibility of noncondensed substructures to MnP (*30*). However, the attack of lignin on the less resistant phenolic b-O-4 linkages by MnP may result in the solubilization of condensed fragments of lignin, thus explaining the reduction of the labeling with the anti-GS_{ZL} after MnP treatment.

In a similar approach, using the appropriate anti-guaiacyl antibodies (*33)* , it could be seen that guaiacyl lignin of the condensed type was largely represented in high yield

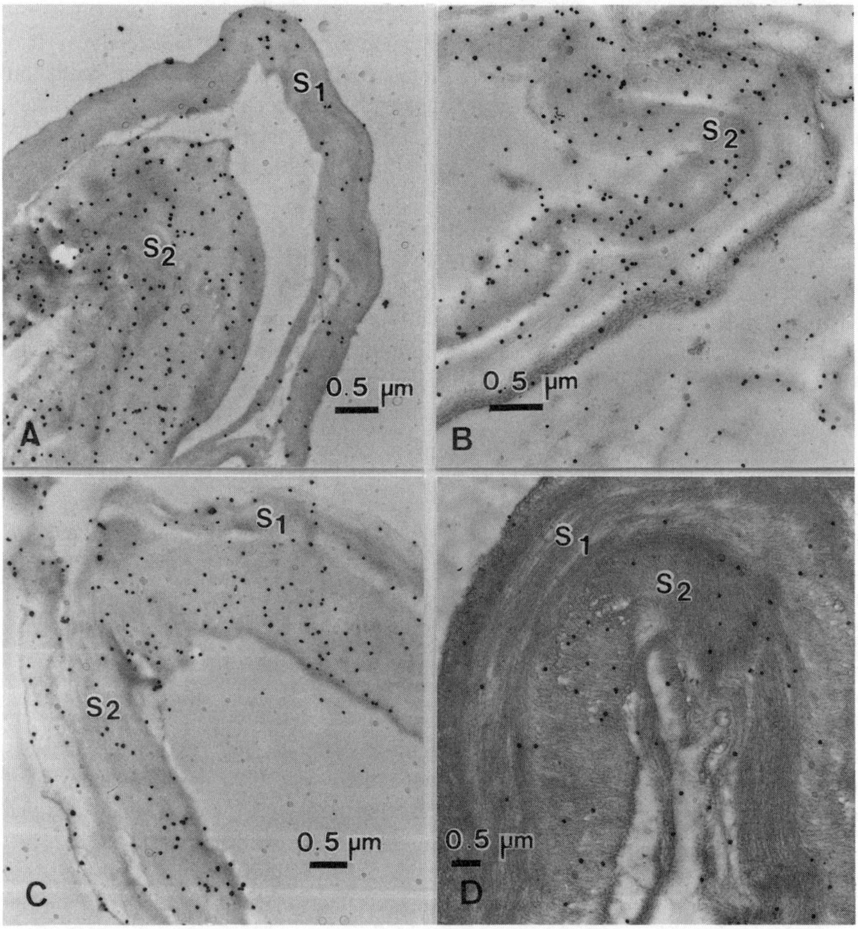

Figure 5. Specific immunogold labeling of lignins in wheat straw R-CMP treated with MnP. (A and B): anti-GS(ZT). A, untreated pulp. The labeling is more abundant in S_2 layer; B, MnP treated pulp. No change in the intensity of the gold labeling; (C and D): anti-GS (ZL). C, untreated pulp; D, MnP treated pulp. A strong decrease in the number of gold particles follows the action of the enzyme; (E and F): anti-G(ZL). E, untreated pulp; F, no diminution of the G(ZL) distribution after MnP treatment.

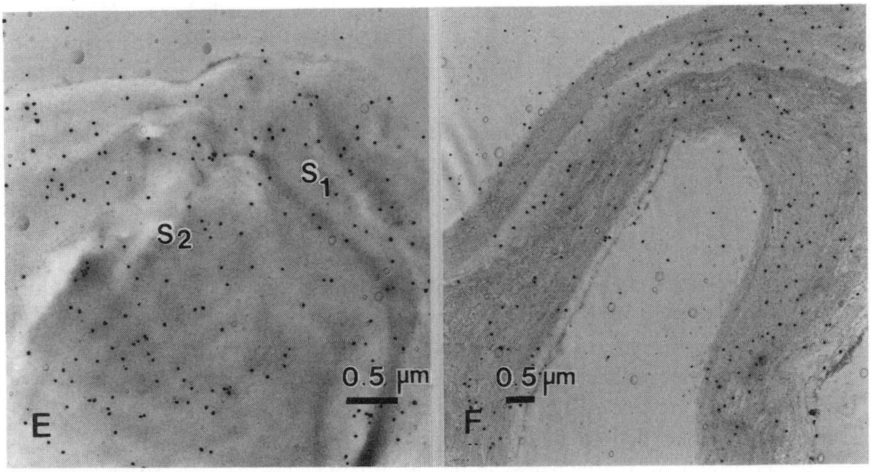

Figure 5. *Continued.*

wheat straw pulp (Fig. 5E), as expected from lignin analysis of wheat straw (*29,36*). The treatment with MnP did not induce significant loss of condensed guaiacyl lignin, as judged by the high intensity of the labeling (Fig. 5F) performed in parallel on the untreated and MnP-treated pulps, with the same antibody dilution. This indicates that G units and particularly the condensed type are more resistant to MnP than GS units, in agreement with studies on synthetic lignins (*30*) which demonstrated that G polymers were less rapidly oxidized than GS polymers. This was also the case when DHPs were degraded by *P. chrysosporium* (*37*).

The above results showing that guaiacyl-syringyl units, of the condensed type, were more degradable by MnP in straw high-yield pulps were confirmed in pulps from poplar wood. Here also, a substantial decrease in the labeling of GS lignins was observed after MnP treatment, whereas labeling of condensed G lignins was not significantly reduced. Interestingly, the treatment of poplar R-CMP pulp with Lac, beside inducing a clear diminution of GS units, resulted in a strong reduction in the labeling of condensed G lignins. This result is difficult to explain since *in vitro* Lac was reported as degrading GS noncondensed substructures more rapidly than G condensed linkages (*15,28*).

Another difference was observed in the labeling of G lignin after treatments with MnP and Lac-ABTS, respectively, showing that more G lignin was removed by Lac than by MnP. Both enzymes readily oxidize free phenolic units but Lac-ABTS can also oxidize nonphenolic units that are unattacked by MnP (*15,32*). This could explain the more extensive removal of guaiacyl moieties by Lac plus redox-mediator.

Conclusions

Transmission electron microscopy combined with immunocytochemical labeling of high yield pulps from wheat straw and poplar wood has shown that treatments with MnP and Lac performed local ultrastructural alterations in the fiber walls. Both enzymes were more effective after secondary refining of the pulps, the refining allowing improved penetration of the enzymes. This results in a softening and swelling of the fiber wall, and in an enhancement of defibrillation. It is important to note that the spatial delignification effects performed by the enzymes is largely oriented by the previous mechanical treatments which create the necessary openings for their penetration. This agrees with previous conclusions by Messner and Srebotnik (*38)* that biopulping with ligninolytic fungi does not remove the bulk of lignin but causes alterations in the cell wall structure.

Our observations show that both MnP and Lac affect the secondary wall structure in a similar way, although MnP does so more intensely . However, Lac showed a particular tendency to degrade lignin in middle lamellae and cell corners. This was more pronounced with wheat straw pulp, possibly because of the different nature of grass lignin in this cell wall area. The modification and removal of lignin by Lac in the middle lamellae could be a crucial factor leading to improved cell separation during biopulping. From the results of this study it seems doubtful that single isolated enzymes could perform extensive delignification of high-yield pulps at the first refining stage. However, if the enzyme treatment is placed after a second alkaline refining of CMP, the action of ligninolytic enzymes is more effective on fiber ultrastructure. The relationship between the nature of lignins, their ultrastructural localization and their susceptibility to lignin oxidative systems, when more clearly understood, should allow better uses of ligninolytic enzymes in pulp manufacture.

Acknowledgments

The authors wish to thank SAICA (Spain) and SICEM-SAGA (Italy) for the gift of pulps. This research was supported in part by grants from the EC programs AIR_2-CT_{93}-1219, and AIR_3-CT_{94}-2065.

Literature cited

1. Eriksson, K. E.; Kirk, T. K. In *Comprehensive Biotechnology*; 4, Moo-Young, M., Ed.; Pergamon Press, **1985**, pp. 271-295.
2. Kirk, T. K.; Burgess, R. R.; Konig Jr., J.W. In *Frontiers in Industrial Mycology*, Leatham, G. F., Ed.; Chapman & Hall: New-York, **1992**, p. 99.
3. Messner, K.; Srebotnik, E. *FEMS Microbiol. Rev.* **1994**, 13, pp. 351-364.
4. Akhtar, M. *Holzforschung* **1994**, *48*, pp. 199-202.
5. Leatham, G. F.; Myers, G. C. *Tappi J.* **1990**, *73*, pp.192-197.
6. Hatakka, A.; Mettälä, A.; Härkönen; Paavilainen, L. *Proceedings of the 6th International Conference on Biotechnology in the pulp and paper industry*; Messner, K., Srebotnik, E., Eds; Facultas-Universitätsverlag: Vienna, **1996**, pp. 229-232.
7. Reid, I. D., *TIBTECH*. **1991**, *9*, pp. 262-265.
8. Paice, M. G.; Jurasek, L.; Ho, C.; Bourbonnais, R.; Archibald, F. S. *Tappi J.* **1993**, pp. 217-221.
9. Tran, A.V.; Chambers, R. P. *Appl. Microbiol. Biotechnol.* **1987**, *25*, pp. 484-490.
10. Hirai, H. R.; Kondo, R.; Sakai, K. *Mokuzai Gakkaishi* **1994**, *40*, pp. 980-986.
11. Kondo, R.; Kurashiki, K.; Sakai, K. *Appl. Environ. Microbiol.* **1994**, *60*, pp.921-926.
12. Farrell, R.; U.S. Patent 4,690,895; **1987**.
13. Paice, M. G.; Reid, I.D.; Bourbonnais, R.; Archibald, F.S. *Appl. Environ. Microbiol.* **1993**, *59*, pp. 260-265.
14. Kondo, R.; Harazono, K.; Sakai, K. *Appl. Environ. Microbiol.* **1994**, *60*, pp. 4359-4363.
15. Bourbonnais, R.; Paice, M. G. *Appl. Environ. Microbiol.***1992**, *36*, pp. 823-827.
16. Sigoillot, J. C.; Petit-Conil, M.; Ruel, K.; Moukha, S.; Comtat, J.; Laugero, C.; Joseleau, J. P.; de Choudens, C.; Asther, M. *Holzforschung* **1997**, in press.
17. Katagiri, N.; Tsutsumi, Y., Nishida, T. *Appl. Environ. Microbiol.* **1995**, *61*, pp. 617-622.
18. Laugero, C.; Sigoillot, J. C., Moukha, S.; Frasse, P.; Bellon-Fontaine, M.N.; Bonnarme, P.; Mougin, C. *Appl. Microbiol. Biotechnol.* **1996**, *44*, pp. 717-723.
19. Gross, B.; Yonnet, G.; Picque, D.; Brunevie, P.; Corrieu, G.; Asther, M. *Appl. Microbiol. Biotechnol.* **1990**, *34*, pp. 371-387.

20. Ruel, K.; Goring, D. A. I.; Barnoud, F. *Wood Sci. Technol.* **1978**, *12*, pp. 287-291.
21. Joseleau, J. P.; Ruel, K. In *Biotechnology in the Pulp and Paper Industry*; Kuwahara, M.; Shimada, M., Eds.; UniPublishers CO, Ltd.: Tokyo, Japan, part II, **1992**, pp. 195-202.
22. Ruel, K.; Ambert, K.; Joseleau, J. P. *FEMS Microbiol. Rev.* **1994**, *13*, pp. 241-254.
23. Lobos, S.; Larrain, J.; Salas, D.; Cullon, D.; Vicuña, R. *Microbiology* **1994**, *140*, pp. 2691-2698.
24. Ruel, K.; Barnoud, F.; Eriksson, K. E. *Holzforschung* **1984, 38,** pp. 61-68.
25. Michalowicz, G.; Toussaint, B.; Vignon, M. *Holzforschung* **1991**, *45*, pp. 175-179.
26. Kallavus, U.; Gravitis, J. *Holzforschung* **1995**, *49*, pp. 182-188.
27. Brownell, H. H.; Saddler, J.N. *Biotech. Bioeng.* **1987**, *29*, pp. 228-235.
28. Higuchi, T. In *Plant Cell Wall Polymers, Biogenesis and Biodegradation*, Lewis, N.G., Paice, N. G. Eds.; ACS Symp. Ser., **1989**, *399*, pp. 482-502.
29. Lapierre, C. In *Forage Cell Wall Structure and Digestibility*, Jung, H.G.; Buxton, D.R.; Hatfield, R. D.; Ralph, J., Eds.; American Society of Agronomy Inc.: Madison, **1993**, pp. 133-163.
30. Wariishi, H.; Valli, K., Gold, M. H. *Biochem. Biophys. Res. Comm.* **1991**, *176*, pp. 269-275.
31. Hammel, K. .; Moen, M. A. *Arch. Biochem. Biophys.* **1991**, *242*, pp. 329-341.
32. Fukushima, Y.; Kirk, T. K. *Appl. Environ. Microbiol.* **1995**, *61*, pp. 872-876.
33. Ruel, K.; Faix, O.; Joseleau, J. P. *J. Trace Microprobe Techniques* **1994**, *12*, pp. 247-265.
34. Burlat, V.; Ambert, K.; Ruel, K.; Joseleau, J.P. *Plant Physiol. Biochem.* **1997**, *32*, in Press.
35. Chiang, V. L.; Funaoka, M. *Holzforschung* **1990**, *44*, pp. 309-313.
36. Lawther, J. M.; Sun, M.; Banks, W.B. *J. Agric. Food. Chem.* **1996**, *44*, pp. 1241-1247.
37. Faix, O.; Mozuch, M. D.; Kirk, T.K. *Holzforschung* **1985**, *39*, pp. 203-208.
38. Messner, K.; Srebotnik, E. *FEMS Microbiol. Rev.* **1994**, *13*, pp. 351-364.

Chapter 11

Applications of Enzymes in Paper Deinking Processes

A. L. Moerkbak and W. Zimmermann

Biotechnology Laboratory, Sohngaardsholmsvej 57, 9000 Aalborg, Denmark

Enzymes can be applied for the deinking of wastepaper for the recovery of secondary fibers. Mainly cellulases and hemicellulases have been utilized for the deinking of different types of newsprint and office waste papers. Ink removal efficiencies and brightness gains have been reported in several cases to be equal or superior to conventional chemical deinking procedures. Paper printed with vegetable oil based inks could be deinked efficiently in a deinking procedure including a neutral deinking chemical in combination with lipase, cellulase and xylanase. Treatment with lipase resulted in a higher decrease of residual ink area and dirt count compared to treatments with a combination of the enzymes which gave the highest increase in brightness of the deinked paper.

Worldwide, paper production has increased over the last ten years by 4% annually and is projected to further increase by 2% per year until year 2000 to about 300 million tons. Recycled paper will by then contribute more than 40% of the raw material supply for paper making while in the U.S., Japan and Germany, the recovery rates are expected to be well above 50%. Utilization of the secondary fiber from recycled paper requires effective removal of impurities including the printing inks in a deinking process. The printing inks are removed from the fibers using chemical and mechanical treatment to displace ink particles from the surface of the fibers. Conventional deinking chemicals can have a negative impact on both the environment and the quality of the recycled paper. On the other hand, treatments with enzymes have been shown to be effective in assisting or replacing the chemical deinking processes in addition to improving paper quality such as strength and dewatering properties (*1*). The ease with which the ink can be removed from the fibers is influenced by the type of inks and drying method used, the printing process and the age of the paper.

The ecological and qualitative demands on ink products have resulted in a continuous development of more environmentally friendly products. Inks based on vegetable oils (VOI) represent such advanced products, since they reduce VOC (volatile organic components) emission and are made from renewable resources unlike petroleum based inks. However, limited information is available on the properties of paper printed with these inks in the enzymatic deinking processes.

In this paper, the factors affecting the deinking behavior of recycled paper are summarized, some of the recent research on the enzymatic deinking of wastepaper is described and results from the enzymatic deinking of VOI printed paper are presented.

Paper Printing Processes

The paper printing processes used require different types of inks thereby influencing the deinking behavior of recycled paper (*2*). The two major printing techniques used are impact and nonimpact printing. Impact printing includes flexography, gravure, offset and letterpress which is often used for printing of newsprint, while nonimpact printing methods include laser printing and photocopying. Nonimpact-printed papers, which make up an increasing proportion of recycled paper, are generally more difficult to deink than impact-printed paper (*3*).

Ink Types and Deinking Behavior

Inks consist of pigments to supply color, and a vehicle to carry the pigment and bind it to the paper. Ink formulations also contain small amounts of modifiers such as drying aids or lubricants which, however, are not considered to affect the deinking behavior (*2*). The effectiveness of ink removal is to a large extent influenced by the type of vehicle and drying method used. In newsprint inks the vehicle consists of mineral or vegetable oil dispersing the pigment and alkyd resins or other hydrocarbon resins attaching the pigments to the surface of the fiber. The resins used depend on the drying mechanism by which the ink attaches to the paper: I) absorption into the paper, ii) evaporation of volatile components in the vehicle, iii) oxidation, where the solvents penetrate the sheet leaving a film of drying oil and resin forming a solid polymer and iv) radiation curing using ultraviolet or electron beam radiation forming a solid polymer by polymerization. Absorption and evaporation type inks are relatively easy to disperse or emulsify while the ink films formed by oxidation and radiation curing are more difficult to deink.

Efforts to replace the petroleum-based conventional inks beginning in the 1970s have resulted in the development of inks based on vegetable oils, mainly soybean oils. Already in 1992, 75 % of the daily newspapers in the U.S. were printed with VOI (*4*). VOI inks made from renewable and biodegradable resources are considered environmentally friendly since they facilitate waste treatment in recycling and reduce the emission of VOC (*5*). The so-called non-drying mineral oil based inks release VOC as they dry by evaporation. In contrast, the "semi-drying" VOI based inks are drying by oxidation with very little evaporation of organic compounds. The cross-linked ink film formed by oxidation of VOI based inks has been considered to result in a poorer deinking behavior (*2*). However, other studies have not confirmed a negative effect of VOI on the deinking process (*6*) and even stated that VOI printed papers could be deinked faster resulting in recycled fibers with improved quality (*7, 8*).

Conventional Wastepaper Deinking Methods

A number of processes in different combinations are used for recycled fiber deinking, with the main steps being pulping and chemical treatment for ink dispersion and emulsification followed by washing and/or flotation for separation of the inks from the fibers. If required, a bleaching step is added (*9*).

Chemical Deinking. The chemicals commonly used in conventional alkaline deinking processes include sodium hydroxide for ink dispersion and saponification of resins, and fatty acid soaps, polyphosphate, silicate and nonionic surfactants to promote ink dispersion and emulsification. Hydrogen peroxide, sodium hydrosulfite or chlorine compounds are used in the bleaching step (*9*). Some of these chemicals can create waste disposal problems and have a negative effect on the quality of the recycled paper such as yellowing following deinking at high pH values which makes a bleaching step necessary.

Neutral Deinking. This deinking process is performed at neutral pH with non-ionic detergents, for example fatty alcohol ethoxylates. The formation of alkali-solubilized contaminants and the yellowing of the fibers caused by high pH is prevented which makes the addition of bleaching chemicals such as hydrogen peroxide redundant. This results in a considerable reduction in chemical costs (*10*). However, lower brightness values of the recycled fibers are obtained compared to conventional deinking which may require a separate high consistency bleaching step or assistance by another deinking method.

Enzyme Assisted Deinking Methods

Previous work has shown that enzymes can be favorably used in the deinking process resulting in a reduction of residual ink areas and dirt count, brightness increases, and improvement of paper quality (*1*). Different approaches have been applied for enzymatic deinking of recycled paper. Cellulases, hemicellulases, peroxidases and laccases have been used to modify fiber surfaces resulting in the release of ink particles which can subsequently be removed in washing and flotation steps. On the other hand, esterases such as lipases can be used to degrade ink components directly thereby releasing pigments from the fibers.

Deinking Effect of Cellulases, Hemicellulases, Peroxidases and Laccases on Impact Printed Paper. The early research on enzymatic deinking of recycled paper was concentrated on the application of cellulases for the deinking of old newspaper (Table I). Deinking of offset print with an alkaline cellulase in combination with surfactants resulted in varying brightness gains (*11*) depending on the surfactant used. Other alkaline active cellulases were tested on newspaper in combination with deinking chemicals. The increase in brightness obtained was higher after sequential treatment compared to simultaneous cellulase and chemical treatment and could be further increased when the pH was shifted from alkaline to acidic after flotation (*12, 13*). Treatment of newspaper with cellulases from *Trichoderma reesei* at a low pH resulted in an increase in brightness which could not be raised further by increasing the enzyme concentration. This could be explained by excessive ink particle size reduction (*14, 15*). Increasing the pH or extending the reaction times resulted in a brightness decrease of the deinked paper. In contrast, an increase in brightness was obtained by increasing enzyme concentrations and reaction times in deinking of old newsprint with cellulases from *Trichoderma viride* and xylanases from *Aspergillus niger* (*16*). Deinking with a combination of cellulase and xylanase exceeded brightness gains obtained by chemical deinking.

Cellulase and xylanase preparations were also tested in low pH deinking processes of colored offset (*17*) and black and white letterpress printed newsprint (*18*). The highest brightness gains of colored offset newsprint were obtained after treatment with a mixture of cellulases and xylanases whereas the lowest residual ink area determined by image analysis was obtained after treatment with an enzyme preparation mainly containing xylanases (*17*). In contrast, black and white letterpress showed the highest brightness values after treatment with a xylanase mixture and the lowest residual ink area after treatment with a mixture of cellulases and xylanases (*18*) indicating that the deinking results can differ depending on the type of paper used and evaluation method applied. Flexographic-printed newspaper could be deinked with the same enzymes resulting in a release of the water-base ink and brightness levels well above those obtained with conventional deinking indicating that enzymes could be favorably employed in deinking of water-base ink printed wastepaper (*19*).

Table I. Enzymes used in deinking of impact and nonimpact printed paper.

Printing type	Enzymes	Reference
Impact:		
Newsprint	Cellulases	11 - 16
Newsprint, Magazines	Laccase	20, 26
Newsprint	Peroxidase	26
Newsprint	Cellulases/xylanases	16 - 19
Magazines	Cellulases	13
Newsprint	Lipases/esterases	27 - 31
Nonimpact:		
Xerographic print	Cellulases	23
Laser and Xerographic print	Cellulases	24, 25
Laser and Xerographic print	Cellulases/xylanases	24

In a different approach to enzymatic ink removal with hydrolase-type enzymes like cellulases and hemicellulases, oxidoreductive enzymes involved in lignin degradation have been tested and comparable results to conventional deinking have been reported by treatment of recycled paper with laccases (*20*).

Deinking Effect of Cellulases, Hemicellulases, Laccases and Peroxidases on Nonimpact Printed Paper. Non-impact printed paper such as recovered laser and xerographic printed office paper is difficult to deink by conventional deinking methods due to cross linking and strong binding of the thermoplastic ink. The toner particles need to be broken up and reduced in size after release from the fibers to be efficiently removed in the flotation step which, when using chemical deinking, can only be achieved by strong mechanical action detrimental to the fiber quality (*21*). Enzymatic deinking of this type of wastepaper has shown promising results (Table I). Laser and xerographic printed paper treated with cellulase (*22, 23*) or cellulase and xylanase combinations (*24*) resulted in reduced residual ink areas. However, increases in brightness compared to chemical deinking could not be obtained. On the other hand, an increase in brightness and an improvement of the dewatering properties of the pulp fiber (freeness) has been reported after treatment of laser and xerographic wastepaper with a pure alkaline cellulase (*25*).

The deinking effect of peroxidases and laccases has also been tested on a blend of nonimpact printed wastepaper. The brightness obtained was comparable to that of conventional deinking (*26*).

Deinking Effect of Lipase on Soybean Oil Based Ink Printed Paper. The enzymatic deinking with esterases or lipases involves a partial degradation of the carrier or binder in oil-based inks (*27-31*). Lipases could therefore be expected to be effective in the deinking of VOI printed wastepaper. The deinking effect of a lipase, cellulase, and hemicellulase preparation was studied in combination with a neutral deinking system consisting of a surfactant and $NaSiO_3$. The deinking effects were evaluated by the use of image analysis and brightness measurements. As shown in Figure 1, the brightness of the deinked paper was increased by by 2.02-5.81 points in all cases of enzyme treatment compared to a treatment with the neutral deinking chemical. The highest brightness increase was obtained by treatment with a combination of cellulase, xylanase and lipase

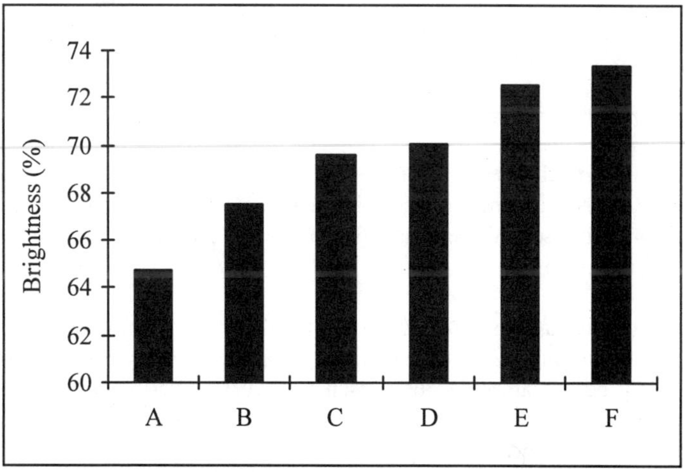

Figure 1. Brightness of VOI printed paper treated with neutral deinking chemicals, (B); with addition of: xylanase, (C); cellulase, (D); lipase, (E); a combination of the three enzymes, (F). Untreated control, (A).

To evaluate the effect of the enzymes added to the neutral deinking chemicals, the dirt count and residual ink areas of enzyme treated paper were compared to those after neutral chemical deinking. The measured dirt count and the residual ink area was increased after treatment with xylanase and cellulase whereas treatment with lipase alone and in combination with the xylanase and cellulase decreased the dirt count and residual ink (Figure 2). The highest decrease in dirt count and residual ink areas was obtained by treatment of the paper with the lipase. As shown in Figure 3, 64% of particles in all size ranges have been removed. In contrast, treatment with the xylanase or cellulase resulted in an increase of the number of particles in the small and medium size range, while the number of particles in the size range from 0.08 - 0.4 mm^2 was decreased resulting in the overall increase in the dirt count seen in Figure 2.

The increase in the small particle size range could also explain the observed increases in residual ink area after xylanase and cellulase treatment since a larger population of small particles increases the total surface area of the particles and thereby the residual ink area. The increase in dirt count and residual ink area of samples treated with xylanase and cellulase, however, did not correlate with the measured increases in brightness.

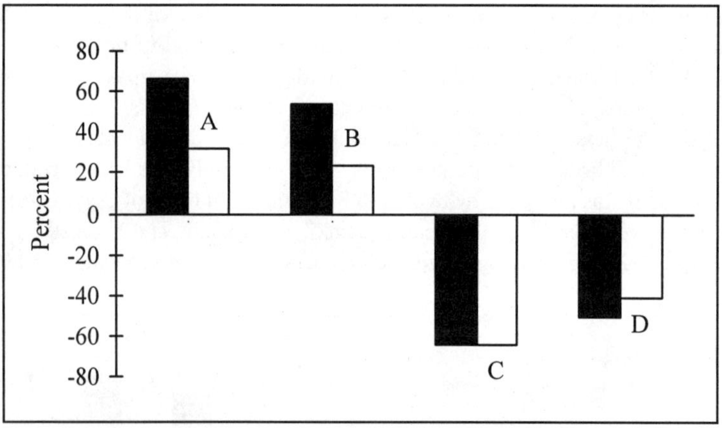

Figure 2. Percent changes in dirt count (black bar) and residual ink area (white bar) of VOI printed paper treated with neutral deinking chemicals and with addition of: xylanase, (A); cellulase, (B); lipase, (C); and a combination of the three enzymes, (D).

Autoclaved commercial lipase only gave a small decrease in dirt count and residual ink area and no brightness increase indicating that the deinking effect of the lipase can not solely be explained by the presence of surface active compounds in the enzyme preparation, as has been suggested recently for cellulase-assisted deinking (*32*) or by surface activity of the lipase.

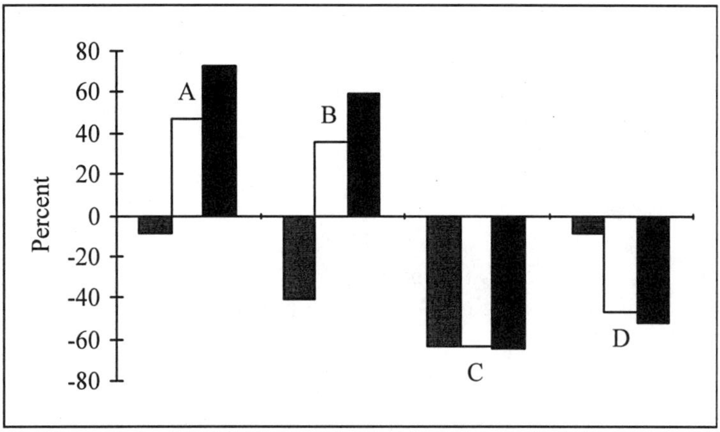

Figure 3. Percent change of particles in the size range 0.08 - 0.4 mm^2 (striped bar); 0.005 - 0.08 mm^2 (white bar); and < 0.005 mm^2 (black bar) on VOI printed paper treated with neutral deinking chemicals and with addition of: xylanase, (A); cellulase, (B); lipase, (C); and a combination of the three enzymes, (D).

Conclusion

The enzymes applied in a deinking process need to be carefully selected since the process conditions such as pH, temperature, pulp consistency, and mechanical stress have a large effect on their activity and stability (*33, 34*). The results from enzymatic deinking trials also indicate that the enzymes used in the enzymatic deinking process should be selected according to ink, printing and wastepaper type.

Generally, enzymes seem to effect a reduction of ink particle size, independent of the type of paper or ink used (*1*). The reported results concerning increases in brightness of enzyme-treated wastepaper with cellulases and hemicellulases are in some cases contradictory and also difficult to compare, since various deinking procedures have been applied by different authors (*1, 32*).

Several mechanisms have been suggested to explain the effect of cellulases in deinking. The enzymes could act by a peeling process of cellulose microfibrils (*35*) or by hydrolysis of fines clogging the paper sheet on the machine wire which could contribute to the often observed increase in dewatering properties of the paper (*36*). A hydrolysis of easily accessible cellulose chains could be supported by mechanical action in the pulper (*37*) or cellulases could act by removal of cellulosic fines and microfibrils (*36*). While xylanases are thought to release ink and lignin from the fiber surface by cleavage of the xylan back-bone releasing xylodextrins and attached lignin (*16*), very little is known about the mechanism by which the oxidoreductive enzymes effect ink removal. A thorough understanding of the mechanism by which the enzymes and a specific type of paper and ink interact is necessary for the selection of the most efficient enzymatic deinking process. This topic needs further research, since only limited information is available confirming general hypotheses.

Esterases and lipases are efficient in deinking due to a partial degradation of the carrier or the binder in oil-based inks. This makes lipases the enzymes of choice in deinking processes for vegetable oil based printed wastepaper. In combination with neutral deinking, they are part of an environmentally friendly deinking process with reduction in chemical costs and waste problems.

Acknowledgment

This work was supported by a grant from the Danish Research Academy to Anne Louise Mørkbak. We thank Kolb AG, Switzerland for providing the surfactant and Akzo Nobel A/S, Denmark for providing paper printed with vegetable oil based ink.

LITERATURE CITED

1. Welt, T; Dinus, R. J. *Progr. in Paper Recycl.* **1995**, *4*, pp. 36-47.
2. Borchardt, J. K. *Progr. in Paper Recycl.* **1995**, *11*, pp. 81-87.
3. Vidotti, R. M.; Johnson, D. A.; Thompson, E. V. *Proc. Tappi Pulping Conf.* **1992**, pp. 643-652.
4. Anon. *Soy Ink Executive Summary*, American Newspaper Publishers Association 1992.
5. Power, J. *NZ Printer Magazine* **1992,** *September/October*, 52.
6. Rosinski, J. *Progr. in Paper Recycl.* **1995**, *4,* pp. 55-62.
7. Anon. *Deinking of Soy Inks During Paper Recycling*. Final Report, Western Michigan University **1991.**
8. Anon. *Newspaper Focus* **1992**, *5*, p. 25.
9. Biermann, C. J. *Essentials of pulping and papermaking;* Academic Press: San Diego, Ca, **1993**, pp. 263-282.
10. Schmid, H. *Wochenblatt fŸrPapierfabrikation* **1995**, *14/15*, pp, 621-626.
11. Urushibata, H. *Jap. Pat. 59 009 299* **1984.**
12. Nomura, Y.; Shoji, S., *Jap. pat. 59 494* **1988.**
13. Baret, J-L . A. G; Leclerc, M.; Lamort, J-P. *US pat. 5 364 501* **1994.**
14. Kim, T. J.; Ow, S. S.-K.; Eom, T-J. *Proc. Tappi Pulping Conf..* **1991**, *2*, pp. 1023-1030.
15. Ow, S. S.-K. and Eom, T.-J. Proc. EUCEPA Symp. Additives, Pigments and Fillers Pulp and Paper Industry, **1990**, pp. 85-94.
16. Paik, K.H.; Park, J.Y. *J. of Korea TAPPI* **1993**, *25*, pp. 42-52.
17. Prasad, D. Y.; Heitmann, J.A.; Joyce, T. W. *Nordic Pulp and Paper Res. J.* **1993**, *2*, pp. 284-286.
18. Prasad, D. Y.; Heitmann, J. A.; Joyce, T. W. *Progress in Paper Recycling* **1992**, *1,* pp. 21-30.
19. Prasad, D. Y.; Heitmann, J. A.; Joyce, T. W. *Parpiripar* **1992**, *36*, p. 122.
20. Call , H. P., Strittmatter, G. *Das Papier* **1992** *46*, pp. V32-37.
21. Woodward, T. W. *PIMA Mag.* **1993**, pp. 49-51.
22. Abubakr; S., Klungness, J. H., Gaumnitz, L. *Progr. in Paper Recycl.* **1994,** *3*, pp. 17-23.

23. Jeffries, T. W.; Klungness, J. H.; Sykes, M. S.; Rutledge-Cropsey, K. R. *Proc. Tappi Recycl. Symp.* **1993**, pp. 183-188.
24. Jeffries, T. W.; Klungness, J. H.; Sykes, M. S.; Rutledge-Cropsey, K. R. *Tappi J.* **1994**, *77*, pp. 173-179.
25. Prasad, D. Y. *Appita* **1993**, *46*, pp. 289-292.
26. Call, H. P.; Von Raven, A; Leyerer, H. *Das Papier* **1990,** *44*, pp. V33-41.
27. Sugi, T.; Nakamura, K. *Jap. pat.* 249,291/91 **1990.**
28. Fukuda, S.; Hayashi, S.; Ochiai, H.; Ihizumi, T.; Nakamura, K. *Jap. pat.* 229,290/90 **1990.**
29. Sharyo, M.; Sakaguchi, H. *Jap. pat.* 2,160,984 **1990.**
30. Guy Vare, J. A.; Lucrelk, M.; Sharyo, M.; Sakaguchi, H. *Jap. pat.* 150,984/90 **1990.**
31. Nakano, J. *J. of Korea Tappi* **1993**, *25,* pp. 85-91.
32. Xia, Z.: Baudry, A.-R.; Bourbonnais, R. *Progr. in Paper Recycl.* **1996** *5*, pp. 46-58.
33. Kochavi, D.; Videbaek, T.; Cedroni, D. *Am. Dyestuff Reporter* **1990**, *79*, pp. 24-28.
34. Daneault, C.; Leduc, C.; Valade, J. L. *Tappi J.* **1994**, *77*, pp. 125-131.
35. Eom, T. J.; Ow, S. S. *Brit. pat.* 2231595, **1990**.
36. Jackson, L. S.; Heitmann, J. A.; Joyce, T. W. *Tappi J.* **1993**, pp. 147-154.
37. Zeyer, C.; Joyce, T. W.; Heitmann, J. A.; Rucker, J. W. *Tappi J.* **1994**, 77, pp. 169-177.

Chapter 12

Characterization of Oxidoreductases, Ferrireductase, and Manganese Peroxidase, from White-Rot Fungi

R. Kondo, H. Hirai, K. Harazono, and K. Sakai

Department of Forest Products, Kyushu University, Fukuoka 812–81 Japan

Manganese dioxide was reduced to Mn(II) in a liquid culture of *Phanerochaete sordida* YK-624 which showed a high bleaching ability of unbleached hardwood kraft pulp. The reduction of manganese dioxide was mediated by Fe(II)-chelator complex, and enhanced when NADPH was added as an electron donor. Fe(III)-chelator, produced when manganese dioxide is reduced to Mn(III) or Mn(II), was reduced by the cell-free extract from the fungus. NADPH-dependent ferrireductase was purified to apparent homogeneity from the cell-free extract. This enzyme was obtained for the first time from wood-rotting fungi. It is concluded that the NADPH-ferrireductase was involved in the reduction of manganese dioxide for production and function of manganese peroxidase (MnP). MnP isoenzymes from liquid culture of white rot fungus *Ganoderma* sp. YK-505 were purified and characterized. They exhibited stabilities against high temperature and pH comparing with MnP from *Phanerochaete chrysosporium.* It is suggested that MnP from *Ganoderma* sp. YK-505 may be useful for biotechnological applications and studies of the relationship between structure and function of MnP isoenzymes.

Several species of white-rot fungi have been studied intensively in recent years because of their abilities to preferentially degrade and remove the lignin from wood. Typical white-rot fungi, *Phanerochaete chrysosporium* (*1,2*) and *Trametes versicolor* (*3,4*) have the ability to bleach kraft pulp. It has been reported recently that newly isolated lignin-degrading fungi such as IZU-154 (*5,6*) *Phanerochaete sordida* YK-624 (*7,8*) and unknown species SKB-1152 (*9*) have shown higher bleaching abilities than typical white-rot fungi, and that the activity of manganese peroxidase (MnP) was dominant in the biological bleaching of kraft pulp with *P. chrysosporium*, *T. versicolor*, and *P. sordida* YK-624 (*1,7,10*). More recently, we have reported that MnP was a key enzyme in the biological bleaching of unbleached hardwood kraft pulp (UKP) with *P. sordida*

YK-624 due to the presence of manganese ions in the UKP which were necessary to bleach the UKP (*11*). Thus, it is assumed that mainly MnP is involved in the biological bleaching of kraft pulp with white-rot fungi compared with lignin peroxidase and laccase. Bleaching of UKP by isolated MnP was first reported by Paice *et al.* (*10*) and Kondo *et al.* (*12*) have found that the bleaching of UKP was conducted successfully with partly purified MnP secreted from *P. sordida* YK-624.

The Mn(II) ion is necessary to bleach UKP with *P. sordida* YK-624 because of the production (*13*) and function (*14*) of MnP. UKP contains about 50 mg/kg pulp of the element Mn (*10,11*), and *P. sordida* YK-624 utilizes this element during the biological bleaching of UKP because this fungus produces MnP and brightened UKP in a culture containing only UKP and water (*7,8*). However, the brightness increase of UKP during the *in vitro* MnP treatment had not been observed without the addition of $MnSO_4$ (*12*). It is expected that the dominant species of Mn in UKP is Mn(IV), such as manganese dioxide, but Mn(II) is present in small quantities. Hence, it is thought that *P. sordida* YK-624 reduces the manganese dioxide present to Mn(II) during the biological bleaching of UKP. One possibility is that some low molecular weight compounds may stoichiometrically reduce manganese dioxide (*15*). Recently, Roy *et al.* have reported that cellobiose:quinone oxidoreductase (CBQase) from *T. versicolor* efficiently catalyzed the reduction of manganese dioxide to Mn(II) and Mn(III) (*16*). In the present study, we have shown the NAD(P)H-dependent reduction of manganese dioxide which is mediated by ferrous chelate in the white-rot fungus *P. sordida* YK-624.

Many ligninolytic fungi produce MnP in various cultures such as synthetic liquid and solid-state media (*17, 18*). Enzymological and kinetic studies of the isozymes could be useful for understanding catalytic characteristics and improving the stability of peroxidases. In a previous study, ligninolytic fungi with high selectivity for lignin were screened and isolated from decayed wood samples, and biological bleaching of hardwood kraft pulp with those fungi was carried out (*7*). Although in most fungi, pulp brightness increases were shown to be related to MnP activities, *Ganoderma* sp. YK-505 deviated from this relationship. Therefore, we expected MnP from *Ganoderma* sp. YK-505 might have different properties when compared to other white rot fungi. In this study, the properties of purified MnPs from liquid cultures of *Ganoderma* sp. YK-505 and *P. chrysosporium* were compared.

NADPH-dependent ferrireductase from *Phanerochaete sordida* YK-624

Manganese dioxide-reduction by *P. sordida* YK-624. Two fungal disks of *P. sordida* YK-624 were added to each 100 ml Erlenmeyer flask containing 10 ml of $MnSO_4$-lacking Kirk and Tien (KT) medium (*19*) supplemented with 1% glucose, 12 mM ammonium tartrate, and 20 mM 2,2-dimethylsuccinate (pH 4.5), and incubated at 308 °C for several days. One ml of sterilized (1218 °C, 15 min.) 23 mmol/ml manganese dioxide suspension was added to each flask on an indicated day. After the incubation, 1 ml of 100 mM ethylenediaminetetraacetic acid · 2Na (EDTA) aq. was added to the reaction mixture, and the concentration of the Mn(II)-EDTA complex was analyzed by HPLC. Fig. 1 shows a change of manganese dioxide-reducing activity by the fungus during the cultivation. The data represents the concentration of Mn(II) accumulated during a 24-hour incubation after the addition of 23 μmol manganese dioxide to 10 ml of the cultures at indicated days. The reduction of manganese dioxide by fungal culture was observed at all cultivation times, and on days 3-7, particularly. In the fungal culture, CBQase activity, which can reduce manganese dioxide (*16*) was not detected in the culture and cell-free extract.

We examined the characteristics of the manganese dioxide-reduction system in washed mycelia of the fungus. Standard reaction mixtures of 10 ml consisted of 20 mM 2,2-dimethylsuccinate buffer containing 0.1% Tween 80 (pH 4.5), 23 μmol manganese dioxide, and the washed mycelium (about 40 mg dry weight). To determine the effect of metal ions or chelator, metal elements based on KT medium (MEK), MEK without nitrilotriacetate (NTA), MEK without $FeSO_4$, MEK without $ZnSO_4$, MEK without $CoSO_4$, or MEK without $CuSO_4$ were added to the standard reaction mixture. Reactions were made at 308C for 24 hours, and then 1 ml of 100 mM EDTA aq. was added to each reaction mixture. Mn(II)-EDTA complex was assayed. Fig. 2 shows the effects of the elements of MEK on manganese dioxide-reduction by the washed mycelia. The addition of MEK stimulated the manganese dioxide-reducing activity of the mycelia. To clarify the elements involved in the reduction of manganese dioxide, the manganese dioxide-reducing activity of the mycelia in MEK without NTA, Fe(II), Zn(II), Co(II), or Cu(II) were determined. No manganese dioxide-reducing activity was detected in the absence of NTA, and decrease of the activity was observed by the lack of Fe(II).

Reduction of ferric chelate by ferrireductase. Mycelial mats grown in KT culture containing 1.7 mM $MnSO_4$ and 0.13 mM $FeSO_4$ were taken from the culture on day 5 and were washed with ice-cold distilled water. Mycelial mats were homogenized with 700 ml of extraction buffer (20 mM phosphate buffer (pH 7.0) containing 0.05% Tween 80 and 0.004% phenylmethylsulfonyl fluoride in a Waring blender at a high speed (15200 rpm) for 5 min (20 sec blending with 15 min intervals). The homogenate was centrifuged at 7000 rpm for 30 min. The supernatant was fractionated by sequential additions of solid ammonium sulfate, with each addition being followed by centrifugation at 7000 rpm for 40 min. The reduction of ferric chelate was measured by monitoring absorbency at 510 nm using 1.5 mM ferrous-specific chelate 1,10-phenanthroline (PHT) (*20*). The reaction mixture contained 20 mM sodium phosphate buffer (pH 7.0), fractionated cell-free extract, 0.1 mM $FeCl_3$, 0.1 mM NTA, 0.1 mM NADH or NADPH, and 1.5 mM PHT. As shown in Table 1, the ferrireductase activities were detected in the cell-free extract, and NADPH-dependent ferrireductase activity was almost the same as NADH-dependent activity. In ammonium sulfate-fractionation, ferrireductase activity was detected in the protein precipitated between 30 and 65% ammonium sulfate saturation. However, NADH-dependent ferrireductase activity, which was also detected between 30 and 65% ammonium sulfate saturation, decreased during the fractionation.

Table 2 summarizes the purification procedure from 260 g (wet weight) of cells grown in the liquid media. After the Mono Q chromatography, SDS- and IEF-PAGE analysis of the active fraction revealed that the protein consisted of single band. The

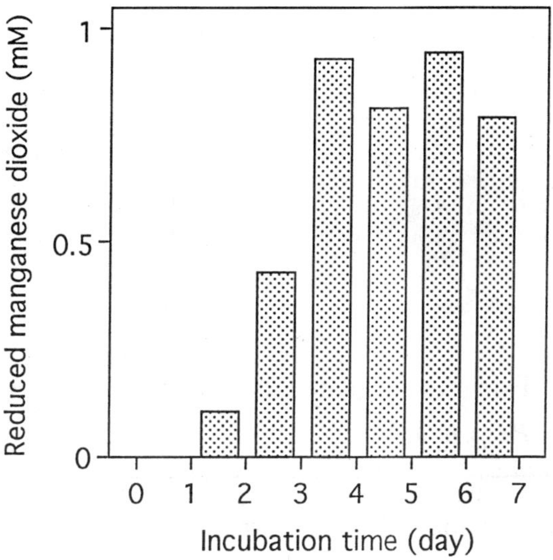

Figure 1. Reduction of manganese dioxide by the fungal culture system. (Reproduced with permission of Hirai et al., Mokuzai Gakkaishi. 43, 249, 1997. Copyright 1997 The Japan Wood Research Society)

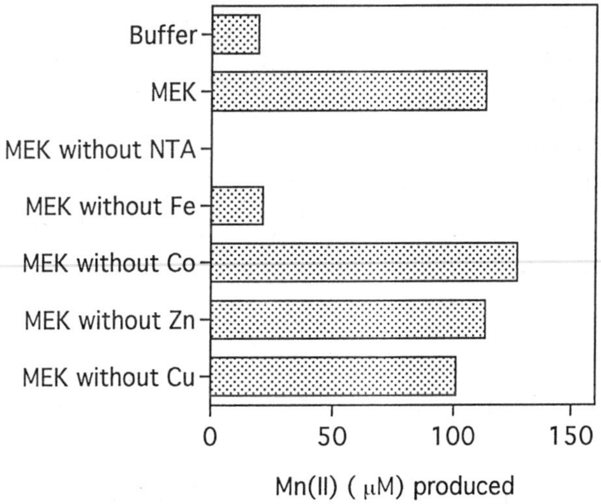

Figure 2. Effects of elements of MEK on the reduction of manganese dioxide by washed mycelia in 20 mM 2,2 -dimethylsuccinate buffer (pH 4.5). (Reproduced with permission of Hirai et al., Mokuzai Gakkaishi. 43, 250, 1997. Copyright 1997 The Japan Wood Research Society)

Table 1. Ferrireductase activity in cell-free extract of *P. sordida* YK-624.

Enzyme solution	Ferrireductase activity (nkat)	
	NADPH	NADH
Cell-free extract	219	243
Ammonium sulfate fractionation		
0-30%	trace	2
30-65%	177	73
65%-	0	0

Table 2. Purification of NADPH-dependent ferrireductase from *P. sordida* YK-624.

Step	Volume (ml)	Protein (mg)	Total act.(nkat)	Specific act.(nkat/mg)	Yield (%)	Purification (fold)
Cell-free extract	830	703	2127	3	100	1.0
$(NH_4)_2SO_4$ precipitate	250	293	1285	4	60	1.5
Phenyl Toyopearl	50	22	360	16	17	5.5
Superdex 75	20	3.7	220	60	10	19.8
Mono Q	7.5	9.8×10^{-2}	102	1041	4.8	346.9

overall enzyme yield was 4.8%, with a concomitant 346.9 -fold purification. Several electron donors and acceptors were tested as substrates (Table 3). When NADPH was used as electron donor, Fe(III)-NTA was significantly reduced by the purified protein, but the reduction hardly occurred with NADH. No reduction of Fe(III)-NTA was observed when cellobiose and succinate were used as electron donors. NADPH oxidation was observed when Fe(III)-NTA, Mn(III)-malonate, and methoxybenzoquinone were used as electron acceptors. Particularly, the enzyme showed a greater rate of NADPH-oxidation for Fe(III)-NTA than for the other electron acceptors. No oxidation of NADPH was observed when 2-hydroxy-1,4-naphthoquinone, phenazine methosulfate, plumbagin, and veratraldehyde were used.

Mechanism of manganese dioxide-reduction by *P. sordida* YK-624. The proposed mechanism for the reduction of manganese dioxide by *P. sordida* YK-624 is shown in Scheme 1. It is suggested that the reduction of manganese dioxide was performed by Fe(II) complex, and that Fe(III) complex produced was reduced to regenerate Fe(II) complex by an intracellular NAD(P)H-dependent ferrireductase system. The proposed mechanism was supported by the following observations: (I) The reduction of manganese dioxide by the washed mycelia was dependent on the Fe ion and NTA; (ii) Manganese dioxide-reducing activity was roughly proportional to the concentrations of Fe(II)-NTA and NADH; (iii) Ferric chelate was reduced by the cell-free extract from *P. sordida* with NAD(P)H, and (iv) NADPH-dependent ferrireductase was purified from the cell-free extract.

The siderophores are biosynthesized by organisms under negative iron control, and they are released to the environment where the ferrie ion-siderophore complexes are produced. These complexes are subsequently taken up by the microorganisms. In bacteria, it is known that this uptake mechanism is often specific and that it involves a number of proteins, which are also produced only under negative iron control. In fungi, this is presumed to be the case as well, but no transport proteins have been identified so far (*21*). As shown in Scheme 1, it is expected that iron such as iron-protein complexes are assimilated and secreted by the fungus. Fe(III)-protein complex would be directly reduced by the ferrireductase, and the produced Fe(II)-protein complex probably secretes to the extracellular region. This reduction system probably is involved in the acquisition of Mn(II) which is important in the production and function of MnP during the biological bleaching of UKP with *P. sordida* YK-624.

Characterization of MnP isoenzymes from *Ganoderma* sp. YK-505.

For MnP preparation, a carbon and nitrogen-limited medium as described by Kondo *et al.* (*12*) was used. *Ganoderma* sp. YK-505 was cultivated in 500-ml Erlenmeyer flask containing 200 ml of liquid culture medium. After 7 days of incubation, the supernatant was separated from the mycelium by filtration through glass fiber. The liquid was frozen at -208C. After removing the extracellular glucan by filtration, the supernatant was dialyzed against 10 mM sodium acetate buffer (pH 4.5). The liquid was concentrated to a volume of about 200 ml by ultrafiltration with nitrocellulose membrane (10-kDa cut

Table 3. Substrate specificity of ferrireductase from *P. sordida* YK-624[a].

Substrate	Relative act.(%)
Electron donor	
NADPH	100.0
NADH	1.4
Cellobiose	0
Succinate	0
Electron acceptor	
Fe(III)-NTA	100.0
Mn(III)-malonate	46.1
Methoxybenzoquinone	35.8
Veratraldehyde	0
2-Hydroxy-1,4-naphthoquinone	0
Phenazine methosulfate	0
Plumbagin	0

[a] The effects of electron donors were determined with 100 μM Fe(III) -NTA, 127 ng purified protein, and 50 μM electron donor.
The effects of electron acceptors were determined with 100 μM electron acceptor, 127 ng purified protein, and 50 μM NADPH.
In the case of Fe(III)-NTA, 1.5 mM PHT was added to the reaction mixtures.

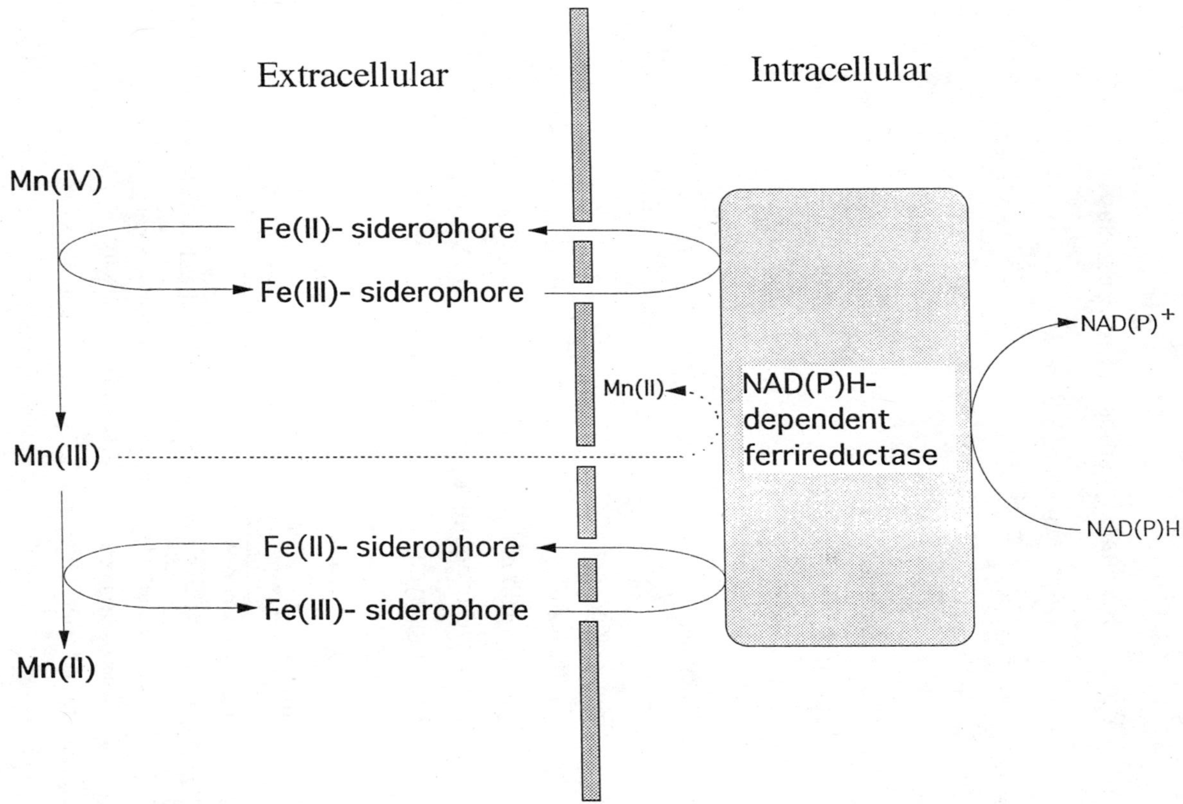

Scheme 1. Proposed mechanism of manganese dioxide-reduction by *P. sordida* YK-624.

off). Polyethylene glycol (average molecular weight, 3000) was added to the liquid to make a 5% solution. The pH was then adjusted to 7.2 with 5 M aqueous NaOH. After the slime was filtered off, the filtrate was subjected to chromatography. MnP isoenzymes were purified using DEAE-Sepharose CL-6B, Sephacryl S-200, phenyl Toyopearl and Mono Q chromatography. Firstly, anion-exchange chromatography with DEAE-Sepharose CL-6B revealed two MnP fractions (G-1 and G-2), one of which contained phenol oxidase (PhO) activity (Fig. 3A). After chromatography with Sephacryl S-200 and phenyl Toyopearl, each MnP fraction were subjected to Mono Q column chromatography (Fig. 3B and C). G-1 and G-2 fractions eluted on Mono Q column chromatography were used for experiments described below. The molecular masses of both MnPs were 43 kDa as determined by SDS-PAGE. The visible spectra of native MnPs included Soret bands near 408, 502 and 640 nm. The spectra of H_2O_2-oxidized

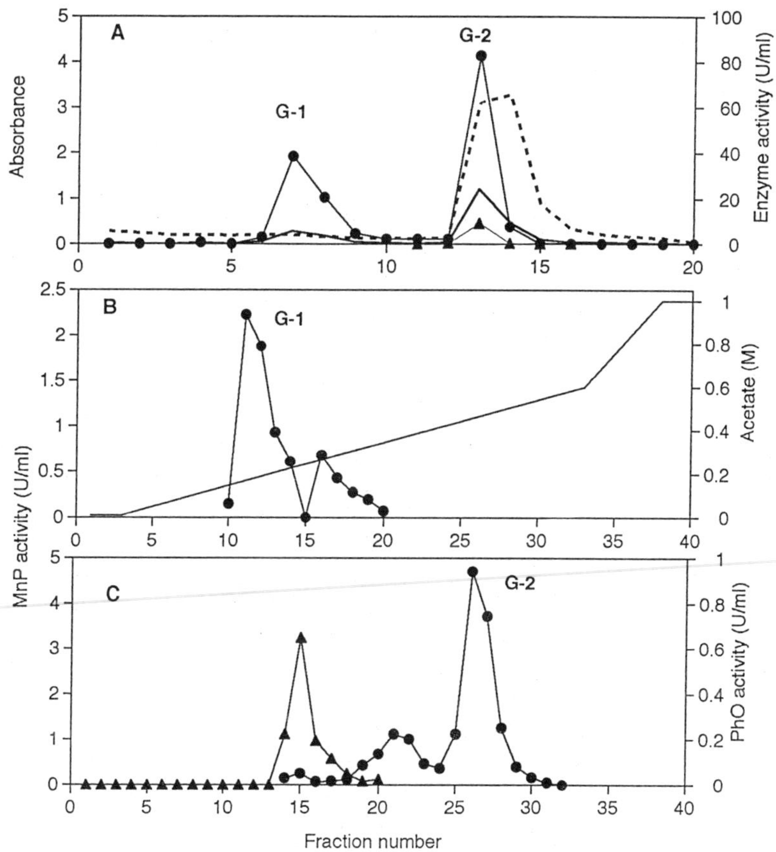

Figure 3. Chromatographies of filtrate of culture of *Ganoderma* sp. YK-505 on DEAE-Sepharose (A) and Mono Q (B and C). Symbols: ●, MnP activity; ▲, PhO activity ; —, absorbance at 405 nm; ---, absorbance at 280 nm.

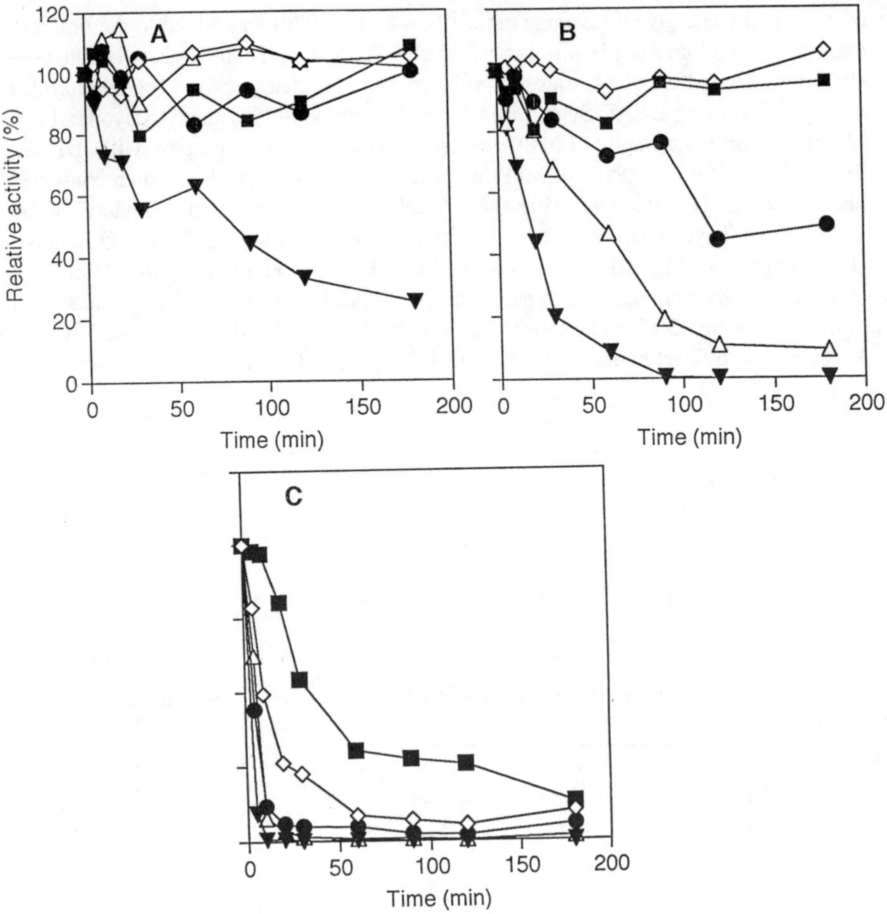

Figure 4. Stability of G-1 (A), G-2 (B) and *P. chrysosporium* MnP (C) during incubation at pH 4.5 and various temperatures. Symbols: ■, 45℃; ◇, 50; ●, 55℃; △, 60℃; ▼, 65℃.

MnPs included bands as follows: compound II, 422, 530 and 557 nm; compound III, 420, 546 and 578 nm. The spectrum of the alkaline pyridine ferrous hemochrome showed maxima at 415, 523 and 556 nm, characteristic of protoporphyrin IX.

MnP isoenzymes were incubated under various temperatures at pH 4.5, and then the remaining activities were measured as given times (Fig. 4). G-1 and G-2 showed much more stability against high temperature than that from *P. chrysosporium*. MnP isoenzymes were incubated under various pH values at 378°C, and then the remaining activities were measured as given times (Fig. 5). The MnP, G-2 was much superior to MnP from *P. chrysosporium*, for stability against high pH. Recently, Sutherland *et al.* (*22*) have reported that MnP from *P. chrysosporium* was very susceptible to thermal inactivation due to the loss of calcium from the enzyme. They demonstrated that when MnP from *P. chrysosporium* is exposed to temperature of 458°C or greater, or to pH 6.5

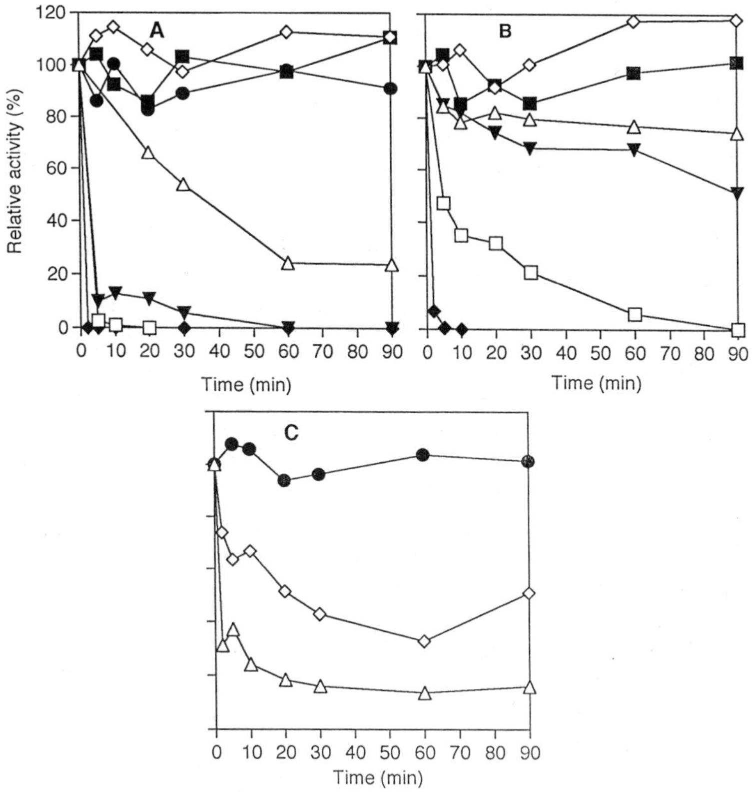

Figure 5. Stability of G-1 (A), G-2 (B) and *P. chrysosporium* MnP (C) during incubation at 37℃ and various pHs. Symbols: ■, pH 5.5; ◇, pH 6.5; ●, pH 7.0; △, pH 7.5; ▼, pH 8.0; □, pH 8.5; ◆, pH 9.0.

or greater, calcium bound on the distal side of heme was released from the enzyme and then structural changes occur in the heme environment of MnP G-1 and MnP G-2 were more stable than MnP from *P. chrysosporium* when exposed to high temperature and pH, suggesting that they have different structures from the extreme MnP from *P. chrysosporium.* It is suggested that MnPs from *Ganoderma* sp. YK-505 may be useful for biotechnological applications and studies of the relationship between structure and function of MnP isoenzymes.

Literature cited

1. Katagiri, N; Tsutsumi, Y.; Nishida, T. *Appl. Environ. Microbiol.,* **1995,** *61*, pp 617-622.
2. Katagiri, N; Tsutsumi, Y.; Nishida, T. *Mokuzai Gakkaishi*, **1995,** *41*, pp 780-784.

3. Paice, M. G.; Jurasek, L.; Ho, C.; Bourbonnais, R.; Archibald, F. S. *Tappi J.,* **1989,** *72(5)*, pp 217-221.
4. Reid, I. D.; Paice, M. G.; Ho, C.; Jurasek, L. *Tappi J.*, 1990,. *73(8)*, pp 149-153.
5. Fujita, K.; Kondo, R.; Sakai, K.; Kashino, Y.; Nishida, T.; Takahara, Y. *Tappi J.*, **1991**, *74(11)*, pp 123-127.
6. Fujita, K.; Kondo, R.; Sakai, K.; Kashino, Y.; Nishida, T.; Takahara, Y. *Tappi J.,***1993**, *76(1)*, pp 81-84.
7. Hirai, H.; Kondo, R.; Sakai, K. *Mokuzai Gakkaishi*, **1994**, *40*, pp 980-986.
8. Tsuchikawa, K.; Kondo, R.; Sakai, K. *Japan Tappi J.*, **1995**, *49*, pp 1332-1338.
9. Iimori, T.; Kaneko, R.; Yoshikawa, H.; Machida, M.; Yoshioka, H. Murakami, K. *Mokuzai Gakkaishi*, **1994,** *40*, pp 733-737.
10. Paice, M. G.; Reid, I. D.; Bourbonnais, R.; Archibald, F. S.; Jurasek, L. *Appl. Environ. Microbiol.***,** **1993**, *59*, pp 260-265.
11. Hirai, H.; Kondo, R.; Sakai, K. *Mokuzai Gakkaishi*, **1995**, *41*, pp 69-75.
12. Kondo, R.; Harazono, K.; Sakai, K. *Appl. Environ. Microbiol.,* **1994**, *60*, pp 4359-4363.
13. Brown, J. A.; Glenn, J. K.; Gold, M. H. *J. Bacteriol.*, **1990,** *172*, pp 3125-3130.
14. Wariishi, H.; Akileswaran, L.; Gold, M. H. *Biochemistry*, **1988**, *27*, pp 5365-5370.
15. Stone, A. T.; Morgan, J. J. *Environ. Sci. Technol.*, **1984,** *18*, pp 617-624.
16. Roy, B. P.; Paice, M. G.; Archibald, F. S.; Misra, S. K.; Misiak, L. E. *J. Biol. Chem.*,**1994**, *269*, pp 19745-19750.
17. Orth, A., Royse, D. J., Tien, M. *Appl. Environ. Microbiol.*, **1993**, *59*, pp 4017-4023.
18. Pel•ez, F., Martìnez, M. J., Martìnez, A. T. *Mycol. Res.*, **1995**, *99*, pp 37-42.
19. Tien, M.; Kirk, T. K. *Methods Enzymol.*, **1988**, *161*, pp 238-249.
20. Khindaria, A.; Grover, T. A.; Aust, S. D. *Arch. Biochem. Biophys.,* **1994**, *314*, pp 301-306.
21. Helm, D.; Winkelmann, G.: "Metal ions in fungi";Winkelman, G.; Winge, D., Ed.; Mycology series No. 11; Marcel Dekker Inc.: New York, **1994**, pp 39-98.
22. Sutherland, G. R. J.; Aust, S. D. *Arch. Biochem. Biophys.*, **1996**, *332*, pp 128-134.

Chapter 13

Isolation and Expression of Genes for Hemicellulases from Extremely Thermophilic Culturable and Unculturable Bacteria

P. L. Bergquist[1,2], M. D. Gibbs[1], D. J. Saul[2], R. A. Reeves[1], D. D. Morris[2], and V. S. J. Te'o[1]

[1]School of Biological Sciences, Macquarie Unviersity, North Ryde, Sydney, New South Wales 2109, Australia
[2]Centre for Gene Technology, University of Auckland, Private Bag 92019, Auckland, New Zealand

Our current research has focused on the isolation of genes coding for thermostable hemicellulolytic enzymes that function under alkaline conditions, and we have isolated xylanase and mannanase genes from both culturable bacteria isolated after enrichment and non-culturable organisms from consortia obtained after enhanced growth. Genomic or biomass DNA is subjected to the polymerase chain reaction (PCR) using consensus primers for xylanases or mannanases that we have designed. The full length sequences are obtained either by genomic walking PCR or by using consensus amino- and carboxy-terminal PCR primers that allow the amplification of the entire gene in a form that allows direct cloning into an expression vector. A number of novel genes have been isolated and cloned. They have been expressed in mesophilic bacteria and fungi (such as yeast and *Trichoderma*), since the mesophilic proteins can be removed by a simple heating step, giving a facile 20-fold purification. Selected candidate enzymes have been tested for their ability to enhance the bleaching of conventional and oxygen-bleached *Pinus radiata* and *Eucalyptus* kraft pulps.

Xylan is a major component of plant hemicellulose and is the second most abundant polysaccharide in Nature. It is a complex polymer consisting of a β-D-1,4-linked xylanopyranoside backbone which can be acetylated or substituted with arabinosyl and glucuronic acid side groups. The complete enzymatic hydrolysis of xylan into assimilable sugars requires a number of enzymes including endo-β-1,4-xylanase (EC 3.2.1.8). Galactoglucomannan is the principal hemicellulose of *Pinus radiata* softwood pulp and comprises a β-1,4-linked heteropolymer of mannose and glucose substituted with galactose and acetyl groups. Low-galactose mannans ('glucomannans') are also present in some hard- and softwood species (*1*). Complete enzymatic hydrolysis of galactoglucomannan also requires the concerted action of several enzymes, including endo-1,4-β-D-mannan mannohydrolase (EC 3.2.1.7) to cleave the mannan backbone.

The enzymes involved in the metabolism of plant carbohydrate polymers have been grouped into 35 different families on the basis of primary and tertiary sequence homologies (*2*). The endo-1,4-β-D-xylanases comprise families 10 and 11. The only similarity between members of these two families is their ability to hydrolyze the acetyl-methylglucuronoxylans of hardwoods and arabinomethylxylans of softwoods, since they are unrelated biochemically and structurally. Like most other cellulolytic and hemicellulolytic enzymes, xylanases are highly modular in structure and may be composed of either a single domain or a number of distinct domains broadly classified as catalytic or noncatalytic. Linker peptides often delineate the individual domains of multidomain enzymes into discrete and functionally-independent units. The catalytic domain of a xylanase determines the hydrolytic activity and hence governs the classification of the enzyme as belonging to family 10 or 11. β-mannanases from a number of bacterial, fungal and plant sources have been isolated and characterised (*3,4*). The nucleotide sequences of 10 β-mannanase genes have been reported (*5-16*). There are two distinct groups of mannanases and some appear to be related to cellulases (*2*).

Thermophilic bacteria, isolated from geothermal areas or other hot environments, produce enzymes that function at high temperatures and are resistant to denaturation. Such enzymes are of potential importance in industrial processes, but usually they are produced in low yields in the laboratory. Kraft pulping is a process widely used in the pulp and paper industry and involves the alkaline sulphate cooking of wood fibres to remove about 95% of the lignin. The remaining 5% gives the pulp a dark brown colour which tends to darken further by exposure to light due to oxidation. The brown colour can be removed by a multi-stage bleaching process using chlorine, chlorine dioxide or oxygen in order to obtain a white pulp for high quality paper. Currently, there is concern regarding the environmental effects of compounds generated from the bleaching process. Enzymes can aid in the removal of lignin from pulp without any harmful chemical side products (*18*). The amount of chlorine or chlorine dioxide needed is decreased by treating the pulp with enzymes prior to bleaching (*19,20*). Xylanases have been the enzymes of choice for aiding bleaching but mannanases have been shown to act synergistically (*21*). Most of the enzymes available for pulp bleaching are derived from mesophilic organisms and have pH optima from 5-7, and consequently, are not ideal for use with the kraft process which operates at temperatures above 100°C under alkaline conditions. Accordingly we have undertaken research with the objective of isolating stable enzymes that have activity closer to the operating temperature and pH of the kraft pulping process in the pre-bleaching stage of pulp manufacture.

Current research reported here has focused on the isolation of genes coding for thermostable enzymes that function under alkaline conditions. Novel genetic techniques have been developed to isolate xylanase and mannanase genes from both culturable organisms isolated after enrichment and non-culturable organisms that cannot be maintained in pure culture. Genomic DNA from either the culturable organism or from biomass enriched *in situ* in a percolation system has been prepared and subjected to the polymerase chain reaction (PCR) using consensus primers that span the active site sequences of family 10 or family 11 xylanases or using consensus mannanase primers that we have designed. The PCR products are cloned and sequenced and novel enzymes are identified. The full length gene sequences are obtained either by genomic walking PCR or by using consensus amino- and carboxy-terminal PCR primers that allow the amplification of the entire gene in a form that allows direct cloning into an expression vector. Using this technique, we have cloned a number of genes for the novel heat-stable enzymes into mesophilic bacteria and fungi (such as yeast), realising a cost-advantage in the preparation of the enzymes in that the mesophilic proteins can be removed by a simple heating step, giving a facile 20-fold purification. Selected candidate enzymes have been tested for their ability to enhance the bleaching of conventional and oxygen-bleached kraft pulps and several function as well as commercially-available enzyme preparations but at alkaline pH and at temperatures up to 40°C higher than mesophilic hemicellulases.

Materials and Methods

DNA preparation and Library construction. All bacteria were from the collection of H.W. Morgan, Waikato University, New Zealand. Culturable bacteria had been isolated by enrichment and screening procedures on hemicelluloses. The following pure cultures were used: *Caldicellulosiruptor saccharolyticus* (22); *Caldicellulosiruptor* strain Rt8B.4 (23); *Caldicellulosiruptor* strain Tok7B.1 (*24*); *Dictyoglomus thermophilum* Rt46B.1 (*25*); *Thermotoga* strain FjSSB.3 (*26*). Unculturable bacteria were isolated *en masse* from enrichment's on xylan or mannan, using an enclosed percolation system at 83-90°C, pH 9.0. Unculturables II and III were grown in the presence of xylan as a carbon source and Unculturables IV with glucomannan (locust bean gum) at pH9.0 and 88°C. Bulk DNA was prepared from the resulting biomass by standard methods (*27*).

DNA preparation, manipulation and digestion with restriction enzymes were performed according to Sambrook *et al.* (*27*) Gene libraries were constructed in λZAPII by the methods of Short *et al.* (*28*). Xylanase-positive plaques were identified from replica plates that were overlaid with oat spelt xylan and stained with Congo Red (*29*) Genomic DNA from either culturable organisms or from biomass enriched *in situ* in the percolation system was prepared and subjected to PCR using consensus primers that span the active site sequences of family 10 or 11 xylanases or forward and reverse mannanase primers that we have designed (Fig. 5). The PCR products were cloned and sequenced and novel enzymes were identified. The full length sequences were obtained either by genomic walking PCR (*11,30*) or by using consensus amino- and carboxy- terminal primers that allow amplification of the entire gene in a form that allows direct cloning into the expression vector pJLA602 (*31*).

Enzyme assays. Assays for enzymes involved in xylan degradation have been described elsewhere (*32*). Quantitative assays were carried out by measuring the release of reducing sugar from oat spelt xylan (*33*). In some cases, purified recombinant enzymes have been assayed for release of reducing sugar and lignin from kraft pulp (*34*).

Enzyme-aided ECF and TCF Bleaching. Two xylanases (from *Dictyoglomus* and *Thermotoga*) and two mannanases (from *Caldicellulosiruptor* Rt8B.4 and *Dictyoglomus*) were tested on *Pinus radiata* kraft-oxygen pulp at commercial levels of enzyme charge at pH 7 and 75°C using a standard DED sequence by Dr. K. K. Y. Wong, PAPRO New Zealand. The *Dictyoglomus* XynB enzyme was tested on *Eucalyptus* pulp by Dr. P. J. Nelson, CSIRO Forestry and Forest Products, Clayton, Victoria, Australia. Oxygen-delignified eucalypt kraft pulp was treated in a D(EO)DD sequence in an ECF bleaching sequence and in a TCF bleaching sequence with hydrogen peroxide.

Other Methods. Polymerase Chain Reactions (PCR) were performed in a Perkin-Elmer Cetus DNA Thermal Cycler. DNA sequencing was carried out using an Applied Biosystems 373A DNA sequencer and Catalyst 800 Robotic Workstation. Sequences were analysed on a Silicon Graphics IRIX workstation using the GCG Software (*35*).

Nucleotide sequence accession numbers. *Dictyoglomus thermophilum* Rt46B.1 *xynA*: L39875; *xynB*: U766545.

Results and Discussion

Cloning and Expression of xylanase genes. Xylanases have been grouped into two unrelated families on the basis of primary sequence homologies (*2*). However, based on the functional clustering of xylanase catalytic properties on various xylan substrates, and the existence of high pI and low pI family G xylanases, it is clear the xylanases can be further subdivided within these two families according to additional functional and physicochemical criteria (*36-38*). Like most other glycosyl hydrolases responsible for the metabolism of plant carbohydrate polymers, β-1,4-xylanases are highly modular in structure, and the catalytic domains of families 10 and 11 xylanases can be associated with a wide range of downstream and/or upstream non-catalytic domains (*39*). Typically, the characterised non-catalytic domains from microbial multidomain xylanases have apparent roles in the propagation of either substrate-binding, cell-wall association or protein-protein interactions. However, other functions including thermal-stabilisation and the initiation of plant nodulation responses have also been reported (*40-42*). A striking feature of many microbial xylanolytic systems is the presence of multiple family 10 and/or 11 xylanases which are produced as discrete gene-products (*43*). Detailed comparisons of the catalytic properties of xylanases isolated from multi-xylanase systems have revealed distinctions in the hydrolytic activities of the individual xylanases present, including differences in the yields, rates of hydrolysis, and hydrolytic products from different xylan substrates of varying complexity. These observations suggest that xylanase multiplicity is a mechanism employed by both bacteria and fungi to enhance their xylanolytic capabilities on complex xylan substrates.

We have described elsewhere the isolation of recombinant plasmids expressing xylanases from expression libraries of genomic DNA (*34*) and the genomic walking PCR (GWPCR) technique utilising consensus primers to conserved sequences at the active sites of families 10 and 11 xylanases (*11,44*). This latter technique has proved to be particularly valuable in revealing the presence of xylanase genes that escaped detection in expression libraries of culturable thermophilic bacteria (*45*). For example, six separate isolates of obligately anaerobic extremely-thermophilic, xylanolytic bacteria that were placed in Clostridial cluster 4 of Rainey *et al* (*46*) b y microbiological and molecular criteria were all found to have three subfamilies of family 10 xylanases. These genes were related as far as their catalytic domains were concerned, but they varied in the number and extent of their non-catalytic domains (*30*). Similar results have been reported for *Thermotoga* xylanases (*44,47*) and for xylanases from *Caldicellulosiruptor* Tok7B.1, where the family 10 xylanase activities are parts of multidomain enzymes with other catalytic activities (*16*).

Recent experiments with bulk genomic DNA from Unculturables II and III have shown that both family 10 and 11 consensus fragments can be amplified and isolated, and these fragments when cloned show sequence variation. Use of N- and C-terminal consensus primers to the three family 10 xylanases and to two family 11 enzymes have shown that entire genes can be amplified, cloned and expressed from the bulk DNA. These genes also show some sequence diversity, but we believe that the actual range of variation in Nature may be limited. Comparison of enzyme families shows that proteins are surprisingly robust to substitution and can maintain catalytic function even after almost total substitution of their amino acid sequences (*48*). For example, Henrissat and his collaborators have shown that the glycosyl hydrolases can be divided into more than 60 families and when a three-dimensional structure is solved, the catalytic residues are conserved in a fold at the active site but other residues are rarely homologous (*48, 49*). It is thought that natural proteins have evolved *in vivo* by the successive fixation of individual mutations. They have evolved for the benefit of organisms whose growth environment may not match the conditions required for biotechnological processes. There is considerable room for improvement in properties for which selective pressure was never applied, and it should be possible to achieve some improvements by engineering a protein's amino acid sequence. Protein engineering has the potential to dramatically enhance the performance of xylanases for bleaching purposes and may supplant screening procedures in natural environments.

Two xylanase genes from *Dictyoglomus* *Dictyoglomus thermophilum* isolate Rt46B.1 is an extremely thermophilic anaerobic bacterium which was isolated from a thermal spring in New Zealand (25). The endoxylanase component of the xylanolytic system of Rt46B.1 comprises two unrelated enzymes which are expressed as distinct gene products: a high MW/low pI family 10 xylanase (XynA) and a low MW/high pI family 11 xylanase (XynB). XynA is a single-domain enzyme, whilst XynB is a multidomain enzyme, consisting an N-terminal family 11 xylanase catalytic domain and a C-terminal domain which may enhance the affinity of the xylanase for xylan (Fig. 1). Both enzymes exhibit optimal xylanase activity at pH6.5, 85°C, and are active over a broad pH range with over 50% activity between pH5.0 and pH9.0. In addition, the two xylanases show more or less equivalent activity on both isolated and fibre-bound xylan substrates, as measured by the release of reducing sugars.

Soluble xylan and kraft pulp hydrolysis by *Dictyoglomus* XynA and XynB Structurally and physicochemically, the two xylanases are quite distinct. Hence, as would be expected, these enzymes exhibited different catalytic properties. One method of assessing the activity of a xylanase on kraft pulp is to measure the reducing sugars released following enzymatic hydrolysis. However, it should be noted that there has not yet been any conclusive evidence showing a direct correlation between the release of reducing sugars during enzyme treatment and the subsequent enhancement of pulp bleaching.

Figure 1 shows the several genetic constructions used in these studies. The *xynB* gene has been modified at its 5'-end to enhance expression and, in some cases, stop codons have been inserted to produce catalytic domain proteins only. XynB3 and B4 have had the leader sequence removed but differ in that B4 is the full length gene product with the noncatalytic domain included. Following a 6 hr incubation at 75°C with 0.4 g of oxygen-bleached *Pinus radiata* kraft, both XynB3 and XynB4 (50 XU/g pulp) released 0.0625 µmoles of net reducing sugar per weight dry gram of pulp. Therefore, the XynB C-terminal domain did not alter the activity of the xylanase domain on kraft pulp. By comparison, an equivalent application of the Rt46B.1 XynA xylanase released 0.0875 µmoles reducing sugar per dry gram weight of pulp. The composition of the reducing sugars released by the xylanases from oat-spelt xylan and kraft pulp were examined using thin-layer paper chromatography and high performance liquid chromatography. The family 10 xylanase produced predominantly xylose and xylobiose following extended hydrolysis of both soluble and fibre-bound xylan substrates. In contrast, the family 11 xylanase produced xylobiose and xylotriose as the lowest molecular weight xylooligosaccharides from the respective substrates under similar conditions. Additionally, XynB yielded a range of high molecular weight xylooligosaccharides (larger than X3) which appeared to correspond to short substituted xylooligosaccharides. These limit xylooligosaccharides resulting from the action of XynB could be further hydrolysed by the XynA enzyme. The xylan cleavage patterns of both xylanases were dependant upon the pH at which the hydrolysis assays were performed.

Binding to kraft pulp Both xylanases bound to kraft pulp in a pH-dependant fashion, most likely mediated though non-specific hydrophobic interactions. The degree of binding of the XynA to kraft pulp was reduced dramatically as the solution pH was increased from pH 4.0 to pH 10.0 (Fig. 2A). Accordingly, the enzyme could be bound to kraft pulp at low pH (pH 4.0) and subsequently eluted upon resuspension of the 229A-treated pulp in a high pH buffer (pH 10.0, see Fig. 2B). The kraft pulp binding profile of XynB under varying pH conditions was different to that observed for the other xylanase, and consequently, no respective binding/elution steps could be performed on this enzyme.

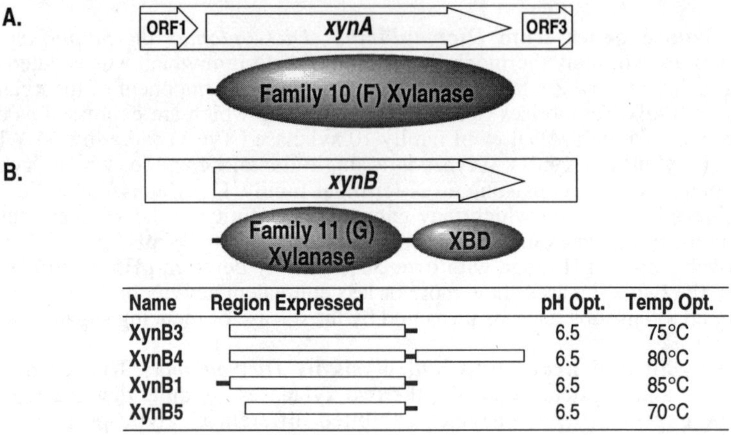

Figure 1. Diagrammatic representation of *Dictyoglomus thermophilum* Rt46B.1 Family 10 (A) and 11 xylanases (B). The lower portion of the figure shows the structure and biochemical characteristics of four genetically-engineered proteins derived from the wild type XynB enzyme.

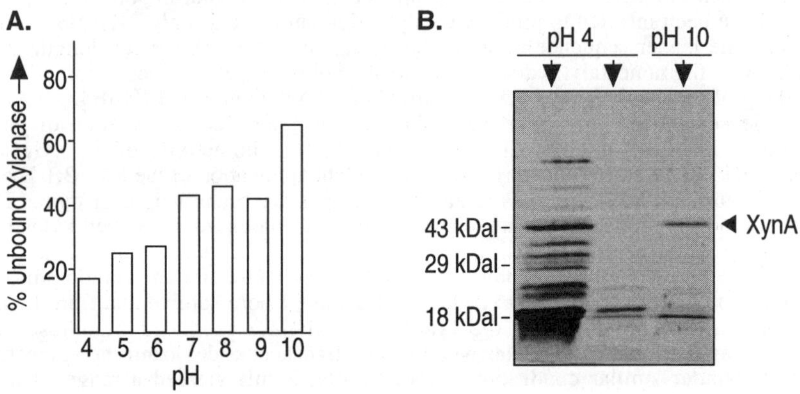

Figure 2. Binding of XynA to kraft pulp. 2A. XynA enzyme activity remaining in solution at various pH values. 2B. SDS-PAGE of XynA remaining in solution after incubation of partially-purified enzyme with kraft pulp at pH 4 (lane 2) and pH10 (lane3). Lane 1 is an *E .coli* extract from a strain producing the recombinant enzyme incubated at pH4 and run on the gel, showing the XynA protein migrating at 43 kDa.

Activity of the enzymes on kraft and oxygen-bleached pulp Small-scale bleaching trials have been performed for both xylanases on *Pinus radiata* and *Eucalyptus* kraft pulps. The high thermal stability of XynB3 at its temperature optimum of 75°C, and the very high yields attainable from the *xynB3* expression plasmid, made this enzyme an ideal candidate for laboratory-scale bleaching trials. Table 1 shows the release of lignin from *Pinus* kraft pulp by the four enzymes and it can be seen that XynB shows the greatest absorbance a 280*nM*. Use of XynB with *Pinus radiata* kraft pulp (at a fixed enzyme dosage of 10XU/g pulp) in an ECF bleaching sequence resulted in a chemical savings of at least 20% at the D stage, and an increase in brightness by 6 ISO points in comparison to the reference pulp. XynB was also able to improve the bleachability of *Eucalyptus* kraft pulp, increasing the ISO brightness by 1.6 points in a TCF bleaching sequence (Fig. 3). The family 10 xylanase, by comparison, effected only slight improvement on the bleachability of *Pinus radiata* kraft pulp, and showed no detectable activity on *Eucalyptus* kraft pulp. This work has been reported previously (*44*).

Table I. Release of chromophore from *Pinus radiata* oxygen-bleached kraft pulp by recombinant thermophilic enzymes

Enzyme	*Relative Absorbance at 280nm*	*Net Absorbance*
Reference	1.64	-
D. thermophilum Rt46B.1 XynB[a]	3.72	2.08
Thermotoga sp. FjSS3B.1 XynA[a]	2.01	0.37
D. thermophilum Rt46B.1 ManA[b]	1.78	0.14
Caldicellulosiruptor sp. Rt8B.4 ManA[b]	1.64	0.00

[a]Endo-β-1,4-xylanase, [b]Endo-β-1,4-mannanase.

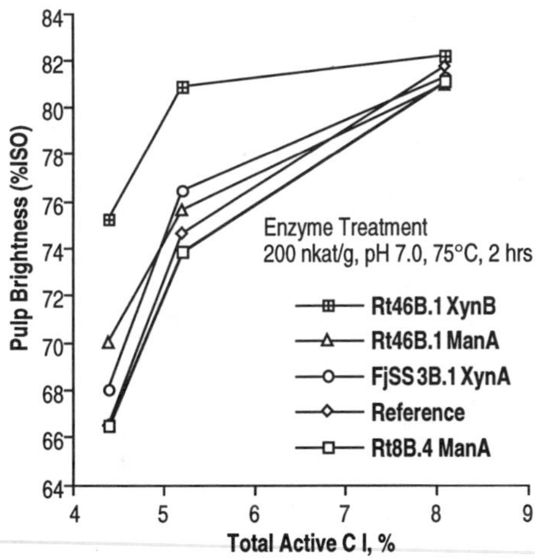

Figure 3. DED bleaching of *Pinus radiata* kraft-oxygen pulp after treatment with thermophilic xylanases and mannanases. Abbreviations as for Table 1.

This brightness increase with XynB is significant,. If a brightness of 90% ISO was required for the eucalypt kraft-oxygen pulp in the absence of a xylanase pre-treatment, the actual bleaching sequence would have to be modified, either by the addition of hydrogen peroxide to the (EO) stage or by the use of a second alkaline extraction stage, such as D(EO)DED. Treatment of eucalypt kraft-oxygen pulp with XynB3 (10 XU/g pulp) enabled a target brightness of 88% ISO to be attained using 3.4% ClO_2 (as active chlorine) in a D(EO)DD sequence compared with 4.0% ClO_2 for untreated pulp (Fig. 4A). The consequent savings in chlorine dioxide was 15% of the chemical charge used for the untreated pulp, or 6 Kg active chlorine per tonne of pulp.

The XynB3 xylanase was also able to enhance the bleachability of eucalypt kraft-oxygen pulp in TCF bleaching. Treatment of the pulp at a xylanase dosage of 10 XU/g increased the final pulp brightness from 82.2% ISO to 86.6% ISO following a Q(PO) TCF bleaching sequence (where Q represents a chelation stage with EDTA, and PO represents high pressure hydrogen-peroxide bleaching).

Comparative bleaching studies were performed to compare the efficacy of XynB pre-treatments on pulp bleachability to the commercial xylanase Irgazyme-40 (Ciba-Geigy Australia Ltd) and a noncommercial xylanase designated DCPX. At a constant enzyme dosage of 10 XU/g pulp, all three xylanases lowered the bleached pulp yield by 3% (based on the yield of unbleached pulp), however, the final pulp brightness attained by each xylanase following a D(EO)DD ECF bleaching sequence ranged from 89.1% ISO for DCPX, 89.4% ISO for Irgazyme-40, and 89.8% ISO for XynB3 (Fig. 4B). Hence at the same enzyme dosage, XynB3 was the most effective of the three xylanases. Furthermore, XynB3 pretreatment was performed for only 2 hours at 75°C, whereas pretreatment of the eucalypt pulp using the Irgazyme-40 and DCPX xylanases required 3 hour incubations at 60°C, and 53°C, respectively.

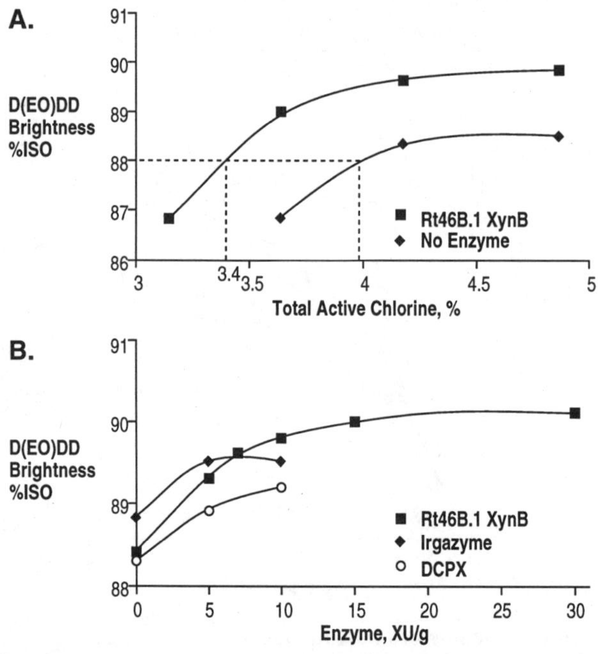

Figure 4. Bleaching with Dictyoglomus XynB. 4A. Effect of *D. thermophilum* Rt46B.1 XynB treatment on D(EO)DD brightness of *Eucalyptus* kraft- oxygen pulp with various total active chlorine charges. 4B. Effect of enzyme dosage on the D(EO)DD brightness of *Eucalyptus* kraft-oxygen pulp treated with various xylanases. Rt46B.1 XynB; *D. thermophilum* Rt46B.1 enzyme; Irgazyme, Ciba-Geigy and Co.; DCPX, non-commercial xylanase.

Expression of *Dictyoglomus* xylanases in fungi *E. coli* produces insufficient recombinant enzyme for mill-scale trials although it is ideal as a host for preliminary characterisation. For this reason, we have transferred our genes to the yeast *Kluyveromyces lactis* and the filamentous fungus *Trichoderma reesei.* These organisms are capable of a far greater production of extra-cellular enzymes as hosts for heterologous gene expression. We have developed expression vectors based on the 2μ-like plasmid pKD1 of *Kluyveromyces drosophilarium* and a plasmid from *Aspergillus* carrying the *amdS* gene for selection. Our *Kluyveromyces* work is currently the most advanced. *xynA* and *B* were fused individually in-frame with a synthetic secretion signal derived from the *K. lactis* killer toxin and expressed under control of the *K. lactis LAC4* (β-galactosidase) promoter. Correctly processed XynA with full biological activity on oat spelts xylan was secreted during shake-flask cultivation of *K. lactis* transformants. One isolate combined high plasmid stability and good yield and has been employed for scaled-up production of XynA in chemostat culture (see accompanying paper by Walsh *et al.*, *this volume*).

Isolation of β-mannanases Unlike xylanases, comparatively few β-mannanase genes have been isolated and sequenced. Two gene families encoding β-mannanases have been identified on the basis of amino acid sequence similarities (*2*). Only two β-mannanase genes were available for comparison at the time the classification was made. ManA from *C. saccharolyticus* was classed as a member of a large group of endoglucanases, although it has low sequence homology with these enzymes when compared using direct alignment methods. This group also has been classified as glycosyl hydrolase family 5 in a classification of cellulases, cellobiohydrolases and xylanases (*49*). A number of mannanase genes have been sequenced since this classification (*2*) and all share homology with either *manA* from *C. saccharolyticus* or the *Bacillus* mannanase gene. They fall into two unrelated groups which correspond to families 5 and 26 defined by hydrophobic cluster analysis (*49,50*). These two families share no detectable amino acid sequence homology. It is of interest in this context that both ManA and ManB from *C. saccharolyticus* are in family 5 whereas ManA from *Caldicellulosiruptor* Rt8B.4 is in family 26, yet on SSU 16S RNA data, these two bacterial strains are closely related (*23*). Apparently, they have acquired their mannanase genes by horizontal transfer from different ancestral micro-organisms.

An alignment of ManA of *D. thermophilum* Rt46B.1 with the other six members of family 26 shows that it has considerable homology with the β-mannanase from *Caldicellulosiruptor* strain Rt8B.4 and to the putative amino acid sequences of two other bacterial mannanases from bacteria, ManA from *Bacillus subtilis* (*13*) and ManA from *Pseudomonas fluorescens* (*51*). As part of this analysis, a previously unidentified gene coding for a homologous enzyme was identified, which presumably is also a mannanase. Matsushita *et al.* (*7*) published a sequence coding for a β-1,4-endoglucanase from *Bacteroides ruminicola.* However, comparison with the mannanase domains from *D. thermophilum* Rt46B.1 and *Caldicellulosiruptor* Rt8B.4 ManA suggests that the *Bacteroides* enzyme is a multidomain enzyme possessing both a mannanase and an endoglucanase domain. The putative mannanase domain of *B. ruminicola* was included in homology comparisons, although there is no enzymatic data available to confirm this inclusion.

The phylogenetic relationships of the seven mannanase domains and the putative *Bacteroides* mannanase were inferred using the criterion of maximal parsimony (*52*) and the *D. thermophilum* Rt46B.1 mannanase domain was seen to be most closely related to the mannanase from *Caldicellulosiruptor* Rt8B.4. Sequence comparisons showed that none of these mannanases share any detectable homology with the mannanase domain of ManA from *C. saccharolyticus*, and that they appear to be members of a unique and unrelated class of mannanases. To avoid possible confusion, homologs of the *Dictyoglomus thermophilum* Rt46B.1 ManA mannanase are referred to as family 26-type mannanases, and *C. saccharolyticus* ManA mannanase domains as family 5-type mannanases.

The mannanase domain of ManA from *C. saccharolyticus* has been classified as a family 5 glycosyl hydrolase (*50*). The majority of members of this group are endoglucanases. A classification system based purely on shared homology is clearly unsatisfactory when the enzymes display completely different catalytic activities. Four homologs of the mannanase domain of ManA were available for analysis. Two of these were from bacterial sources, ManA from *Streptomyces lividans*, and ManB from *C. saccharolyticus* (*9, 11*), and two from fungal sources; ManA from *Trichoderma reesei* and Man1 from *Aspergillus aculeatus* (*15, 10*).

The availability of a number of family 26 mannanase sequences provides the opportunity for the design of PCR primers to highly conserved regions and allowed the identification of strains that carry mannanase genes from this family as well as providing the necessary sequence information for the isolation of the entire genes by genomic walking PCR (*11, 53*). Figure 5 shows an alignment of highly conserved regions in family 26 mannanases and the PCR primers that allow the amplification of a 150 - 200bp fragment that can be sequenced for the identification of novel genes and for the design of primers that allow the sequence of the 5' and 3' ends of the gene to be determined. These primers have been successfully tested against the genes known to be present in genomic DNA from *D. thermophilum* Rt46B.1 and *Caldicellulosiruptor* Rt8B.4 and have revealed several other family A mannanases present in culturable and unculturable hemicellulolytic, thermophilic bacteria (*data not shown; 11*). Further sequence information on family 5 mannanases is required before consensus primers can be designed to allow the identification and isolation of these genes.

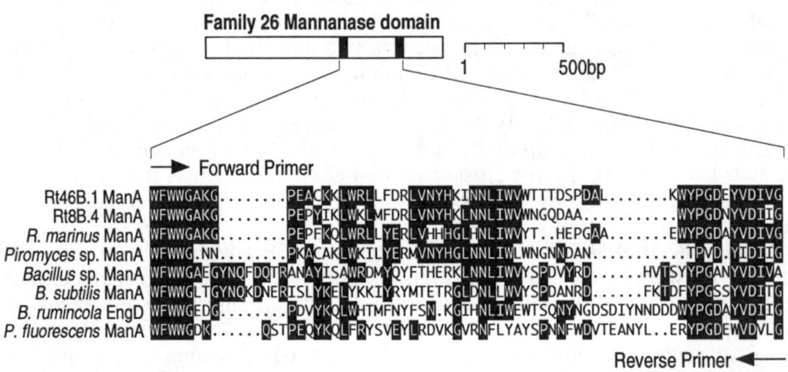

Figure 5. Consensus sequences for PCR primers specific for family 26 β-mannanases from bacteria. Abbreviations: Rt46B.1: *D. thermophilum* Rt46B.1 ManA; Rt8B.4: *Caldicellulosiruptor* Rt8B.4 ManA; B: *Bacillus*; P: *Pseudomonas*; R: *Rhodothermus.*

Bleaching studies with mannanases Samples of both family 5 and family 26 mannanases from *Caldicellulosiruptor* strains have been tested by Drs. Tom Clark, Bob Allison and Ken Wong, PAPRO, Rotorua, New Zealand for their action in pulp bleaching. Allison *et al* (*54*) have reported on comparative studies of a *Trichoderma* mannanase (family 26), an *Aspergillus niger* mannanase (probably family 5) and the catalytic domain of a family 5 mannanase from *Caldicellulosiruptor saccharolyticus* in the bleaching of *Pinus radiata* and MCC kraft pulp using a simple DED sequence. The thermophilic enzyme was less effective than the *Trichoderma* enzyme as judged by increased ISO brightness and kappa number decrease. This result may be attributed to the strong and rapid binding of the enzyme to the kraft pulp (*11*). The family A enzymes from *Caldicellulosiruptor* strain Rt8.B4 and *Dictyoglomus thermophilum* Rt46B.1 were ineffective in bleaching using similar conditions. Just as the family 11 enzymes are markedly superior to the family 10 xylanases in bleaching tests, so the family 5 enzymes appear to be superior to the to the family 26 enzymes in this application. Representative enzymes from neither mannanase family have been tested for synergistic effects in bleaching with xylanases (*21*).

Concluding remarks

The XynB xylanase from *Dictyoglomus thermophilum* Rt46.B1 was able to enhance the bleachability of eucalypt kraft-oxygen pulp to both ECF and TCF bleaching. In view of the high temperature characteristics of XynB, this xylanase appears to be well suited to the high temperature applications encountered during kraft pulping and bleaching procedures.

Some interesting results in basic molecular genetics have arisen from this work. The broad range in both the net production level and thermal characteristics of the different XynB enzymes prepared in this study of the *Dictyoglomus* XynB enzyme highlights some of the considerations which should be addressed during the design of an expression system. Whilst the removal of the N-terminal leader peptide may be essential for high level heterologous expression in *E. coli*, it is important to ensure that the engineered xylanases possess all the mature N-terminal residues for maximal temperature stability. Additionally, as the stability of the immediate N-terminus may effect both the final yield and stability of the enzyme, the use of peptide sequence analysis software to predict the characteristics of the N-terminal peptide may prove invaluable.

We have described the genomic walking PCR approach for the identification and cloning of bacterial family 10 and 11 xylanase and family 26 mannanase genes. The procedure is expedient and highly sensitive, and avoids the construction and screening recombinant gene-libraries. Clearly however, the technique can be applied to almost any bacterial gene-family from which a representative consensus can be deduced. The methodology has been applied to xylanase and mannanase genes from unculturable bacterial sources, which presently represent a largely untapped source of biodiversity. However, using primers developed for the isolation of complete genes from genomic DNA from culturable thermophilic bacteria and applying them to biomass DNA from unculturables showed that many related genes are present in Nature and our study of subsets of genes from family 10 suggests that there may be limits to the genetic variability that can be found. Most enzymes have evolved to function within the intricate network of chemical and physical processes that make up a living system, and consequently, some of their properties may be sub-optimal when removed from their natural context for biotechnological applications. This result is to be expected if there has never been selective pressure for a particular biochemical trait or chemical characteristic. Accordingly, it should be possible to achieve some improvements by engineering an enzyme's amino acid sequence by evolutionary genetic techniques *in vitro* and protein engineering has the potential to dramatically enhance protein performance in a wide variety of technologically-interesting environments.

Our current experimental approach recognises that natural proteins have evolved for the benefit of organisms whose growth environment may not match those required for biotechnological processes. We are endeavouring to alter the pH and temperature optima of selected thermostable xylanases by iterative random mutagenesis to match their activities to industrial conditions. Genetic diversity is being created by *in vitro* evolution involving repeated cycles of random mutation, *in vitro* recombination and selection. By using *in vitro* evolution, we hope to produce a series of mutant proteins with extended pH and temperature optima. Analyses of the three-dimensional structures of these proteins should allow correlation of sequence changes and activities at various pH values and temperatures.

Acknowledgments

This work was supported in part by the University of Auckland Research Grants Committee, Pacific Enzymes Ltd., the Foundation for Research, Science and Technology, and a Macquarie University Research Grant. We thank Dr. Ken Wong, PAPRO, New Zealand, and Dr. Peter Nelson, CSIRO Forestry and Forestry Products, Australia, for the bleaching data.

References

1. Eriksson, K. E. L. In: K. E. L. Eriksson, R. A. Blanchette and P. Ander (eds.), *Microbial and Enzymatic Degradation of Wood and Wood Components.* **1990,** pp.181. Springer-Verlag Series in Wood Science, New York.
2. Henrissat, B. *Biochem. J.* **1991,** *280,* 309.
3. Araujo, A. & Ward, O. P. Jour. Appl. Bacteriol. **1990,** *68,* 253.
4. McCleary, B. V. *Methods Enzymol.* **1988,** *160,* 596.
5. Akino, T., Kato, C. & Horikoshi, K. *Appl. Environ. Microbiol.* **1989,** *55,* 3179.
6. Lüthi, E., Jasmat, N. B., Grayling, R. A., Love, D. R. & Bergquist, P. L. *Appl. Environ. Microbiol.* **1991**, *57,* 694.
7. Matsushita, O., Russell, J. B. & Wilson, D. B. *Jour. Bacteriol.* **1991,** *173,* 6919.
8. Gibbs, M. D., Saul, D. J., Lüthi, E. & Bergquist, P. L. *Appl. Environ. Microbiol.* **1992,** *58,* 3864.
9. Arcand, N., Kluepfel, D., Paradis, F. W., Morosoli, R. & Shareck, F. *Biochem. Jour.* **1993,** *290,* 857.
10. Christgau, S., Kauppinen, S., Vind, J., Kofod, L. V. and Dalbge, H. *Biochem. Molec. Biol. Int.* **1994,** *33,* 917.
11. Morris, D. D., Reeves, R. A., Gibbs, M. D., Saul, D. J. & Bergquist, P. L. *Appl. Environ. Microbiol.* **1995,** *61,* 2262.
12. Fanutti, C., Ponyi, T., Black, G. W., Hazlewood, G. P. & Gilbert, H. J. *J. Biol. Chem.* **1995**, *270,* 29314.
13. Mendoza, N. S., Arai, M., Sugimoto, K., Ueda, M., Kawaguchi, T. & Joson, L. M. *Biochim. Biophys. Acta.* **1995,** *1243,* 552.
14. Politz, O., Krah, M., Borriess, R. Unpublished submission to the GenBank nucleotide Databank. **1995**. Accession No. X90947.
15. Stalbrand, H., Saloheimo, A., Vehmaanpera, J., Henrissat, B. and Penttila, M. *Appl. Environ. Microbiol.* **1995,** *61,* 1090.
16. Gibbs, M. D., Elinder, A. U., Reeves, R. A. & Bergquist, P. L. *FEMS Microbiol. Lett.* **1997**, *141,* 37.
17. Reeves, R. A., Gibbs, M. D., & Bergquist, P. L. *Appl. Environ. Microbiol.* **1997**, submitted.
18. Kantelinen, A., Hortling, B., Sundquist, J., Linko, M. & Viikari, L. *Holzforsh.* **1993,** *47,* 318.
19. Viikari, L., Kantelinen, A., Buchert, J. & Puls, J. *Appl Microbiol Biotechnol.* **1994,** *41,* 124.
20. Viikari,L., Sundquist, J. & Kettunen, J. *Paper and Timber* **1991,** *73,* 384.
21. Buchert, J. Salminen, J., Siika-Aho, M., Ranua, M. & Viikari, L. *Holzforschung.* **1993,** *47,* 473.
22. Rainey, F.A., Donnison,A.M., Janssen, P.H., Saul, D., Rodrigo, A., Bergquist, P.L., Daniel R.M., Stackebrandt E., & Morgan. H.W. *FEMS Microbiol. Lett.* **1994,** *120,* 263.
23. Dwivedi, P. P., Gibbs, M. D., Saul, D. J. & Bergquist, P. L. *Appl. Microbiol. Biotechnol.* **1996,** *45,* 86.
24. Gibbs, M. D., Reeves, R. A. & Bergquist, P. L.**1997**, unpublished.
25. Patel, B.; Morgan, H.; Weigel, J.; & Daniel, R. *Arch. Microbiol.* **1987,** *147,* 21.

26. Saul, D. J., Williams, L. C., Reeves, R. A., Gibbs, M. D. & Bergquist, P. L. *Appl. Environ. Microbiol.* **1995**, 61, 4110.
27. Sambrook, J., Frisch, E.F. & Maniatis, T., Eds.; *Molecular Cloning: A Laboratory Manual,* Cold Spring Harbor Press, Cold Spring Harbor, New York (1989).
28. Short, J. M., Fernandez, J.M., Sarge, J. A., & Huse, W. D. *Nucl. Acid Res.* **1988,** *16,* 7583.
29. Teather, R.M. & Wood, P.K. *Appl. Environ. Microbiol.* **1982,** *43* 777.
30. Morris, D. D., Chin, C. W. J., Koh, M-H., Wong, K. K. K., Allison, R. W., Nelson, P. J. & Bergquist, P. L. Appl. Environ. Microbiol. **1997,** submitted
31. Schauder, B., Blöcker, H., Frank, R. & McCarthy, J. E. G. *Gene* **1987,** *52,* 279.
32. Lüthi, E., Love, D. R., McAnulty, J., Wallace, C., Caughey, P. A., Saul, D. J. & Bergquist, P. L. *Appl. Environ. Microbiol.* **1990,** *56,* 1017.
33. Lever, M., **1973**, Biochem. Med., *7,* 274-281.
34. Gibbs, M. D., Reeves, R. A. & Bergquist, P. L. *Appl. Environ. Microbiol.* **1995,** *61,* 4403.
35. Devereux, J., Haeberli, P. & Smithies, O. *Nucl. Acids Res.* **1984,** *12,* 387.
36. Biely, P., Kluepfel, D., Morosoli, R. & Shareck, F. *Biochim. Biophys. Acta* **1993,** *1162,* 246.
37. Maringer, U., Wong, K.K.Y., Saddler, J.N. & Kubicek, C.P. *Biotechnol. Appl. Biochem.* **1995,** *21,* 49.
38. Zhu, H., Paradis, F.W., Krell, P.J., Phillips, J.P. & Forsberg, C.W. *Jour. Bact.* **1994,** *176,* 3885.
39. Gilkes, N. R., Henrissat, B., Kilburn, D. G., Miller, M. C. & Warren, R. A. J. *Microbiol. Reviews.* **1991**, *55,* 2303.
40. Black, G. W., Hazlewood, G. P., Millward-Sadler, S. J., Laurie, J. I. & Gilbert, H. J. *Biochem. J.* **1995.** *307,* 191.
41. Clarke, J., Davidson, K., Gilbert, H.J., Fontes, C.M.G.A. & Hazlewood, G.P. *FEMS Microbiol. Lett.* **1996,** *139,* 27.
42. Fontes, C.M., Hazelwood, G. P. , Morag, E., Hall, J., Hirst, B. H. & Gilbert, H. J. *Biochem. Jour.* **1995,** *307,* 151.
43. Wong, K.K.Y., Tan, L.U.L. & Saddler, J.N. *Microbiol. Rev.* **1988,** *52,* 305.
44. Bergquist, P. L., Gibbs, M. D., Saul, D. J., Te'o, V. S. J., Reeves, R. & Morris, D. *Biotechnology in the Pulp and Paper Industry: Recent Advances in Applied and Fundemental Research.* Ed. Srebotnik, E. Facultas-Univ.-Verl., Vienna, Austria. **1996,** p497 - 502.
45. Bergquist, P. L., Gibbs, M. D., Saul, D. J., Reeves, R. A., Morris D. D. & Te'o, V. S. J. In: *Enzymes for Pulp and Paper Processing,* Ed T. J. Jeffries and L. Viikari. *American Chemical Society Symposium Series* **1996,** *655,* 85. Washington, D. C.
46. Rainey, F., Ward, N., Morgan, H., Toalster, R. & Stackebrandt, E. *J. Bacteriol.* **1993,** *175,* 4772.
47. Winterhalter, C., Heinrich, P., Candussio, A., Wich G. & Liebl, W. *Mol. Microbiol.* **1995,** *15,* 431.
48. Davies, G. and Henrissat, B. *Structure* **1995,** *3,* 853.
49. Henrissat, B., Callebaut, I., Fabrega, S., Lehn, P., Mornon, J.-P. & Davies, G. *Proc. Nat. Acad Sci. USA* **1995,** *92,* 7090.
50. Henrissat, B. and Bairoch, A. *Biochem. Jour.* **1993,** *293,* 781.
51. Braithwaite, K. L., Black, G. W., Hazelwood, G. P., Ali, B. R. S. & Gilbert, H. J. *Biochem. Jour.* **1995,** *305,* 1005.
52. Swofford, D. Smithsonian Institute (publ.), **1993,** Washington, D. C., U. S. A.
53. Morris D. D., Gibbs, M. D., & Bergquist, P. L. In: *Enzymes for Pulp and Paper Processing,* Ed T. J. Jeffries and L. Viikari. *American Chemical Society Symposium Series,* **1996,** 101. Washington, D. C.
54. Allison, R. W., Clark, T. A. & Suurnäkki. *Appita* **1996,** *49,* 18.

Chapter 14

Expression and Secretion of Bacterial Thermophilic Hemicellulases in *Kluyveromyces lactis*

D. J. Walsh[1], M. D. Gibbs[2], and P. L. Bergquist[1,2]

[1]Centre for Gene Technology, University of Auckland, Private Bag 92019, Auckland, New Zealand
[2]School of Biological Sciences, Macquarie University, North Ryde, Sydney, New South Wales 2109, Australia

The yeast *Kluyveromyces lactis* has been developed as a host for extracellular production of thermophilic hemicellulases employing expression vectors based on the 2μ-like plasmid pKD1 of *Kluyveromyces drosophilarium.* A β-1,4-xylanase gene (*xynA*) from *Dictyoglomus thermophilum* strain Rt46B.1 was fused in-frame with a synthetic secretion signal derived from the *K. lactis* killer toxin and expressed under control of the *K. lactis LAC4* (β-galactosidase) promoter. Correctly processed xylanase enzyme with full biological activity on Oat Spelt Xylan was secreted during shake-flask cultivation of *K. lactis* transformants. Yield was found to be dependent on the strain and the composition of the growth medium. The transcriptional activity of the *LAC4* promoter dramatically affected mitotic stability of the expression vector under non-selective conditions. However, one isolate combined higher plasmid stability and good yield and has been employed for scaled-up production of XynA and other thermostable hemicellulases in chemostat culture. Similar results have been obtained for expression of a fusion of the *xynA* gene of *Thermotoga* strain FjSS3.B1 cloned into the same secretion vectors.

The utilization of hemicellulases in the pulp and paper industry is one of the most important new fields of bulk industrial enzymology. In the kraft process (the major pulping method worldwide), about 95% of the lignin present in wood fibre is removed by alkaline sulphate cooking. Residual lignin gives pulp a brown colour and is removed by bleaching with a variety of chemical agents, mainly chlorine and chlorine dioxide. Effluents contain toxic chlorinated organic compounds and must be treated before discharge.

Enzymatic pretreatment of kraft pulp with hemicellulases, especially endo-β-1,4-xylanases, allows paper of comparable brightness to be produced using up to 30% less chlorine (*1,2*) and is now used in several mills worldwide. However, the mesophilic enzymes currently used require cooling and pH adjustment of the pulp prior to addition. Ideally, enzymes used in kraft mills should be active above 60°C and pH >7. In collaboration with the Thermophile Unit, University of Waikato, we have been successful in isolating a number of thermophilic, alkaline-active hemicellulases from microorganisms found in thermal areas (*3,4*), and work by the Pulp and Paper Research Organisation of New Zealand (PAPRO) demonstrates that some of these do improve the bleachability of *Pinus radiata* pine kraft pulp (*5*). We now aim to produce recombinant enzymes in sufficient quantity for large-scale mill trials.

Heterologous production of candidate enzymes is necessary due to low yields from the native organisms which are themselves difficult to grow in the laboratory, generally requiring anaerobic conditions and high temperatures. In order to simplify purification, it is preferable to engineer recombinant thermophilic enzymes for secretion into the growth medium, allowing removal of contaminating mesophilic proteins by simple heat treatment.

We have employed the yeast *Kluyveromyces lactis* as a host organism due to its capacity for high level secretion of foreign proteins (*6*) and have developed episomal vectors for hemicellulase gene expression based on the 2μ-like pKD1 plasmid of *Kluyveromyces drosophilarium* (*7*). In our system, xylanase genes are amplified by PCR from bacterial genomic DNA and cloned in-frame with the signal sequence of the *K. lactis* killer toxin a-subunit for expression under control of either the *S. cerevisiae* PGK promoter or the *K. lactis* LAC4 promoter. The presence of a kanamycin resistance cassette on vectors containing the entire pKD1 sequence extends the spectrum of accessible host strains to all wildtype isolates of *K. lactis.*

Here we illustrate some aspects of our work on expression of thermostable bacterial xylanases from *Dictyoglomus thermophilum* strain Rt46B.1, originally isolated from a hot pool in Rotorua, New Zealand, and *Thermotoga* sp. strain FjSS3B.1, which was enriched from a hot spring on Savu-Savu beach in Fiji.

Materials and Methods

Strains and media. The *E. coli* strain used for routine cloning manipulations was JM101 [F' *traD36 lacI^{q}Δ(lacZ)M15 proA$^+$B$^+$/supE thiΔ(lac-proA,B)*]. Bacterial strains were grown at 37°C in Luria broth, supplemented with ampicillin (60 μg/ml) or kanamycin (30 μg/ml) where required for plasmid selection. *K. lactis* strains used as hosts for gene expression were MD2/1 (*MATa uraA, argA, lysA,* K^+, pKD1$^+$) and wild-type strains CBS1065 and CBS2359. Strain MD2/1 was transformed to uracil prototrophy by the lithium acetate procedure (*8*). Transformants were selected on SD minimal agar medium (0.67% yeast nitrogen base without amino acids, 40 mg/ml L-lysine, 40 mg/ml L-arginine, 2% glucose and 2% agar) and grown at 30°C in shake-flasks containing three different media: YNB, YNB supplemented with 1% Casamino acids (YNB-C) and YEP (1% yeast extract, 2% bactopeptone and 2% glucose or galactose, as indicated). Wild-type strains were transformed via selection for resistance to G418 (Geneticin, Life Technologies, Gaithersburg, MD) as described by Bianchi *et al.* (*9*).

Plasmid constructions. The structures of the *K. lactis* shuttle vectors used in this study are shown in Figure 1. Plasmid pSPGK1 contains the replication origin of the *Kluyveromyces* plasmid pKD1 (*10*) and a promoter-terminator cassette derived from the *S. cerevisiae PGK* gene, as described by Mellor *et al.* (*11*). A synthetic secretion signal derived from the pre-region of the *K. lactis* killer toxin a-subunit (*12*) is situated downstream of the *PGK* promoter and is followed immediately by a unique *Eco*RI site. Plasmid pCXJ-kan1 (*9*) contains the entire sequence of pKD1 and the kanamycin resistance gene (Kan) of Tn*903*, which confers resistance to G418 in yeast. Plasmid pCXJ1 was derived from pCXJ-kan1 by recircularization following removal of the

Kan gene on a *Sal*I fragment. Plasmid pNZ2931 (*4*) was used as the source of the *xynA* gene of *Dictyoglomus thermophilum* strain Rt46B.1. The nucleotide sequence encoding the mature xylanase (codons 30 to 353) was amplified by PCR with the introduction of *Eco*RI sites (underlined) at each end of the gene, as shown in Figure 2. Following digestion with *Eco*RI, the PCR product was inserted into *Eco*RI-digested pSPGK1 and recombinant plasmids were digested with *Hin*dIII and *Nde*I in order to identify those containing *xynA* in the required orientation. The resulting plasmid was denoted pSPGK-xyn. An expression cartridge in which the promoter and terminator regions of the *K. lactis LAC4* gene flank the secretion signal-*xynA* fusion region from pSPGK-xyn was constructed in pUC19 and transferred to pCXJ1 according to the strategy summarized elsewhere (Walsh and Bergquist, submitted). The completed expression cartridge was recovered from pLAC-3 on a 2.7 kb *Sal*I/*Hin*dIII fragment and inserted into *Sal*I/*Hin*dIII-digested pCXJ1 to create pCXJ-xyn (Figure 1B). Finally, the Kan gene was inserted on a 1.3 kb *Sal*I fragment to give plasmid pCXJK-xyn. Plasmids pSPGK-xyn and pCXJK-xyn were the *xynA* expression vectors used to transform *K. lactis*.

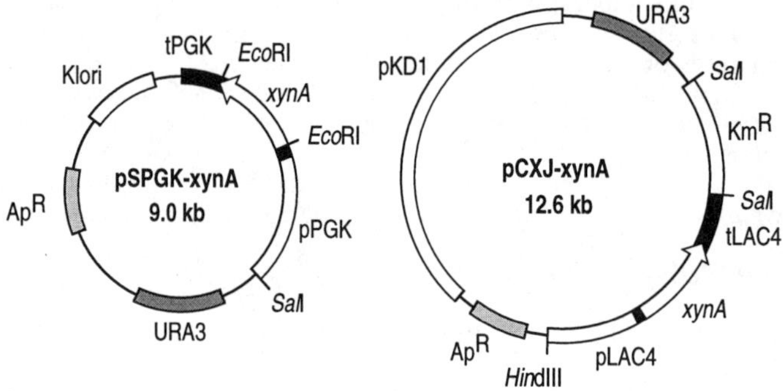

Figure 1. Recombinant plasmids expressing *Dictyoglomus* xylanase. The leader sequence of the *K. lactis* killer toxin α-subunit gene is shown as the filled portion of the circle in front of the *xynA* gene. Plasmid pSPGK-xynA was introduced into *K. lactis* strain MD2/1 (a, uraA, lysA, argA, pKD1$^+$) via selection of Ura$^+$ transformants. Plasmid pCXJ-xynA was introduced into *K. lactis* strain MW98-8C (a, uraA, lysA, argA, pKD1^o) via selection of Ura$^+$ transformants and into *K. lactis* wildtype strains CBS1065, CBS2359 and CBS683 via selection for resistance to the antibiotic G418.

The xynA gene of *Thermotoga FjSS3.B1* was amplified by PCR from plasmid pNZ2824 (*3*) using the upstream primer TF1: 5'-GGTTCA-ACGCGTGTTCCCTTGAGAGTG-3' and the downstream primer TF2: 5'-CGCTG-ACGCGTTTATCTTTCCTTCAG-3', so introducing *Mlu*I sites (underlined) at each end of the gene. Following digestion with MluI the PCR product was inserted into *Mlu*I-digested pCXJ1 to generate plasmid for transformation into *K. lactis* essentially as described above for the *Dictyoglomus xyn* gene. The secretion signal/xylanase junctions are shown in Figure 2.

Sodium dodecyl sulphate-polyacrylamide gel electrophoresis (SDS-PAGE) and zymogram analysis. Samples were subjected to electrophoresis in SDS-12% (wt/vol) polyacrylamide gels according to Laemmli (*12*). Protein bands were visualised by staining with Coomassie brilliant blue G-250 and sized by comparison with high range protein molecular weight standards (200-14.3 kDa, Life Technologies) and with purified *Dictyoglomus* XynA expressed in *E. coli*. Protein bands with xylanase activity were localized *in situ* in 12% polyacrylamide gels containing 0.1% (wt/vol) oat spelts xylan (Sigma Chemical Co., St. Louis, MO). Following electrophoresis, SDS was removed by washing the gel in three changes of 25% isopropanol, 25 mM sodium acetate, pH 6.0, each for 30 min at room temperature. Proteins were renatured by soaking the gel overnight at 4°C in 25 mM bis Tris propane, adjusted to pH 6.5 at 85°C. The gel was then immersed in a fresh sample of the same buffer and incubated in a sealed container at 85°C for 30 min. Zones of xylanase activity were then visualized by the Congo Red assay of Teather and Wood (*13*).

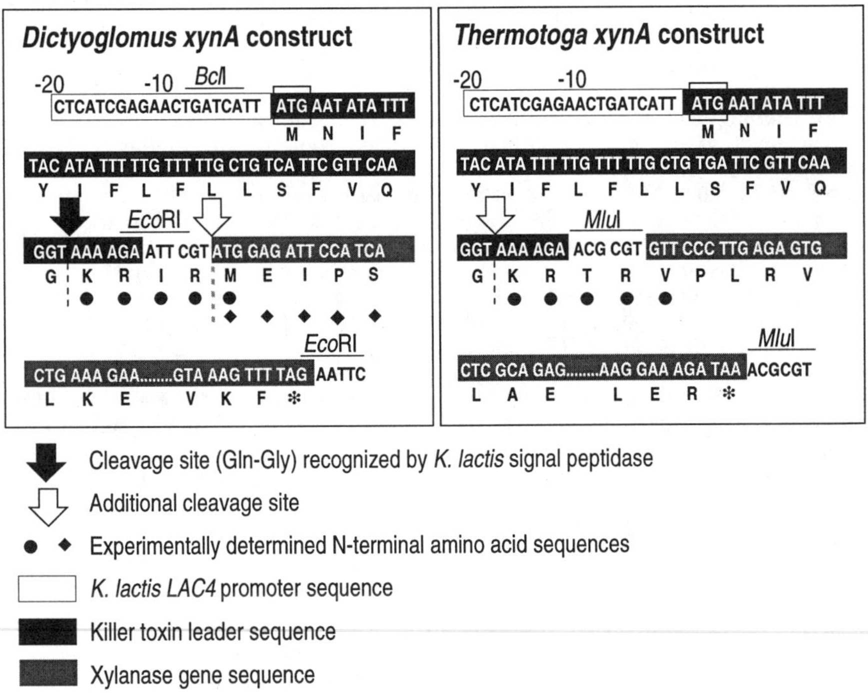

Figure 2. Secretion signal/xylanase gene junctions.

Protein microsequencing. The N-terminal sequence of the *Thermotoga* xylanase secreted by *K. lactis* was determined using supernatant from a culture of CBS1065[pCXJK-xyn]. A sample was subjected to electrophoresis on a SDS-12% polyacrylamide gel and proteins were electrotransferred to ProBlott™ membrane (Applied Biosystems Inc., Foster City, CA) and stained with Coomassie Blue according to Matsudaira (*14*). The XynA band was excised and sequenced using an Applied Biosystems Procise protein sequencer.

Plasmid stability determination. The fraction of plasmid-containing cells in cultures of *K. lactis* transformants was measured by comparing growth on non-selective and selective plates. For MD2/1 cultures, individual colonies were transferred from YPD plates to ura⁻ minimal plates. For CBS1065 and CBS2359 cultures, colonies were transferred from YEP glucose plates to YEP glucose containing 200 μg/ml G418. Two hundred colonies were transferred to selective medium for each culture sample analysed.

Assays for xylanase activity. The presence of xylanase activity in *K. lactis* culture supernatants was detected by plate assay on 0.8% agarose (wt/vol) containing 0.3% (wt/vol) oat spelts xylan, followed by incubation at 80°C for 1 h. Zones of clearing, indicating xylan hydrolysis, were visualised by the Congo Red assay. The enzymatic release of reducing sugar was determined by a modification of the method of Lever (*15*). Enzyme activity is expressed in Xylanase Units (XUs) where 1 XU is the amount of enzyme required to release 1 mmole of reducing sugar from xylan per minute (2). Temperature optima of yeast-derived recombinant xylanases and their thermal stabilities were determined as described previously (*3,4*).

Results

Construction of xylanase expression vectors. The rationale behind the construction of the secretion vectors is described for the *xynA* gene of *Dictyoglomus*. Similar considerations held for the *xynA* gene from *Thermotoga* FjSS3.B1. Plasmid pSPGK-xyn contains an in-frame fusion between the *Dictyoglomus xynA* structural gene without the DNA coding for the leader sequence and the secretion signal of the *K. lactis* killer toxin. The fusion was created at codon Met-30 in the full length XynA amino acid sequence in order to exclude the predicted bacterial signal sequence encoded by residues 1-29 (15). The *K. lactis* secretion signal in pSPGK-xyn is followed by the dipeptide Lys-Arg, which is a potential cleavage site for the Kex1 endopeptidase of *K. lactis* (*17*), and an Ile-Arg pair partly encoded by the *Eco*RI cloning site. In this construction, *xynA* expression is under the control of the *S. cerevisiae PGK* promoter, which also functions in *K. lactis* (*18, 19*). Plasmid pSPGK-xyn contains the replication origin of pKD1 and was accordingly introduced into a *K. lactis* pKD1$^+$ strain, MD2/1.

Plasmid pCXJK-xyn was constructed to allow regulated expression of *Dictyoglomus xynA* under the control of the promoter from the *K. lactis LAC4* gene, which is induced up to 100-fold by galactose and lactose (*20*). In order to ensure inclusion of all essential regulatory elements from the 5' flanking region of *LAC4*, we amplified a 1.1 kb fragment containing all three UAS elements which potentially interact with the *trans*-acting LAC9 protein (*21*). The expression cartridge was designed as a *Sal*I/*Hin*dIII fragment to facilitate direct introduction into the corresponding sites in pCXJ1. The Kan gene was added to allow plasmid selection in wildtype strains and the final recombinant plasmid was introduced into *K. lactis* CBS1065 and CBS2359.

Effect of media composition on xylanase secretion and plasmid stability. The influence of the growth medium on xylanase secretion and plasmid stability was evaluated for recombinant *K. lactis* strains in shake-flask cultures. For strain MD2/1[pSPGK-xyn] three media were utilised: a selective minimal medium (YNB), a richer semidefined medium (YNB plus 1% Casamino acids) which is still selective for the plasmid *URA3* marker, and a rich non-selective medium (YEP). Xylanase activity was detected in cell-free supernatants obtained from all three culture media after 24 hrs of growth (data not shown), consistent with constitutive expression of *xynA* from the *PGK* promoter. Samples obtained at daily intervals thereafter were tested for xylanase activity. Maximal cell density was reached after 24 to 40 h, depending on the growth medium, but xylanase activity continued to increase after this point and reached maximal levels of 3.1, 8.6 and 10.4 XU/ml in YNB, YNB plus Casamino acids and YEP media respectively after 5 days of growth. No xylanase activity was detected by either assay in a culture of MD2/1 harbouring the parent vector pSPGK1. The proportion of plasmid-containing cells remaining at the end of the cultivation period was 46% in SD, 42% in YNB plus Casamino acids but only 5% in YEP, indicating that pSPGK-xyn is mitotically unstable in the absence of selection.

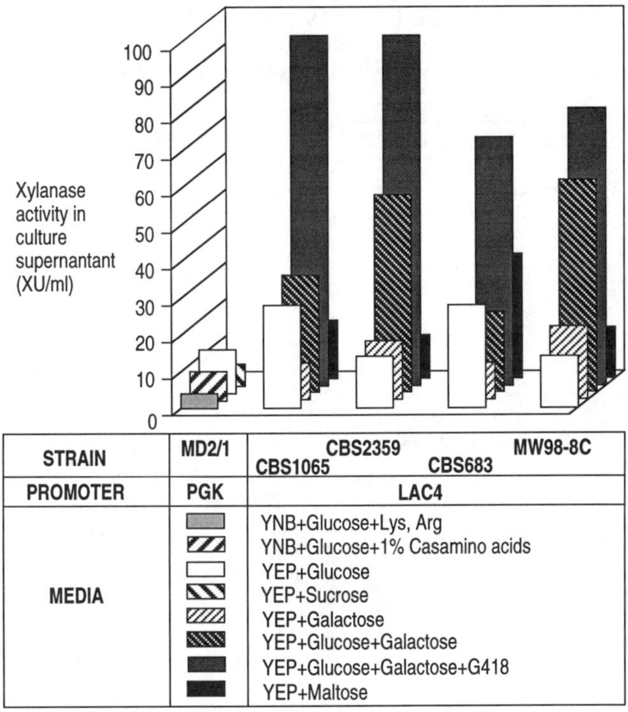

Figure 3. Effect of promoter and growth medium on xylanase expression and secretion.

K. lactis strains harbouring pCXJK-xyn were grown in YEP medium containing either glucose or galactose as the sole carbon source in order to investigate the effect of *LAC4* promoter strength on xylanase production and plasmid stability. For strains CBS1065[pCXJK-xyn] and CBS2359[pCXJK-xyn] the maximal levels of extracellular xylanase activity reached after 5 days of growth in YEP were 28.3 and 14.0 XU/ml respectively. Surprisingly, enzyme activity levels reached only 10.3 XU/ml for both strains grown in YEP medium plus galactose. Therefore, we examined the proportion of plasmid-containing cells remaining in YEP plus glucose and YEP plus galactose stationary phase cultures, which had undergone approximately 10 generations of growth. As shown in Figure 3, plasmid stability in CBS1065 was 92% on glucose but only 26% on galactose. In CBS2359 the plasmid was also highly stable on glucose (88%) but was completely lost during growth on galactose (Figure 4). Plasmid stability in YEP plus glucose medium was monitored over three further culture cycles, representing approximately 40 generations in total, by performing successive back-dilutions of stationary phase cultures in fresh medium. The pCXJK-xyn expression vector was retained by 92% of CBS1065 cells and 68% of CBS2359 cells, compared to a stability of 98% for the parent plasmid pCXJ-kan1 in both strains (Figure 4).

The stability of pCXJK-xyn after 10 generations of growth in YEP plus glucose and galactose was 72% in CBS2359 and 48% in CBS1065 (see Figure 3). We also found that xylanase activity was higher in this medium than in either YEP plus glucose or YEP plus galactose. The maximal levels attained in this medium were 32 and 55 XU/ml in CBS1065 and CBS2359 respectively. The addition to YEP plus glucose and galactose of 200 μg/ml G418 for plasmid selection promoted a further increase in xylanase activity to 95 XU/ml in both strains (Figure 3).

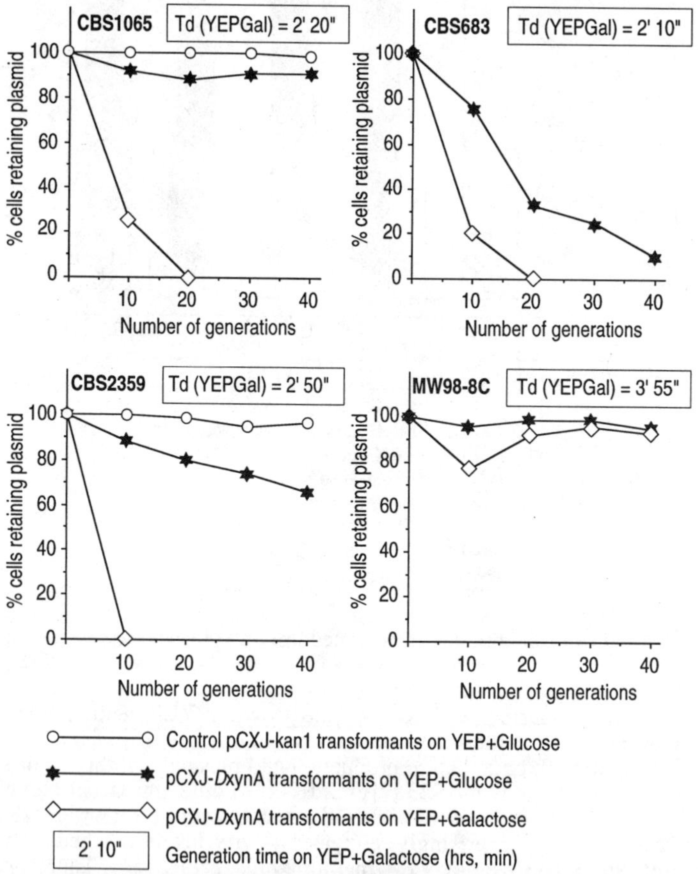

Figure 4. Stability of pCXJ-DxynA recombinant expression plasmids during non-selective growth.

Analysis of *K. lactis* culture supernatants by SDS-PAGE. Proteins secreted from recombinant *K. lactis* strains were visualized by Coomassie blue staining following SDS-PAGE of culture supernatants, as shown in Figure 5. Media used for cultivation of MD2/1[pSPGK-xyn] and CBS1065[pCXJK-xyn] (Figure 5, lanes 2 to 5) contained a prominent protein band with an apparent molecular mass of 35 kDa which co-migrated with *Dictyoglomus* XynA expressed in *E. coli* (Figure 5 lane 1) but which was not produced by the control strain CBS1065[pCXJ-kan1] (not shown). Densitometric scanning of the gel revealed that XynA comprised over 90% of total extracellular proteins in the CBS1065[pCXJK-xyn] YPDG+G418 culture (data not shown). The concentration of the recombinant xylanase was estimated to be 130 µg/ml by comparison with a known amount of *Dictyoglomus* XynA expressed in *E. coli.* This figure is in good agreement with an estimate of 150 µg/ml based on the level of xylanase activity in the CBS1065[pCXJK-xyn] YEP glucose plus G418 sample (95 XU/ml) and the specific activity of the *E. coli*-derived enzyme (630 XU/mg), which indicates that the *E. coli*- and *K. lactis*-derived xylanase preparations have approximately the same specific activity. Two bands can be seen in lanes 3 to 6 in Figure 5. These bands can be accounted for by the two cleavage sites in these constructions, the second, alternative one resulting in a lower molecular mass product was introduced in the cloning procedure as shown in Figure 2. The *Thermotoga* enzyme gives a single band of the correct molecular mass since it has only the Gln-Gly cleavage site recognised by the *K. lactis* signal peptidase (Figure 2; Figure 5, lanes 7 and 9).

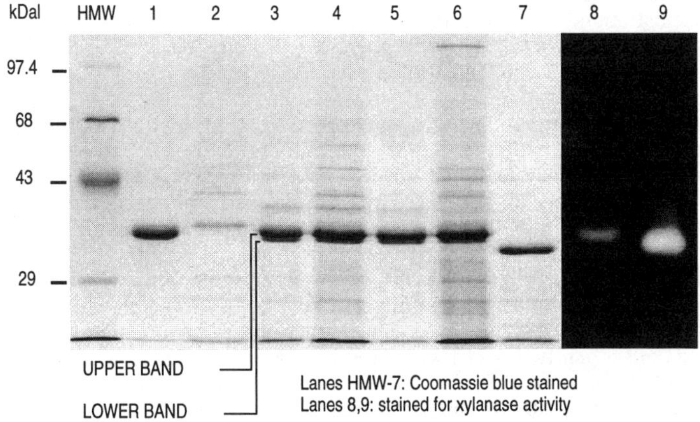

LANE	STRAIN	MEDIUM	XynA [µg/ml]
1.	FPLC-pure *Dictyoglomus* XynA produced in *E. coli*		100
2.	MD2/1 [pSPGK-*DxynA*]	YEP+Glucose	10
3,8.	CBS1065 [pCXJ-*DxynA*]	YEP+Glucose +Galactose+G418	110
4.	CBS2359 [pCXJ-*DxynA*]	YEP+Glucose +Galactose+G418	140
5.	CBS683 [pCXJ-*DxynA*]	YEP+Glucose +Galactose+G418	110
6.	MW98-8C [pCXJ-*DxynA*]	YEP+Glucose +Galactose+G418	130
7,9.	CBS1065 [pCXJ-*TxynA*]	YEP+Glucose +Galactose+G418	80

DxynA/*TxynA* denotes *Dictyoglomus* and *Thermotoga xynA* constructs

Figure 5. Analysis of culture supernatants after growth of *K. lactis* strains carrying recombinant expression plasmids by SDS-PAGE. pCXJ-DxynA = *Dictyoglomus xynA* recombinant plasmid: pCXJ-TxynA = *Thermotoga xynA* recombinant plasmid.

When proteins secreted from CBS1065[pCXJK-DxynA] were stained for xylanase activity (Figure 5, lane 8), we detected a zone of intensive xylan hydrolysis corresponding to the major 35 kDa XynA protein arising from expression of the *Dictyoglomus* gene and a zone of less intense activity which coincided with a minor 38 kDa component detected by Coomassie blue staining (Figure 5, lane 6). The equivalent recombinant strain carrying the *Thermotoga xynA* gene (CBS1065[pCXJK-TxynA]) showed a single zone of clearing (Figure5, lane 9).

Effect of temperature on XynA activity. The *Dictyoglomus* and *Thermotoga* xylanases secreted by *K. lactis* showed maximal activity at 85°- 90°C and over 50% of maximal activity between approximately 68 and 98°C. We observed no or minimal reductions in enzyme activity after culture supernatants diluted in 25 mM bis Tris propane, pH 6.5, had been incubated at 85°C in the absence of substrate for 24 h. Both enzymes were inactivated relatively rapidly at 95°C.

Discussion.

K. lactis strains harbouring pKD1-based xylanase expression vectors efficiently secrete recombinant xylanases at high levels. All strains examined were able to promote secretion of a *Dictyoglomus* xylanase with the same electrophoretic mobility as the control enzyme produced in *E. coli* (approx. 36 kDa). Recombinant xylanase is by far the most abundant protein present in transformed *K. lactis* culture supernatants, since the *Dictyoglomus* and *Thermotoga* XynA proteins comprise over 95% of total proteins secreted from strain CBS1065.

Several parameters influence xylanase secretion levels, including plasmid architecture and composition of the growth medium. Significantly higher levels of extracellular enzyme were attained through LAC4-directed XynA expression on full pKD1 sequence vectors as compared to expression from the PGK promoter on a pKD1 ori plasmid. The carbon source influences secretion levels by modulating promoter strength, which in turn affects plasmid mitotic stability. In non-inducing glucose medium, LAC4-directed *xynA* expression is incompletely repressed in all strains tested, although plasmid stability remains high. Full induction of the LAC4 promoter on galactose results in a drastic reduction in plasmid stability in wild-type strains, and consequently, lower levels of xylanase secretion than on glucose. Plasmid stability remains high on galactose in MW98-8C, probably due to the much longer generation time of this auxotrophic strain.

Our results show that the killer toxin secretion signal is processed by *K. lactis* signal peptidase. The yeast-derived *Dictyoglomus* XynA enzyme comprises two species, differing slightly in MW. N-terminal amino acid sequencing revealed the major, slightly larger, product to result from signal peptidase cleavage after Gln-Gly in the killer toxin signal sequence. The minor product is a truncated variant lacking four N-terminal amino acids (Walsh and Bergquist, submitted). This species may arise via Kex1 endopeptidase cleavage two residues downstream of its putative Lys-Arg recognition sequence (*17*).

Recombinant enzymes produced in *K. lactis* are biologically active as shown in the activity gel included in Figure 5, and are not hyperglycosylated, as we found previously for expression of thermophilic enzymes in *Saccharomyces cerevisiae* (*22*). The *Dictyoglomus* and *Thermotoga* XynA proteins produced in *K. lactis* have optimal activity at 90°C and half-lives at 85°C of approx. 13 hrs and >24 hrs respectively. These figures are comparable to the results reported for these enzymes produced in *E. coli* (*3,4*).

The next steps involve the scale-up of the expression system in fermentors for the production of substantial amounts of enzyme for X-ray crystallography and other basic research studies and for larger-scale bleaching trials. Strains CBS2359 and MW98-8C combine good secretion levels on non-selective media (60-80 µg/ml) with high plasmid stability on YEP Glucose and are thus suitable for scaled-up production of recombinant xylanases in chemostat culture. It will be necessary to develop an appropriate induction strategy to minimize the effects of plasmid instability.

Acknowledgments

We wish to thank Prof. H. Fukuhara for the gift of *K. lactis* vectors pSPGK1 and pCXJ-kan1. Financial support for this work was obtained from the Foundation for Research, Science and Technology, New Zealand and the University of Auckland Research Grants Fund.

References

1. Trotter, P.C. *Tappi Jour.* **1990,** 73, 198.
2. Fleming, B. *Pulp and Paper Canada,* **1991** 92, 13.
3. Saul, D. J., Williams, L. C., Reeves, R. A., Gibbs, M. D. & Bergquist, P. L. *Appl. Environ. Microbiol.* **1995**, 61, 4110.
4. Gibbs, M.D., Reeves, R.A. and Bergquist, P.L. Appl. Environ. Microbiol. **1995,** 61, 4403.
5. Allison, R.W., Wong, K.K.Y., Bergquist, P.L. and Morris, D.D. *PAPRO New Zealand Report* **1996,** No. C623.
6. Fleer, R., Yeh, P., Ammellal, N., Maury, I., Fournier, A., Bacchetta, F., Baduel, P., Jung, H., L'Hôte,H., Becquart, J., Fukuhara, H. and Mayaux, J. F. *Bio/Technol.* **1991,** 9, 968.
7. Ito, H., Fukada,Y., Murata, K. and Kimura, A. *Jour. Bacteriol.* **1983,** 153, 163.
8. Bianchi, M.M., Falcone, C., Chen, M. Wésolowski-Louvel, X.J., Frontali, L. and Fukuhara, H. *Curr. Genet.* **1987,** 16, 95.
9. Chen, X.J., Saliola, M., Falcone, C., Bianchi, M.M. and Fukuhara, H. *Nucleic Acids Res.* **1986,** 14, 4471.
10. Mellor, J., Dobson, M.J., Roberts, N.A., Tuite, M.F., Emtage, J.S., White, S., Lowe, P.A., Patel, T., Kingsman, A.J. and Kingsman, S.M. *Gene* **1983,** 24, 1.
11. Laemmli, U.K. *Nature* 1970, 227, 680.
12. Stark, M.J.R., and Boyd, A. *EMBO J.* **1986,** 5, 1995.
13. Teather, R.M., and Wood, P.J. *Appl. Environ. Microbiol.* **1982,** 43, 777.
14. Matsudaira, P. *Jour. Biol. Chem.* **1987,** 262, 10035.
15. Lever, M. 1973. *Biochem. Med.* **7,** 274.
16. Bailey, M.J., Biely, P. and Poutanen, K. *Jour. Biotechnol.* **1992,** 23, 257.
17. Tanguy-Rougeau, C., Wésolowski-Louvel, M. and Fukuhara, H. *FEBS Lett.* **1988,** 234, 464.
18. Fleer, R., Chen, X.J. Amellal, N., Yeh, P., Fournier, A., Guinet, F., Gault, Faucher, N. D., Folliard, F., Fukuhara, H., and Mayaux, J.-F. *Gene* **1991,** 107, 285.
19. Macreadie, I.G., Castelli, L.A., Ward, A, C., Stark, M.J.R. and Azad, A.A. *Biotechnol. Lett.* **1993,** 15, 213.
20. Dickson, R.C., and Markin, J.S. *Jour. Bacteriol.* **1980,** 142, 777.
21. Leonardo, J.A., Bhairi, S.M. and Dickson, R.C. *Mol. Cell. Biol.* **1987,** 7, 4369.
22. Donald, K.A.G., Carle A., Gibbs, M.D. and Bergquist, P.L. *Appl. Microbiol. Biotechnol.* **1994,** 42, 309.

TEXTILES

Chapter 15

Processing Textile Fibers with Enzymes: An Overview

Artur Cavaco-Paulo

Departamento de Engenharia Têxtil, Universidade do Minho, 4800 Guimarães, Portugal

Advances in biotechnology have made available for the textile industry a new range of auxiliaries - enzymes. In the last decade most of the enzyme based processes and products came out to the market and since then intensive development and research have been done. An overview on the research and successful applications in this "new" textile field is presented.

Enzyme proteins are catalysts for most reactions in living organisms and advances in biotechnology have led to their use for industrial processing of natural products. The main advantages of enzyme catalyzed reactions are: mild temperatures of processing, absence of by-products and most of the enzyme processes are claimed to be environmental friendly. In textile processing of natural fibers there are some stages where enzymes could be used. Fiber components like fats, waxes, starch, hemicelluloses, cellulose, proteins among others are ideal substrates for different classes of enzymes.

The microbial retting of bast fibers was the first biological process used in the textiles from BC times to actuality *(1)*. The first use of enzymes in textile process was made in the 1857 when starch sized cloth was soaked for several hours with water liquor containing barley. In 1900 this process was slightly improved with the use of malt extract, but it was only in 1912, with the use of animal and bacterial amylases, that the process became effective and started to be applied to the industry *(2)*.

In the last 10 years, new enzyme based effects and processes have been introduced such as the aged look in Denim garments and free pills cotton fabrics among others. Nowadays, some enzymes are very successful textile finishes. However, very little is known about the various processing mechanisms. This is mainly due to the lack of biochemical knowledge of textile chemists and to the lack of understanding of textile process by biochemists and biotechnologists.

In the last years, this area has suffer a rapid growth and several reviews and overviews have been written on the subject *(3-7)*. In this paper it is intended to give an overview of the existing processes and related research.

Wool

Wool Composition and Processing. Proteins (about 97 %) are the main material of dry wool, but structural lipids (about 2%) and other compounds (about 1%) like inorganic, nucleic acids and carbohydrates are also present *(8)*. Wool fibers are covered with the "flat-scale" cells usually known as the cuticle and the main bulk of the fiber is the cortex *(2)*. Cross linking is the major characteristic of wool proteins due to its high content of cysteine residues forming disulphide bonds between chains, especially present in the cuticle. The hydrophobic character of wool fiber is mainly due to the lipidic out layer covering the cuticle *(9)*.

Raw wool contains impurities like vegetal matter, suint (from the animal perspiration), grease, excrements, skin residues, dirt among others. Most of these impurities are removed after shearing, by soaking the fibers with detergent and by scouring with neutral or mild alkaline pHs. The vegetable soils are eliminated by treatment with sulfuric acid in a further process call carbonization and the cellulosic impurities residues digested are removed from the fibers by brushing and suction.

Most of the wet processing is made with very mild agitation due to the tendency of wool to felt. This tendency is a result of the presence of the "scales" of the cuticle in wool surface. The removal of this scales confers a anti-shrinking behavior and this is usually done with oxidation products and further polymer application. Most of these oxidation products (halogens derivatives) are environmentally harmful and intensive research has been made for more environmental friendly processes *(3)*.

Anti Shrink Finishing With Enzymes. The search for enzyme processes for anti shrink finishing of wool dates back to 1910, when trypsin and pepsin were used for cleaning skin scales *(10)*. The first studies showed that, the pre swelling of the fiber could determine the extent of proteolysis. It was also stated that, if the cystine disulphide bonds remain intacted, the proteolysis is slow. But, if some of the cross-links are broken, the reaction rate is increased. Several processes (already patents from the 40's *(10)*) based on a oxidative process and further proteolysis had been suggested, but none were applied to the industry, because of the enzyme cost and the unacceptable weight loss obtained *(3)*. Recent reports *(11, 12)* about the use of papain and commercial proteases after oxidative treatment, showed a good "descalating" effect but a high fiber damage.

The use of protein disulphide isomerase has been reported to improve the shrinkage behavior of wool fabrics. This enzyme rearranges disulphide bonds with the aid of a cofactor in a reduced form, like glutathione or dithiothreitol *(13)*. The use of transglutaminase has also been reported to improve shrink proof of wool, by a rather different mechanism, with the formation of new-cross links (N^6-(5-glutamyl)-lysine)) with the liberation of ammonia *(14)*.

Other Enzyme Applications and Effects. The wool proteolysis is claimed to improve the handle *(10)* and the luster of the fibers *(11)*. The improvement of appearance and handle was also verified in wool/cotton blended fabrics by the use of cellulases and proteases enzymes *(15)*. New protease products have been introduced for pilling removal of wool fabrics, that have been previously treated with anti-shrinking agents *(16)*. After those protease treatments, wool seems to have a better dyeability *(17, 18)* but the dye fastness is similar when compared to untreated wool *(18)*.

The presence of commercial proteases in the treatment liquor seems to aid wool bleaching with hydrogen peroxide *(19)*. The carbonization of wool has been reviewed by Heine & Hocker *(3)* and described by Sedelnik *(20)*, where a range of enzymes have been used to remove vegetable matter by reducing the amount of sulfuric acid used. However, the enzymatic degradation of vegetable materials by hydrolases like cellulases and hemicellulases, is a slow process which may never turn carbonization in a industrial enzymatic process.

The removal of the lipidic out layer of wool could improve the fiber hydrophilicity, however, selected lipases didn't produce any effects in wool surface *(21)*. These select lipases were active towards palmitol serine which is a model compound of the wool bound layer of 18-metyleicosanoic acid *(9)*. Only a mixture of lipases with minor proteases quantities produces a higher damage and a slightly increase of water absorption *(21)*.

Final Remarks. It is common sense for wool researchers *(3, 12)* that, at the present stage, the way is the search for specific proteases for the cuticle. However, just a few isolated reports have appeared about protein disulphide isomerases and transglutaminases. The heterogeneity of wool proteins makes it impossible to control the protease hydrolysis during an industrial process. A recent study *(17)* suggest that enzymes affect mainly the inner part of wool, confirming that the enzymes seem to diffuse inside the fiber *(3, 10)*"retting" it (fig. 1).

Silk

Silk fibers are constituted mainly by a double filament of fibroin involved by a layer call sericin *(1)*. Both sericin and fibroin are proteins that after hydrolysis, yield almost 0% in cystein residues. The amount of sericin, in terms of weight loss, after degumming varies, between 17-38 %. The composition of sericin is mainly serine (33 %), aspartic acid (17 %), glycine (14 %) and minor components of other residues. Fibroin is mainly constituted by glycine (44 %) and alanine (29 %). Sericin is more accessible to chemicals than fibroin and it is removed during preparation. An ideal degumming agent would attack specifically peptide bonds near by serine residues. However, several methods have been developed for degumming silk *(22, 23)* like extraction with water, boiling-off with detergent, with alkalis, acids and enzymes. The degumming with commercially available bacterial enzymes is more effective than the use of trypsin and papain. It is also reported that some bacterial proteases work better than the conventional process. Proteases can also be used to change silk fibroin surface, to give the aged look *(24)* in a similar fashion of enzyme washing of denim garments. It has also been recently reported that, proteases may induce improved softness and wetability in silk fabrics *(25)*.

Cotton

Cotton Composition and Processing. Cotton fibers are constituted largely by cellulose (about 94 % in raw state) and other minor materials like pectins, waxes, fats, proteins and minerals. Most of those non-cellulosic materials are on the out-layers (cuticle and primary wall) of the cotton fiber. The non-cellulosic materials are efficiently removed after scouring and bleaching, and cellulose content can go up to 99 %. Scouring nowadays is done in alkaline pH with detergents and a sequestering agent. Bleaching is done with hydrogen peroxide or hipocloride (in decline due to environmental problems). Those processes are done on yarns, fabrics or garments. In woven fabrics a further impurity must be removed; the size that has been added to the warp yarns to prevent breaks during weaving. The most usual sizes for cotton are starch or their derivatives and, they can be removed by amylase enzymes or some combined process using detergent, alkali, sequestering agent and an oxidation product *(2)*.

Desizing of Cotton. The use of α-amylases for desize starch and their derivatives from woven fabrics comes from the beginning of this century, as mentioned above. The enzymes used are mainly of bacterial origin and especially from *Bacilus subtilis.* Advances on bacteria engineering, permits the use, nowadays, of a range of enzymes acting at different temperatures ranges from 20 to up 115°C. The optimum pH is between 5 and 7, depending on the enzymes used, although amylases perform better in hard waters. All kinds of methods can be used ranging from padding to exhaustion methods *(26).* The desizing with amylases is especially done on fabric with yarns previously dyed.

Scouring and Bleaching. Scouring and bleaching of cotton with enzymes is one of the areas where a lot of research efforts have been made in recent years *(3-6, 27-30 and this book).* In these studies, lipases, pectinases, cellulases and their mixtures were used to improve cotton properties. However, previous work indicates that an enzyme based scouring and bleaching is slow, uneffective and therefore uneconomic *(7).* A recent new bleaching process has been proposed, based on the use of enzymatic desizing bath, for further controlled production of hydrogen peroxide by glucose-oxidase *(30).* The resulting gluconic acid was used *(30)* as a sequestering agent of metal ions (Fe III).

Removal of H_2O_2. Catalases have been successfully introduced recently, for removal of hydrogen peroxide after bleaching and prior to dyeing. The high rate of enzymatic decomposing of hydrogen peroxide allows the reduction of the water consumption during the washing of the bleached cotton, and prevents problems in further dyeing. The mechanism of catalase action for textile proposes, were reviewed by Hall *(31)* and the industrial application is described by Schmit *(32).*

Cellulase Finishing. Cellulases are the most successful enzymes used for textile processing. Apparently the same class of enzymes are used for obtaining aged and the renewed look of cotton fabrics *(33-40).* Cellulases are multicomponent enzymes and are mainly constituted by Endoglucanases (EGs), Cellobiohydrolases (CBHs) and Cellobiases. For ageing effects EG or EG rich mixtures are used, while for renewal and depilling effects complete mixtures can be used. Commercially available cellulases are mainly produced from the fungi *Humicola insolens* (maximal activity at pH=7) and *Trichoderma reesei* (maximal activity at pH=5).

Advances in biotechnology of the fungi and bacteria allows the availability in the market of cellulase mixtures enriched with certain components and monocomponents *(41-44).* New genetically designed proteins with EG activity have been recently developed having a specified aminoacid sequence from the conserved regions of the amino acid sequence of natural occurring EGs *(44).* Several studies had been made with CBH and EG enriched components and several compositions have been suggested for better depilling and ageing effects *(33, 34, 37, 38, 40, 43, 45-47).*

Depilling/Cleaning Effects. The most likely mechanism of depilling/cleaning effects seems to be due to the attack/adsorption to the more accessible pills (or fibrils) at the surface of a fabric (or fiber), rather than attack/adsorption on the main fabric (or fiber) structure. The pills become weaker after some cellulase hydrolysis extent and they are removed, by mechanical action, from the fabric (figure 2). Support for this mechanism is provided by the fact that depilling effects only take place with higher levels of mechanical agitation. It is also shown that enzymatic degradation on a high pilled fabric with high levels of mechanical agitation is mainly on the pills itself rather than on the fabric main structure *(33).*

Ageing Effects. Ageing effects are obtained mainly with EG or EG rich crudes. It was showed that enzyme action plus mechanical agitation, simultaneously *(33, 34)* or sequentially *(37)*, will abrade fiber surfaces, releasing cotton powder and defibrillation at the surface (figure 2) *(34)*. On denim fabric the dyed yarns after the enzymatic abrasion will release the "trapped" dye into the liquor in a form of cotton powder aggregations or just the insoluble dye itself. This produces in areas of higher mechanical abrasion, a higher contrast in the blue color. The fibrillation produced during the ageing process is a result of the synergistic action of the cellulases and mechanical action, and therefore, the ageing look is produced with less abrasive methods than the traditional washing with pumice stones. This is the main overvalue of the enzymatic washing process.

Key Features During Cellulase Processing. In both applications mechanical agitation is a very important issue and it seems to create more sites for cellulolytic attack *(38)*. Either due to the increased diffusion into the fabric structure (kind of spinning, knit or woven) or to the increased surface area due to defibrillation.

Other key features, like previous treatments such mercerization *(39, 48)*, scouring and bleaching *(49)* seem to increase further cellulase hydrolysis. Previous direct and reactive dyeing will inhibit cellulase action while, vat dyeing seems to not affect greatly the enzyme hydrolysis *(36, 38, 50)*. Water hardness and high ionic strength buffers reduces cellulases performance, i.e., an acetate buffer seems to be better than a citrate *(38)*. Confirming this behavior, ionic surfactants also inhibit cellulase activity *(50)*.

Dyeability and moisture recovery are not expected to change after cellulase treatment, since no changes were verified in crystallinity of cellulase treated cotton *(33)*. However, due to defibrillation, the water retention has been shown to increase *(51)*. Some times, slight deeper shades are apparently obtained after cellulase hydrolysis due to the enzyme treatment which will clean fiber surfaces giving this sensation.

Experimental evidence using complete crude mixtures and enriched compositions suggest that strength loss is mainly produced by EG activity *(33, 34, 36, 37)*.

Indigo Backstaining During Enzymatic Washing. The reposition of Indigo dye during washing denim garments with cellulases seems to be the major problem. Experimental evidence indicates that, *Humicola insolens* enzymes (neutral) gives less dye reposition than *Trichoderma reesei* enzymes (acid), suggesting that the pH was the major factor to control backstaining *(52)*. Recent studies indicate that, the ability of cellulase enzymes to be strongly adsorbed on the cotton substrates, seems to be the main reason that causes backstaining *(53)*. This confirms why some detergents that desadsorb cellulases reduce backstaining. Further experiments indicate that the nature of the enzyme used, it is an important issue, because Indigo dye seems to stain better in *Trichoderma* than on *Humicola* enzymes *(53)*. The first indication that cellulase binding process plays a major role during backstaining, was suggested by Genencor Patents, where indigo staining levels were reduced by the use of proteases that prevent the enzyme binding during washing *(54)*. The use of some cellulase components with low binding behavior also showed low Indigo staining levels *(54)*.

After Stone-Washing Treatment. The removal of the blue color of Indigo is usually done with NaOCl after the stone washing process. Indigo dye is a substrate for laccase enzymes and this enzyme was recently introduced commercially to replace NaOCl for this after treatment *(55)*.

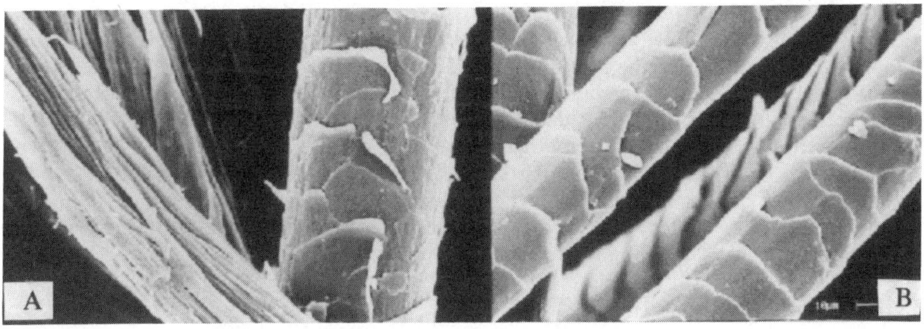

Figure 1- SEM photos of treated (A) and untreated wool (B) fibers with proteases. (offered by Dr. J.Shen, De Montfort University, Leicester, UK)

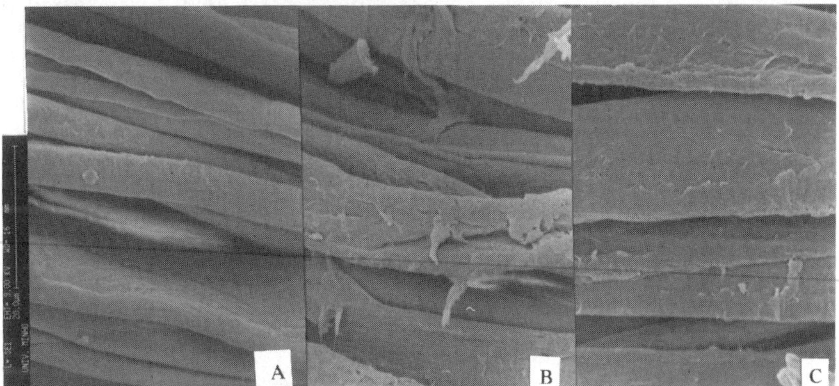

Figure 2 - SEM photos of control (A) EG treated (aged) (B) and further treated with a crude mixture (C). All treatments were done at high levels of mechanical agitation (adapted from reference 34).

Man Made Cellulose Fibers

Man-made cellulose fibers are obtained mainly from wood pulp, dissolved in a solvent and spun to a filament. This process may include, chemical derivatization of cellulose before solvent dissolution and further regeneration. Solubilization is a main issue for production of cellulose man-made fibers. Some studies have been done using cellulases for reduction of cellulose DP, to enhance solubility for fiber production *(56-58)*.

Several studies on a more basic level had been done in order to investigate the changes of several cellulosic fibers caused by cellulases *(47, 47, 60)*. Lyocell is a luxurious and high tenacity cellulosic fiber, produced by a recently developed extrusion process *(57, 58)*. This fiber is the main substrate for cellulases in the area of man-made fibers due to the high pilling tendency of their fabrics after processes with strong mechanical agitation *(60-62)*. Lyocell producers (62) suggest that cellulases are very important finishes, when used in a processing route to obtain the peach-skin feeling: after a process with strong mechanical action, the so called primary fibrillation is produced (with raised longer fibers and fibrils). Then, cellulases can be used to clean fabric and fiber surfaces; after that a new process with high mechanical action is followed and a secondary and uniform fibrillation is produced with very short fibrils, giving the peach-skin feeling.

Lignocellulosic Fibers

Bast fibers (flax, hemp, jute, kenaf and others) are based in cellulose (over 50%), hemicelluloses, lignin, pectins, fats, waxes among others *(1)*. Bast fibers are extracted from the plant stem by a process call "retting" as mentioned before. The objective of retting is just the partial degradation of the fiber materials, in such way that fibers can be obtained from the plant stems. Former retting process of flax were based in combined bacterial action and moisture (the stem in a open grass field) or in water (immersing the stem in slow rivers); nowadays retting is more often done in tanks in water at 30°C *(2)*. Despite of being an old process, much attention has been recently given to retting *(63-65)*. The use of enzymes like hemicellulases and pectinases for retting allows, a more controlled degradation of the fibers and a less environmental charged effluents *(63, 64)*.

The up-grading of jute fibers based on the use of enzymes has been studied by several Indian scientists *(66-68)*. The up-grading of this fiber is based on the use of cellulases for cleaning and softening, making easier further processing *(66, 67)*. The pretreatment of jute fibers with enzymes like cellulases and xylanases enhances further bleaching, in such way, that the jute enzyme treated has increased capacity to retain peroxide and larger surface area of lignin accessible to the oxidant *(68)*. In this study it is also claimed a softening benefit to enzymatic treated jute, due to the reduction of the bulk torsional rigidity and increase of transverse swelling *(68)*. Several softening and appearance benefits are also reported in the literature for fabrics based in bast fibers *(51, 69)*.

Synthetic Fibers

Enzymes, despite being natural catalysts, can also catalyze the modification of synthetic polymers. It is patented the modification of polyacrylonitrile with enzymes (nitrile hydratase) with the aim of increasing the content of amidic groups in the fibers *(70)* with improved dyeability and hydrophilicity. The same enzymes are also used for production of precursors of polyamide 6,6 *(71)*. This could be the beginning of application of enzymes in synthetic textile fibers and it shows that enzymes are not as specific as it was though, being widely used in synthetic organic chemistry *(72)*.

Others Applications

Detergents. Detergents are one of the most important markets for industrial enzymes. The function of the enzymes is to help on the removal of natural soils, by breaking them in smaller parts, that can be wash off better with the detergent action *(73)*.

Proteases are used since the late 60s in fabric washing products and since there are a wide variety of proteic soils, unspecific enzymes are used. This implies that wool and silk fabrics can't be apparently washed with detergent formulations containing proteases, however detergent manufactures just recommend not to soak in these products. It seems that, under mild washing conditions, little or no damage is produced until the proteic articles have been washed repeated times *(73)*.

Lipases have also been introduced in detergent formulations with the function of hydrolyzing fats, improving detergency of fat soils; however it seems very difficult to show the efficiency of lipases in detergents *(73)*. It also seems that lipases adsorb in fat soils and degradation is done between the washes, given complete removal in the next wash *(73)*.

Amylases are also part of a few detergents to remove starch soils (73).

Cellulases are claimed to aid detergency *(74)* during fabric washing and for more than 20 years, the use of cellulases was proposed to the removal of fibrillation produced after multiple washes *(74)*. But from the cellulases available at the time none were enough active at alkaline pHs of wash. Nowadays, these alkaline mixtures are available *(75)* and they are used in detergents renewing cotton article by removing pills and improving color brightness *(74)*. Further studies indicate that, detergent compositions enriched in CBH components and reduced in EG components showed good pill cleaning of cotton garments with reduced degradation *(76)*.

Recent studies have shown that L-amino acid oxidase enzymes in detergents compositions could inhibit the transfer of dyes between differently colored fabrics *(77)*.

REFERENCES

1-Neves, J., *Tecnologia Textil,* LLS Ed., Porto, Portugal, 1982, Vol. 1

2-Peters, R., *Textile Chemistry,* Elsevier Ed., Amsterdam, London, New York, 1963, Vol.1-2

3-Heine, E., Hocker, H., *Rev.Prog. Coloration,* **1995**, *25*, 57-63

4-Weltrowski, M., *L'Industrie Textile,* **1995**, *Jan.*, 52-58

5-Cegarra, J., *JSDC,* **1996**, *112*, 326-329

6-Cegarra, J., *Rev. Ind. Textil,* **1996**, *Dec.*, 42-50

7-Hamlyn, P., *Textiles Mag.*, **1995**, *3*, 6-10

8-Zahn, H., Knott, J., *Chemical Methods for Characterization of Wool at Different Stages of Processing,* Eurotex Ed., Guimarães, Portugal, 1992

9-Negri, A., Cornell, H., Rivett, D., *Text. Res. Jour.*, **1993**, *63*, 109-115

10-Nolte, H., Bishop, D., Hocker, H., *J. Tex. Inst.*, **1996**, *87*, 212-226

11-Levene, R., Sakkour, G., *JSDC,* **1995**, *111*, 352-359

12-Levene, R., Cohen, Y., Barkai, D., *JSDC,* **1996**, *112*, 6-10

13-Patent EP 0276547A1 (1988)

14-Patent JP3213574 (1991)

15-Chikkaki, S., Khan, S., Mehta, R., *Text. Res. Jour.*, **1995**, *65*, 564-569

16-Smith, S., Jackson, J., Ramazio, F., Nilsson, T., Dybdal, L., Anderson, M., *Int. Dyer,* **1995**, *180(9)*, 10-15

17-Yoon, N., Lim, Y., Tahara, M., Takagishi, *Text. Res. Jour.*, **1996**, *66*, 329-336

18-Riva, A., Cegarra, J., Prieto, R., in *15th IFATCC Conference-Book of Papers*, Lucerna, Switzerland, June 13-16, 1990
19-Cegarra, J., Gacen, J., Cayuela, D., Bernardes, A., *Tinctoria*, **1995**, *92(1)*, 32-36
20-Sedelnik, N., *Przeglad Wlokienniczy*, **1993**, *47(12)*, 294-279
21-Monlleo, D., Julia, M., Pinazo, A., Erra, P., in *16th IFATCC Conference*, Maastrich, Holland, June 9-11, 1993
22-Gulrajani, M., *Rev.Prog. Coloration*, **1992**, *22*, 79-89
23-Ghopra, S., Gulrajani, M., *Asian Textile Journal*, **1994**, *2(3)*, 34-46
24-Cheng, K., Yeung, K., *Textile Asia*, **1993**, *Aug.*, 32-35
25-Gulrajani, M., Gupta, S., *Indian Journal of Fibre & Textile Research*, **1995**, *20(4)*, 192-195
26-Cavaco-Paulo, A., Almeida L., *Nova Textil*, **1994**, *Abril*, 34-37
27-Rossner, U., *Melliand Textilber.*, **1993**, *2*, E63-E65
28-Li, Y., Hardin, I., *In Book of Paper AATCC Conference 1996*, Nashville, Tn, pp 87-100
29-Lange, N., *In Book of Paper AATCC Conference 1996*, Nashville, Tn, pp 101-108
30-Schacht, H., Kesting, W., Schollmeyer, E., *Textilveredlung*, **1992**, *30(11-12)*, 237-243
31-Hall, G., *In Book of Paper AATCC Conference 1995*, Atlanta, Ga, pp 581
32-Schmidt, M., *In Book of Paper AATCC Conference 1995*, Atlanta, Ga, pp 248-255
33-Cavaco-Paulo, A., Almeida L., Bishop, D., *Tex. Chem. Col.*, **1996**, *28(6)*, 28-32
34-Cavaco-Paulo, A., Almeida L., Bishop, D., *Text. Res. Jour.*, **1996**, *66*, 287-294
35-Cavaco-Paulo, A., Almeida L., *J. Tex. Inst.*, **1996**, *87*, 227-233
36-Cavaco-Paulo, A., Almeida L., *Biocatalysis*, **1994**, *10*, 353-360
37-Cavaco-Paulo, A., Cortez, J, Almeida L., *JSDC*, **1997**, *113*, 218-222
38-Cavaco-Paulo, A., Almeida L., Bishop, D., *Text. Res. Jour.*, in press
39-Almeida, L., Cavaco-Paulo, A., *Melliand Textilber.*, **1993**, *74*, 404-407
40-Cavaco-Paulo, A., Rios, M.,*In Book of Paper AATCC Conference 1996*, Nashville, Tn, pp 129-136
41-Patent WO 1992/06209
42-Patent WO 1993/24621
43-Stewart, C., *In Book of Paper AATCC Conference 1996*, Nashville, Tn, pp 212-217
44- Patent WO 1996/29397
45-Miettinen-Oinonen, A., Pere, J., Elovainio, M., Ostmam, A., Paloheimo, M., Suominen, P., *In 77th TI Word Conference*, Tampere-Finland, 1996, Vol. I, pp 197-207
46-Khahorst, S., Kumar, A., Mullins, M., *Tex. Chem. Col.*, **1994**, *26(2)*, 13-18
47-Kumar, A., Yoon, M., Purtell, C., *In Book of Paper AATCC Conference 1995*, Atlanta, Ga, pp 238-247
48-Buschle-Diller, G., Zeronian, S., *Tex. Chem. Col.*, **1994**, *26(4)*, 17-24
49-Cavaco-Paulo, A., Almeida, L., *In Proceedings of a FEMS Symposium: Lignin Biodegradation and Transformation*, Ed. JC Duarte, MC Ferreira, P Ander, Forbitec, April 1993, Lisbon, pp 101-102
50- Koo, H., Ueda, M., Wakida, T., Yoshimura, Y., Igarashi, I., *Text. Res. Jour.*, **1994**, *64*, 219-222 and 615-618
51- Buschle-Diller, G., Zeronian, S., Pan, N., Yoon, M., *Text. Res. Jour.*, **1994**, *64*, 270-279
52- Kochavi, D., Videbeak, T., Cedroni, D., *Am. Dyestuff Rep.*, **1990**, *Sept.*, 24-28

53- Cavaco-Paulo, A., Morgado, J., Almeida, L., Killburn, D., *Text. Res. Jour.*, in press
54- Patents WO 93/25655, WO 94/29426, WO 94/21801
55- Biotimes, No 1, **1997** (Novo Nordisk)
56- Vehvillainen, M., Nousiainen, P., Struszczyk, H., *In 77th TI Word Conference*, Tampere-Finland, 1996, Vol. I, pp 349-354
57- Struszczyk, H., *In 77th TI Word Conference*, Tampere-Finland, 1996, Vol. I, pp 315-322
58- Eibl, M., Eichinger, D., Lotz, C., *In 77th TI Word Conference*, Tampere-Finland, 1996, Vol. I, pp 343-348
59- Raahkmo, L., Viikari, L., Buchert, J., Vehvillainen, M., Nousiainen, P., *In Book of Paper AATCC Conference 1996*, Nashville, Tn, pp 436-441
60- Kumar, A., Purtell, C., Lepola, M., *Tex. Chem. Col.*, **1994**, *26(10)*, 25-28
61- Kumar, A., Purtell, C.,Yoon, M., *In 77th TI Word Conference*, Tampere-Finland, 1996, Vol. I, pp 177-187
62- Tencel, Technical Overview, Courtaulds Fibres, September 1996, 111 West 40thSt., New York
63- Patent WO94/13868
64- Schunke, H., Sanio, C., Pape, H., Schunke, U., Matz, C., *Melliand Textilber.*, **1995**, *76*, E101-E104
65- Ramaswamy, G., Ruff, C., Boyd, C., *Text. Res. Jour.*, **1994**, *64*, 305-308
66- Ghosh, B., Dutta, A., *J. Tex. Inst.*, **1980**, *71*, 108-116 and **1983**, *74*, 83-91
67- Ramurthy, V., Sinha, S., Kothari, R., Biotec. Lett., 1987, 9(8), 593-596
68- Kundu, A., Ghosh, B., Chakrabarti, S., *Text. Res. Jour.*, **1991**, *61*, 720-723 and **1993**, *63*, 451-454
69- Chaulam, C., Chingyee, L., Kowkwing, Y., *J. China Tex. Univ.*, **1995**, *12*, 10-16 (English Ed.)
70- Patent EP 0662540-1995
71- Galzy, P. *et al*, in *Abstracts of European Symposium Biocatalysis*, Sept. 12-17, 1993, Graz, Austria, P-70
72- Wong, C., Whitesides, G., Enzymes in Synthetic Organic Chemistry, 1994, Pergmon, Oxford New York, Tokyo
73- Bishop, D., in Chemistry of the Textiles Industry, 1996, Ed., C. Carr, Chapman & Hall (Blackie AP), London, Weinheim, New York, Tokyo, Madras, pp.125-172
74- Murata, M., Hoshino, E., Yokosuka, M., Suzuki, *JOACS*, **1993**, *70(1)*, 53-58
75- Patents EU 92105956.4, EU 92200101.1, EU 91202882.6
76- Patent WO 91/0584
77- Patent WO 94/25574

Chapter 16

Cotton Wall Structure and Enzymatic Treatments

Ian R. Hardin[1], Yonghua Li[1], and Danny Akin[2]

[1]**Textiles, Merchandising, and Interiors, The University of Georgia—Athens, 305 Dawson Hall, Athens, GA 30602–3622**
[2]**Russell Research Center, U.S. Department of Agriculture, Athens, GA 30605**

The application of pectinases and cellulases dramatically improved cotton water absorbency. Microscopy observations, staining tests, water absorbency tests, fiber weight loss analysis and nitrogen content analysis were conducted on raw cotton, and chemically and biochemically treated cotton to both qualitatively and quantitatively determine the nature of changes in the surfaces of cotton caused by the enzymatic treatments. The cotton surface structure was modified by digestion of the pectins in cotton cuticle and the amorphous primary wall cellulose under the cuticle. The destruction of the hydrophobic cotton cuticle under mild enzymatic treatment conditions occurs simultaneously with achievement of adequate water absorbency of the cotton. More severe enzymatic treatments completely remove the cuticle from the cotton fiber surfaces.

The cotton fiber has a multilayered structure. This structure has been studied and characterized by scientists for nearly a century. The structure of the primary wall, in particular, has an influence on textile manufacturing processes, and the properties and uses of textile products made from the cotton fiber. Scouring is a chemical process used to make cotton absorbent. It removes non-cellulosic substances in the primary wall and thus directly modifies that wall structure. Scouring of cotton is conventionally done with alkaline chemicals, which have potential to be environmentally harmful. Part of the reason for examining enzyme systems for scouring of cotton fibers has been the pursuit of environmentally friendly alternatives for scouring. Enzyme systems have been found to be very effective in making cotton water absorbent, and in the process changing the structure of the primary wall (1, 2, 3, 4). The research described herein focuses on the effects of enzymes used for scouring on the structure of the cotton fiber's primary wall.

The Cotton Wall Structure

The general state of knowledge of the cotton cell wall structure has been characterized, from the outside of the fiber to the inside, as cuticle, primary wall, secondary wall and lumen. The details of this layered structure has been defined by a series (5, 6, 7, 8, 9) of articles. The important points related to enzymatic treatments can be re-stated as the following:

1. Non-cellulosic substances located in the cuticle are mainly wax, proteins and pectins. They account for only a very small amount of the fiber weight (about 2.5%) (10). They exist in an amorphous state, which means they are much more amenable to enzymatic hydrolysis than crystalline substances.

2. Waxy materials on the outer surface, as well as esterified pectic substances within the primary wall, are responsible for the non-absorbent characteristics of raw cotton.

3. Primary wall cellulose accounts for only 2.5 percent of the fiber weight, about 30 percent of which is crystalline (6). This primary wall material is more susceptible to enzymatic hydrolysis than the highly crystalline secondary wall cellulose.

Previous Studies of Effects of Enzymes on Cotton Surface Structures

Previous work by this laboratory was conducted by using pectinases, cellulases, proteases, and lipases from different sources to investigate the effects of these enzymes on the properties and structure of cotton (1, 2, 3, 4). Controls were run by treating the fibers with buffer solutions only, without enzymes. Both cellulases and pectinases were found to be very effective in making cotton fiber absorbent. The following structural changes were revealed by staining tests, scanning electron microscopy observations, fiber weight losses and water absorbency tests:

1. Results (1, 2, 4) from Ruthenium Red staining showed that pectic substances and proteins in cotton surface structure were removed by cellulase treatments or pectinase treatments. Ruthenium Red stains pectic substances and proteins by virtue of affinity for carboxyl groups present in the molecules (11).

2. Results (1, 2, 3) from water absorbency tests and fiber weight losses showed that non-absorbent substances such as waxy substances and methylated pectins were removed by cellulase pectinase treatments.

3. Micrographs (1, 2, 4) from scanning electron microscopy work showed that the enzyme-treated fibers have surface characteristics different from those of raw cotton and chemically-extracted cotton. Raw mature cotton has a microscopical appearance characterized by smooth parallel ridges and grooves (12). These features were present in the control fibers and the ethyl alcohol-extracted cotton. Both enzymatic treatments and alkaline treatments changed the fiber surfaces. The characteristics of enzyme-treated fibers were concave groove-dominated surfaces with visible fibrils, shallow cavities and polished faces.

4. Water absorbency tests and microscopy work (1, 2, 4) showed that mixtures of pectinases and cellulases changed the surface structure of cotton more than either species of enzyme alone.

Procedures, Materials and Methods

Raw Cotton Fibers. Raw cotton (Southeastern, 1.08" staple length) fibers from

Thomaston Mills, Thomaston, Georgia, were used. Cotton fibers were used to avoid contamination or damage during fabric making that would obfuscate observations of the fiber surface changes caused by enzymatic treatments.

Enzymatic Treatments. A sample weight of 0.2 grams was used. Twenty milliliters of enzyme solution were used for each sample. Solutions were buffered at a pH of 4.0 with 0. 1 M citric acid and sodium citrate. The solutions were maintained at a temperature of 50° C. Freesol non-ionic surfactant (alkoxylated alkylphenol) was used as an assistant at a concentration of 0.1 percent in solution. All treatments were conducted with mild mechanical agitation. Please see Table I for dosage and treatment time for each sample.

Control Processes for Enzymatic Treatments. The control was obtained by treating the raw cotton in the buffer solution without the enzyme, but under the same conditions as in the enzyme treatments.

Microscope Observations. A JEOL scanning electron microscope was used in this study. All the specimens were glued to the stubs by Duco Cement. Care was taken not to subject the specimen to mechanical damage while gluing. The specimens were coated with palladium using a Samsputter-2A Sputter Coater. The machine setting was level 4.

Nitrogen Determinations. A FP-2000 nitrogen/protein determinator (LECO Corporation) was used. Each specimen was combusted in a furnace and the nitrogen then detected by thermal conductivity. The samples used for the tests were pectinase-treated fibers (P12/60 in Table I) and control fibers (Control in Table I). The nitrogen standard used for the determination was soil standard with 0.040% nitrogen. Each run of the determinator used about 0.2 grams of cotton.

Results and Discussion

Micrographs. Many individual fibers were observed in order to reach accurate conclusions about the characteristics of a sample. One standard adopted for judgement of a treatment effect was that a particular characteristic of the control sample may still appear to some degree in some of the treated specimens, but the particular characteristic of the treated sample should not exist in the control specimens. A characteristic of the treated sample was ascertained by its constant appearance in individual specimens.

The micrographs of control specimens show that the buffer solution (without enzymes) did not change the characteristic of smooth parallel ridges seen in the raw cotton fiber. Figure 1 is a micrograph of the surface of a raw cotton fiber, while Figure 2 is a micrograph of the surface of a control fiber after 20 minutes of treatment with buffer solution. Figure 3 is the micrograph of the surface of a fiber from a control specimen after 180 minutes of treatment. Figures 2 and 3 exhibit micropores or cracks in the surfaces, a phenomenon which was not observed in our previous study (1, 2, 4) and has not been observed in the surfaces of raw cotton. This newly reported observation supports hypotheses discussed later for the enzyme functions.

Cellulase Treatment. Specimens subjected to at least 20 minutes of treatment with a cellulase solution with of four units strength (see Figure 4) showed the emergence of

Table I. Dosages and Treatment Times for Treatments of Cotton

*Cellulase * Treatments*			*Pectinase* Treatments*			*Control*	
Sample	Enzyme Units**	Time in Minutes	Sample	Enzyme Units**	Time in Minutes	Sample	Time in Minutes
C4/20	4	20	P4/20	4	20	Con20	20
C4/60	4	60	P4/60	4	60	Con60	60
C4/120	4	120	P4/120	4	120	Con120	120
C4/180	4	180	P4/180	4	180	Con180	180
C8/20	8	20	P8/20	8	20	-	-
C8/60	8	60	P8/60	8	60	-	-
C8/120	8	120	P8/120	8	120	-	-
C8/180	8	180	P8/180	8	180	-	-
C12/20	12	20	P12/20	12	20	-	-
C12/60	12	60	P12/60	12	60	-	-
C12/120	12	120	P12/120	12	120	-	-
C12/180	12	180	P12/180	12	180	-	-

*Enzymes are from Sigma Chemical Company (Cellulase C1184 and Pectinase P9179)
**Enzyme Units are determined by product specified procedures

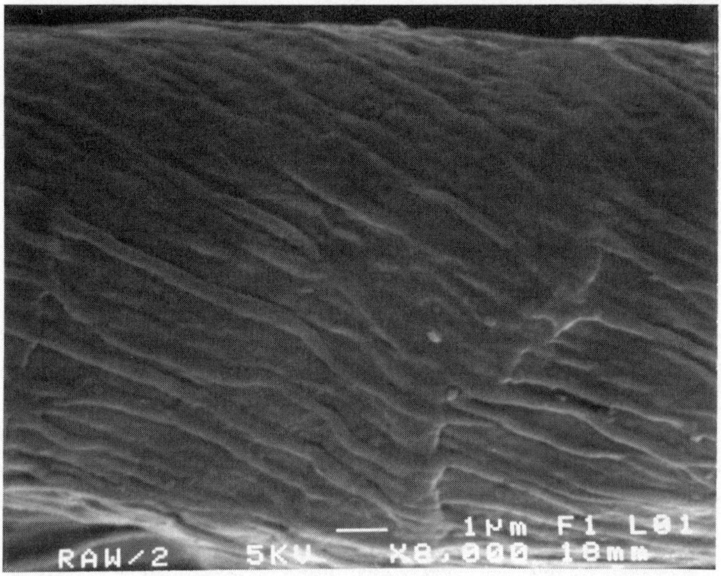

Figure 1. Scanning Electron Micrograph of the Surface of Raw Cotton Fiber.

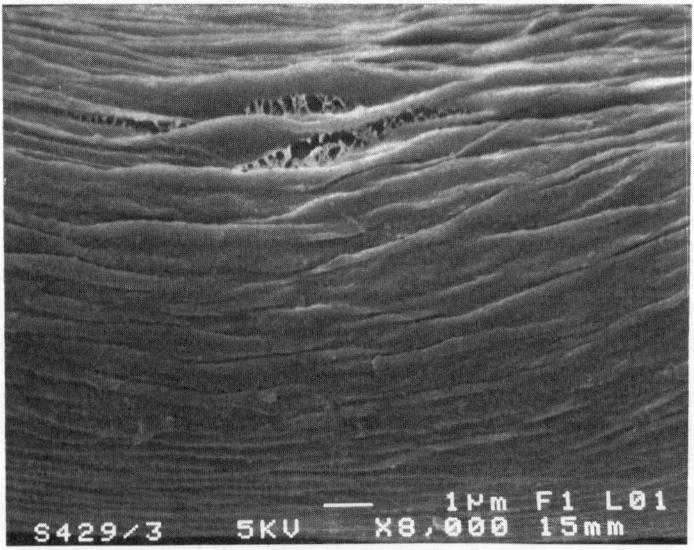

Figure 2. Scanning Electron Micrograph of the Surface of a Fiber from the Control Sample (Treated with Buffer Solution for 20 Minutes) Showing Micropores

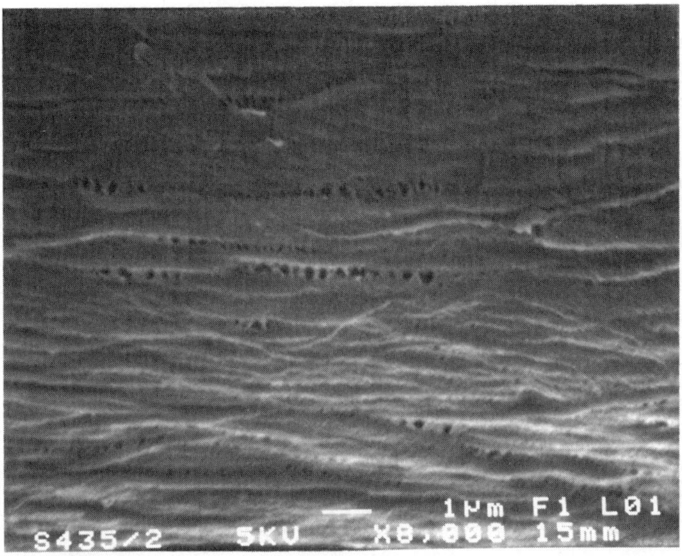

Figure 3. Scanning Electron Micrograph of the Surface of a Fiber from the Control Sample (Treated with Buffer Solution for 180 Minutes) Showing Micropores

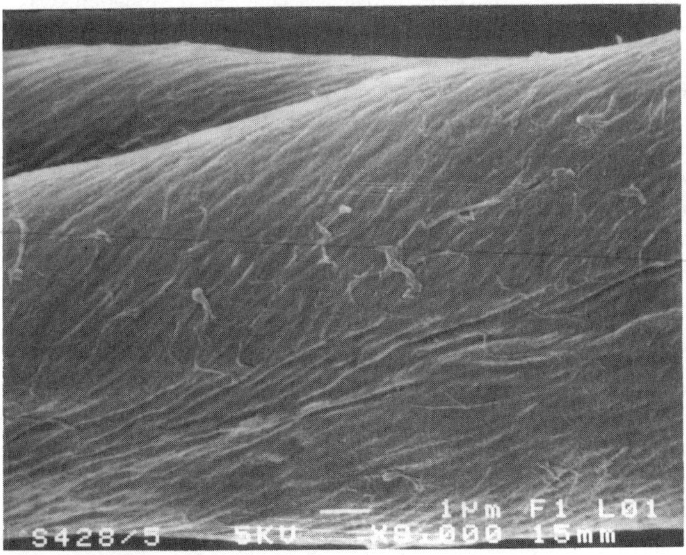

Figure 4. Scanning Electron Micrograph of the Surface of a Fiber from a Cellulase Treated Sample (sample C4/20, see Table I).

microfibrils on the surface of the fiber. These microfibrils were exposed over wide areas of the surfaces of fibers treated with either 4-units solutions for 180 minutes, or 8-units solutions for 60 minutes, both vigorous treatments. Figure 5 shows an example of these observations. As the strength of the cellulase solution was increased (12 units), the surface morphology showed the characteristic microfibrils after 20 minutes, and very clearly exhibited these after 120 minutes. Figure 6 typifies the surface morphology after a thorough cellulase scouring even though, as we show later, the weight loss of the fiber is less than four percent. Such a weight loss should correspond to a loss of water extractables and the waxes, proteins and pectins.

Pectinase Treatment. Treatments with pectinase caused changes in the smooth surface appearance of the raw cotton fiber after 20 minutes with solutions of 4-units strength (Figure 7). Microfibrils were exposed on the fiber surfaces after treatments with 4-units solutions for 180 minutes or 8-units solutions for 60 minutes. These microfibrils were characteristic features of all fiber surfaces after treatment for 60 minutes with 12-units solutions (Figure 8). Figure 9 exemplifies the latter and is typical of the pectinase-treated specimens after longer treatments. The similarities between the specimens in Figure 9 and Figure 6 (cellulase treated) support our hypothesis that the cotton cuticle (waxes, proteins and pectins) was broken down and removed by both the cellulase and pectinase treatments.

Weight Losses. The weight losses varied from sample to sample, even under the same treatment conditions. Statistical analysis was performed to obtain accurate estimations of the significance of the weight losses. Two conditions were chosen for examination. One was 4 units of enzymes/20 minutes of treatment time, and the other one was 12 units of enzymes/60 minutes of treatment time, for both cellulases and pectinases. The former condition was chosen because adequate cotton water absorbency (see references 1,2) was obtained at this level of treatment. The latter condition was chosen because the characteristic enzyme-modified appearances of the cotton fiber (see Figures 6,9) were prevalent at this level. Five duplicate treatments were conducted for each condition. Original weight losses (x_i) and averages (x_j) are listed in Table II.

The Pooled t-test (13) for the difference between two population means was conducted on the following independent pairs:

C4/20 versus Control 20, and P4/20 versus Control 20
C12/60 versus Control 60, and P12/60 versus Control 60
Control 20 versus Control 60
C12/60 versus C4/20, and P12/60 versus P4/20

At a confidence level of 97.5% (α=0.025), the following was concluded:

1. The weight loss value for C4/20 was significantly higher than Control 20, that for C12/60 was significantly higher than Control 60, and that for C12/60 was significantly higher than C4/20. The cellulase treatments on cotton caused significant weight losses of cotton fibers and the more severe the conditions, the higher were the weight losses.

2. The weight loss value for P4/20 was significantly higher than Control 20, that for P12/60 was significantly higher than Control 60, and that for P12/60 was significantly higher than P4/20. The pectinase treatments on cotton caused significant weight losses of cotton fibers and the more severe the conditions, the higher were the weight losses.

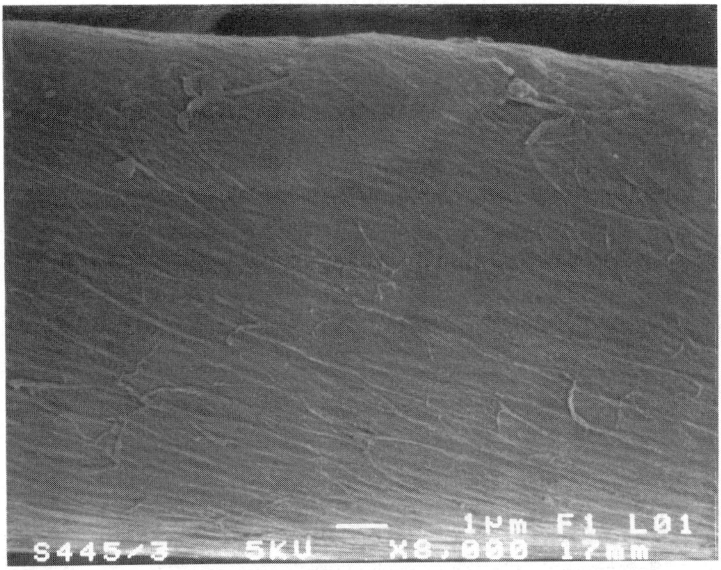

Figure 5. Scanning Electron Micrograph of the Surface of a Fiber from a Cellulase-Treated Sample (sample C8/60, see Table I).

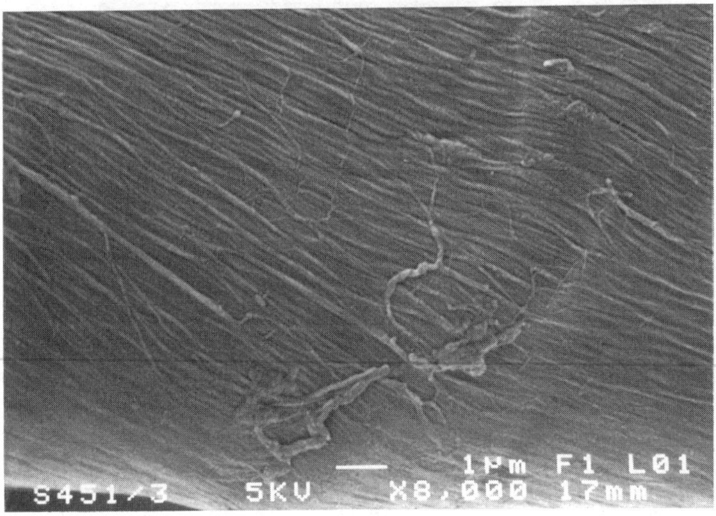

Figure 6. Scanning Electron Micrograph of the Surface of a Fiber from a Cellulase-Treated Sample (sample C12/120, see Table I).

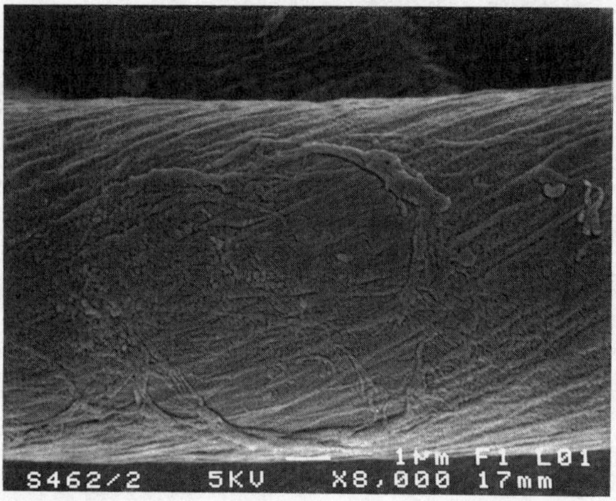

Figure 7. Scanning Electron Micrograph of the Surface of a Fiber from a Pectinase-Treated Sample (sample P4/20, see Table I).

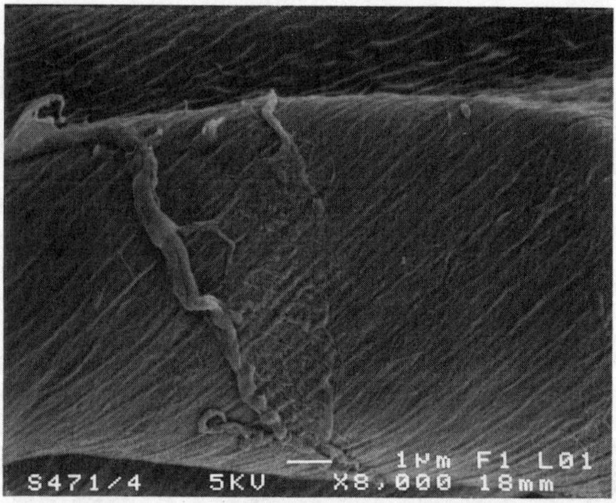

Figure 8. Scanning Electron Micrograph of the Surface of a Fiber from a Pectinase-Treated Sample (sample P12/60, see Table I).

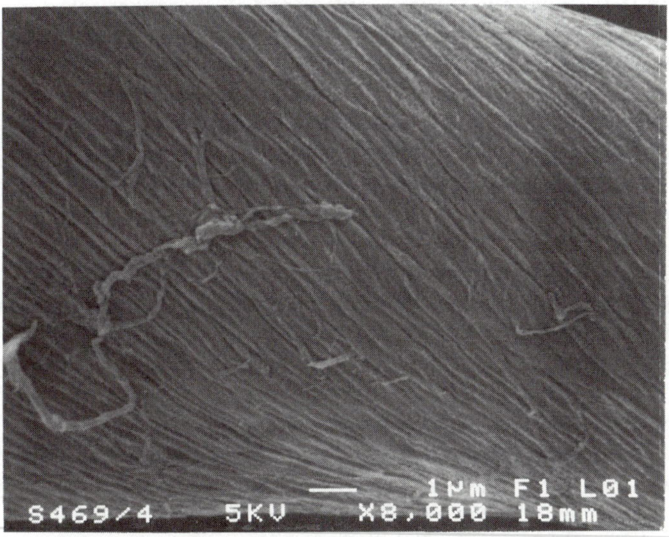

Figure 9. Scanning Electron Micrograph of the Surface of a Fiber from a Pectinase-Treated Sample (sample P8/180, see Table I).

3. The weight loss results for Control 60 did not significantly differ from those for Control 20. The longer treatments of cotton in the buffer solution did not cause higher weight losses.

If the weight losses from the control fibers are looked upon as water-extractable substances (average of 1.16% for 60 minutes of treatments), the non-water-extractable substances removed by enzyme treatments averaged about 2.2 percent (3.36-1.16=2.2, see Table II) for cellulase treatments of 60 minutes, and an average about 1.73 percent (2.89-1.16=1.73, see Table II) for pectinase treatments of 60 minutes. These two percentages (2.2 and 1.73) are smaller than 2.5 percent, the estimated percentage for the cotton cuticle weight. The conclusion is that these treatments removed a large percentage of the waxes, proteins and pectins of the cuticle.

Nitrogen Determinations. Several kinds of non-cellulosic substances exist in cotton. The weight losses caused by enzymatic treatments in these experiments seem to account for the removal of these non-cellulosic substances in total. Quantitative determinations of the changes in amounts of the chemical constituents separately will offer more understanding of the effects of enzymes on the cotton structure. These materials, waxes, proteins, pectins, sugars, and minerals, exist in cotton in very small percentages (2, 10, 14, 15), so additional work is needed to carefully quantify the amounts of these substances and, even more importantly, the location of these in the cuticle and primary wall.

Nitrogenous substances exist in both the cuticle and the lumen (5, 8, 14, 15). Pectinase or cellulase treatments will not affect the nitrogen content in the lumen. But, nitrogen content in the surface of cotton will be affected if the pectinases or cellulases remove the cuticle, or if nitrogenous substances are located in the cuticle between the layer of waxes and the layer of pectins, or with pectins themselves. Table III shows the results of the nitrogen determinations.

The Pooled-t test was conducted to determine if the nitrogen content in P12/60 was lower than in Control 60, or not. At a confidence level of 97.5% (α=0.025), the nitrogen content in P12/60 was significantly lower than in Control 60, even though the absolute difference was small. Pectinase treatments do remove some of the nitrogen substances. Despite the statistical significance of the difference in nitrogen content, the very small real difference probably raises more questions of interpretation than it answers. Whether or not the nitrogen content after treatment includes residual enzyme has not yet been determined.

Cotton Water Absorbency. Our previous studies (1, 2, 3, 4) and follow-up work have consistently showed that cellulases and pectinases can dramatically improve cotton water absorbency. The standard method of the American Association of Textile Chemists & Colorists for fabric wettability is a water-drop test (see AATCC Test Method 39-1980: Evaluation of Wettability). The disappearance of a water drop into the fabric within a second signifies adequate absorbency of the fabric for dyeing, printing and finishing [16]. In this research, adequate absorbency of cotton was obtained by using very low dosages of enzymes (for example, 0.05% owg for P9179 in one case) and in a very short treatment time (for example, 10 to 20 minutes in our case). It must be emphasized that control of

Table II. Cotton Weight Losses in Enzymatic Treatments

Enzyme Units	*Time in Minutes*	*Weight Losses*					
		Readings (x_i)					*Average (X_i)*
C1184=4	20	1.09	1.36	1.50	1.98	1.58	1.50 (C4/20)
P9179=4	20	1.64	1.15	1.51	1.24	1.87	1.48 (P4/20)
Control	20	0.96	1.08	1.11	1.39	0.99	1.11 (Control 20)
C1184=12	60	3.20	3.90	3.79	2.74	3.17	3.36 (C12/60)
P9179=12	60	3.01	2.95	2.60	3.26	2.65	2.89 (P12/60)
Control	60	0.88	1.15	1.30	1.33	1.14	1.16 (Control 60)

Table III. Nitrogen Content in Treated Cotton

Sample	*Nitrogen Readings*					*Averages*
P12/60	0.0893	0.0961	0.0933	0.0906	0.0932	0.0925
Control 60	0.101	0.106	0.108	0.101	0.0964	0.10248

treatment conditions, the nature of the enzyme mixtures, and specific enzyme activities have great influences on the treatment efficiency for adequate cotton absorbency.

Conclusions on the Modifications of Cotton Wall Structure by Enzymatic Treatments

The cumulative evidences from microscopy observations, water absorbency tests, staining tests, weight losses, and nitrogen content determination, indicate the following as working hypotheses for the enzymatic surface modification of raw cotton:

1. Enzymes penetrate the cuticle in aqueous solutions through micropores and make contact with their substrates. Pectinases destroy the cotton cuticle structure by digesting the pectins in the cuticle. Cellulases destroy cotton cuticle structure by digesting the more amorphous primary wall cellulose under the cuticle. Waxes and proteins are physically removed by such enzymatic actions because of the breakdown of the connecting structure holding them on the fiber surface.

2. Enzymatic modification of cotton surface structure with pectinases and cellulases is a progressive process related to enzyme dosages and treatment durations. At preliminary stages of enzymatic functions, the non-absorbent covering on the cotton surface is broken and extensively perforated. This explains the success of relatively mild treatment conditions (dosages and times) in improving cotton water absorbency. Cuticle substances are removed gradually as the treatments involve higher enzyme dosages and longer treatment durations.

3. Any factors in helping enzymes to contact their relative substrates and assisting their catalysis will speed up the process of enzymatic modification of the cotton surface. Examples are uses of surfactant and mechanical agitation in enzymatic treatments. Surfactant lowers the surface tension and promotes penetration of enzymes to the active sites for attack. Mechanical agitation provide extra kinetic energy for molecules to enter the active sites..

4. Work in the future will concentrate on better elucidating the nature of the structural attack by enzymes on the cotton wall. Of particular interest will be the incorporation of chelating agents into pectinase mixtures.. One of us, (DA) and collaborators (17), have shown dramatic increases in pectinase effectiveness on flax fiber with the addition of chelating agents. Attack on calcium bridges included in the pectins is indicated.

Acknowledgment:

Sandy Silver, USDA Russell Research Center, Athens, Georgia

Literature Cited

1. Li, Y. H. and Hardin, I. R., Enzymatic Scouring of Cotton, Book of Papers, *AATCC International Conference & Exhibition* **1996**, pp.87 - 100.
2. Li, Y. H. and Hardin, I. R., Enzymatic Scouring of Cotton: Effects on Structure and Properties, *Textile Chemist and Colorist* **1997**, in press.
3. Hardin, I. R. and Li, Y. H., A Comparison between Enzymatic Scouring and

Alkaline Scouring of Cotton, *Proceedings of Beltwide Cotton Technology Conference* **1997**, pp.745 - 747.
4. Li, Y. H. and Hardin, I. R., Changes in the Cuticle and Primary Wall of Cotton Treated with Cellulases and Pectinases, *Textile Research Journal*, accepted for publication.
5. Tripp, V. W. and Rollins, M. L., Morphology and Chemical Composition of Certain Components of Cotton Fiber Cell Wall, *Analytical Chemistry* **1952**, *vol 24*, pp.1721.
6. Boylston, E. K. and Hebert, J. J., The Primary Wall of Cotton Fibers, *Textile Res. J.* **1995,** *vol 65(7)*, pp.429- 431.
7. Jeffries, R., Jones, D. M., Roberts, J. G., Selby, K., Simmens, S. C. and Warwicker, J. O., Current Ideas on the Structure of Cotton, *Cellulose Chem. Technol.* **1969**, *vol 3*, pp.255-256.
8. Rollins, M. L., The Cotton Fiber, in *The American Cotton Handbook*, Hamby, D. S., Ed., Interscience Publishers, NY, 1965, Vol. I; pp.65, 71, 72, 73.
9. DeGruy, I. V., Carra, J. H., Goynes, W. R., The Fine Structure of Cotton: An Atlas of Cotton Microscopy, Marcel Dekker, Inc. New York, 1973, pp.156-158.
10. Tripp, V. W., Moore, A. T., and Rollins, M. L., Some Observations on the Constitution of the Primary Wall of the Cotton Fiber, *Textile Res. J.* **1951**, *vol 21(12)*, pp.886-894.
11. Rollins, M. L. and deGruy, I. V., Light Microscopy in the Study of Cellulose, in *Instrumental Analysis of Cotton Cellulose and Modified Cotton Cellulose*, O'Connor, R. T., Ed., Marcel Dekker, INC, NY, 1972, pp.145-147.
12. Tripp, V. W., Moore, A. T., and Rollins, M. L., The Surface of Cotton Fibers Part I: Native Fibers, *Textile Res. J.* **1957**, *vol 27*, pp.419-425.
13. Ott, R. L., An Introduction to Statistical Methods and Data Analysis, Fourth Edition, Wadsworth Publishing Company, Belmont, Ca., 1993, pp.262-268.
14. Guthrie, J. D., The Chemistry of Lint Cotton A. Composition, in *Chemistry and Chemical Technology of Cotton*, Kyle Ward, Jr, Ed., Interscience Publishers, NY, 1955, pp.4, 5, 6.
15. McCall, E. R. and Jurgens, J. F., Chemical Composition of Cotton, *Textile Res. J.* **1951**, *vol 21*, pp.19-21.
16. Freytag, R. and Donze, J.-J., Alkali Treatment of Cellulose Fibers, in *Fundamentals and Preparation Part A, Handbook of Fiber Science and Technology: Volume 1, Chemical Processing of Fibers and Fabrics*, Lewin, M. and Sello, S. B., Ed., Marcel Dekker, Inc, New York and Basel, 1983, pp.111-132.
17. Hendriksson, G., Akin, D. and Eriksson, K.-E., The Influence of Chelating Agents and Mechanical Pretreatment on Enzymatic Retting of Flax, *Textile Res. J.* **1997**, in press.

Chapter 17

Bioscouring of Cotton Fabrics with Cellulase Enzyme

E. Csiszár[1], G. Szakács[2], and I. Rusznák[1]

[1]Departments of Organic Chemical Technology and [2]Agricultural Chemical Technology, Technical University of Budapest, H-1521 Budapest, Hungary

When traditional alkaline scouring of desized cotton fabrics was preceded by cellulase enzymatic treatment, two benefits were observed. Beside significant increase in whiteness of fabrics, enhanced removal and alkaline degradation of seed-coat fragments were achieved. Commercial cellulase enzyme was applied before the caustic scouring process, in 1-10 g/l concentration for 0.5-8 hours at pH 5.0 and at 50°C. Enzyme treatment alone resulted in 14-21 % increase in Berger-whiteness. When consecutive cellulase treatment and conventional alkaline scouring were combined, the increase in whiteness was even more significant. Applying cellulase enzyme in concentrations of 1, 5 and 10 g/l, respectively, in biotreatment followed by conventional caustic scouring, a maximum of 18, 25, and 29 % increase in whiteness was observed. Cellulase pretreatment also allowed the reduction of the hydrogen-peroxide consumption in the chemical bleaching step.

Interest in the application of enzymes by textile industry has increased considerably during recent years. Cellulolytic enzymes have emerged quite recently with potential applications in processing of cotton and other cellulosic fabrics. Cellulase enzymes are used for biostoning and fading of jeans (*1*), finishing of cotton and cotton-blended fabrics for improved softness and enzymatic removal of fuzz and pills (*2, 3*). Though the possible application of cellulase, pectinase and lignin degrading enzymes before or during the scouring process has been indicated in the literature (*4*), no detailed study has been reported to use cellulase enzymes for bioscouring and biobleaching of raw cotton fabrics.

We demonstrated recently (Csiszár, E.; Szakács, G.; Rusznák, I. *Textile Res. J.*, in press.) that cellulase enzymatic treatment prior to the alkaline scouring process enhanced the penetration of scouring chemicals into the seed-coat fragments facilitating an easier removal of these impurities from the cotton fabrics. As an additional benefit significant increase in whiteness of fabrics was also observed. The "enzymatic pre-bleaching" is a well-known concept in the pulp and paper industry. Although the mechanism of this process is not completely understood, xylanases and other enzymes have been used commercially since 1990 (*5, 6*). In the present study a combined biological and chemical approach is detailed for bleaching of raw cotton fabrics. The effect of "enzymatic pre-bleaching" (cellulase treatment before traditional caustic scouring) on the consumption of bleaching chemicals has also been studied. The elaboration of environmentally milder bleaching process has been another aim of this work.

Materials and Methods

Greige cotton print cloth fabric (122 g/m^2) obtained from Testfabric Inc., NJ. was used after amylase enzymatic desizing. Commercial cellulase enzyme from *Trichoderma sp.* was kindly provided by Environmental BioTechnologies, Inc., Santa Rosa, California. The following enzyme activities were measured from the product using internationally recognized methods: Filter paper cellulase activity (FPA), according to Ghose (*7*): 108 FPU/ml; β-glucosidase (BG) activity, according to Kubicek (*8*): 81 BGU/ml; 1,4-β-endoglucanase (EG) activity, according to Bailey and Nevalainen (*9*):72,500 EGU/ml; xylanase activity, according to Bailey et. al. (*10*): 12,800 IU/ml.

Cotton fabrics were treated as follows: The conditions for the cellulase treatment were pH 5.0 (0.05 M sodium-acetate-acetic acid buffer), liquor ratio 1:100, enzyme concentrations 1, 5 and 10 g/l, respectively, time of treatment 0.5, 1, 2, 4 and 8 h, respectively, temperature 50°C, concentration of nonionic surfactant 1 g/l. After enzymatic treatment the substrate was washed in hot distilled water twice to deactivate the enzyme and air dried thereafter. Caustic scouring was carried out with pad-steam method at 100 °C for 20 min. in a solution containing 50 g/l sodium-hydroxide and 1 g/l surfactant. After scouring the fabrics were further bleached with hydrogen-peroxide at 100°C for 15 min by pad-steam process. The hydrogen-peroxide (30%, w/w) concentrations were 5, 10, 15, 20, 25 and 30 g/l, respectively, and the concentration of a stabilizer was 1-3 g/l, depending on the hydrogen-peroxide concentration.

X, Y, Z tristimulus values in CIELAB color space (*11*) were measured with ICS Texicon (D65/10°) color-measuring instrument after cellulase treatment, scouring and bleaching processes, respectively. Berger-whiteness (Whiteness), ISO Brightness (Brightness), color (ΔE_{ab}^*)- and chroma (ΔC_{ab}^*) -differences were calculated.

Results and Discussion

Bleaching Effect of Cellulase Pretreatment. The effect of cellulase pretreatment on desized cotton fabric at different enzyme concentrations and times of treatment has been evaluated by whiteness, brightness, CIELAB color- and chroma-differences. These data are summarized in Table I. Figure 1 shows the whiteness of fabrics measured after enzyme treatment with cellulase concentrations of 1, 5 and 10 g/l, respectively, and for different durations.

A 30-min long treatment with different cellulase enzyme concentrations did not cause significant changes in whiteness and brightness. Biotreatment at 10 g/l cellulase concentration resulted in approximately 7 % increase in whiteness. At 5 g/l enzyme concentration identical whiteness value was reached after 2 hours, and at 1 g/l concentration approximately 4 hours treatment time was needed to reach the same whiteness. Cellulase treatment alone resulted in a 14-21 % increase in whiteness and approximately 5 % increase in brightness.

Table I. Whiteness, brightness, CIELAB color- (ΔE_{ab}^*) and chroma-differences (ΔC_{ab}^*) of cotton fabrics measured after cellulase enzymatic treatment at different enzyme concentrations for different durations. Reference (control): desized cotton fabric

Cellulase [g/l]	*Parameter*	*Time of Cellulase Treatment [hour]*					
		0	*0.5*	*1*	*2*	*4*	*8*
	Whiteness	31.9	31.1	31.5	32.0	33.9	36.5
1	Brightness	56.9	57.1	57.3	57.2	58.1	59.1
	ΔE_{ab}^*	-	0.4	0.4	0.2	0.6	1.1
	ΔC_{ab}^*	-	0.3	0.2	0.1	-0.3	-0.8
	Whiteness	31.9	31.3	32.1	34.4	34.3	38.8
5	Brightness	56.9	57.1	57.6	58.6	58.4	60.1
	ΔE_{ab}^*	-	0.3	0.5	0.9	0.8	1.6
	ΔC_{ab}^*	-	0.2	0.1	-0.3	-0.3	-1.2
	Whiteness	31.9	31.2	34.2	34.5	34.5	38.2
10	Brightness	56.9	57.1	57.8	58.3	58.6	59.7
	ΔE_{ab}^*	-	0.4	0.5	0.7	0.8	1.5
	ΔC_{ab}^*	-	0.3	-0.4	-0.4	-0.4	-1.2

Negative chroma-difference values indicate the decolorization effect (Table I., Figure 2.). The higher the enzyme concentration the shorter was the time of treatment required for decolorization of the fabrics. Decolorization effects on fabrics were observed at 1g/l enzyme concentration after 4 hours, at 5 g/l concentration after 2 hours and at 10 g/l enzyme concentration after 1 hour, respectively.

Bleaching Effect of Consecutive Cellulase and Caustic Scouring Processes. After cellulase treatment the cotton fabrics were scoured with caustic soda solution. The color characteristics of the cotton fabrics measured after consecutive enzymatic and scouring processes are shown in Table II.

As a result of cellulase pretreatment, Berger-whiteness values of scoured cotton fabrics increased significantly (Figure 3). After consecutive bio- and chemical treatments, whiteness of the textiles increased by maximum 18 % at 1 g/l, by 25 % at 5 g/l and by 29 % at 10 g/l cellulase concentration, respectively. Chroma difference values were negative at all cellulase concentrations and for all the studied times of

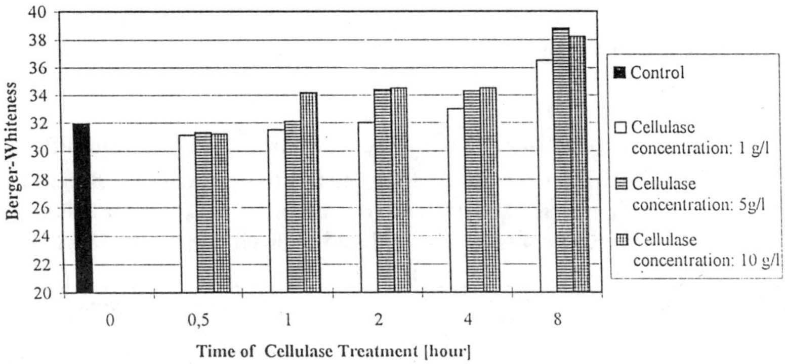

Figure 1. Effect of cellulase concentration and time of treatment on whiteness of cotton fabrics measured after cellulase enzymatic treatment

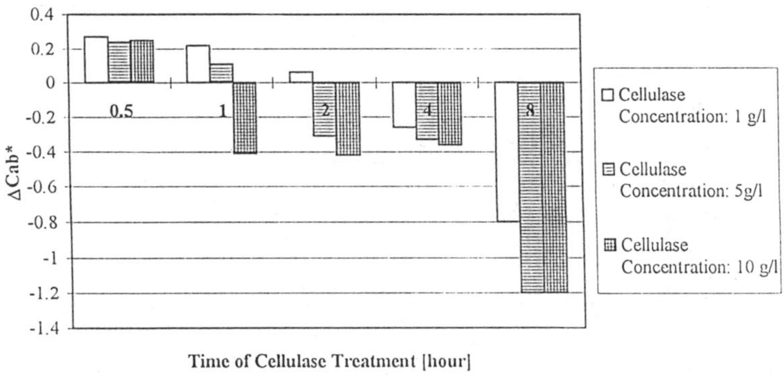

Figure 2. Effect of cellulase concentration and time of treatment on chroma-difference values (ΔC_{ab}^*) of cotton fabrics measured after cellulase enzymatic treatment in comparison with desized cotton fabric

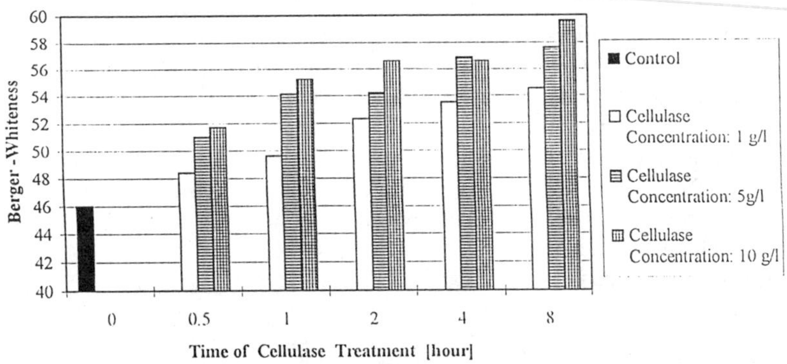

Figure 3. Effect of cellulase concentration and time of treatment on whiteness of cotton fabrics measured after consecutive enzymatic and scouring processes

biotreatment (Figure 4). The decolorization of the fabrics during the consecutive cellulase and alkaline procedures was significant.

Table II. Whiteness, brightness, CIELAB color- (ΔE_{ab}^*) and chroma-differences (ΔC_{ab}^*) of cotton fabrics measured after consecutive cellulase treatment and alkaline scouring process at different enzyme concentrations for different durations. Reference (control): Desized and scoured cotton fabric without cellulase pretreatment

Cellulase [g/l]	*Parameter*	*Time of Cellulase Treatment [hour]*					
		0	*0.5*	*1*	*2*	*4*	*8*
	Whiteness	46.0	48.4	49.6	52.3	53.5	54.5
1	Brightness	63.0	62.4	63.9	64.6	65.9	65.8
	ΔE_{ab}^*	-	1.1	0.8	1.3	1.5	1.7
	ΔC_{ab}^*	-	-0.8	-0.7	-1.3	-1.2	-1.6
	Whiteness	46.0	51.0	54.1	54.2	56.9	57.6
5	Brightness	63.0	64.0	65.7	65.2	67.1	67.0
	ΔE_{ab}^*	-	1.1	1.6	1.6	2.2	2.3
	ΔC_{ab}^*	-	-1.1	-1.5	-1.6	-1.9	-2.1
	Whiteness	46.0	51.7	55.2	56.6	56.6	59.6
10	Brightness	63.0	64.5	66.6	66.7	66.7	68.4
	ΔE_{ab}^*	-	1.2	1.9	2.1	2.1	2.8
	ΔC_{ab}^*	-	-1.2	-1.5	-1.9	-1.9	-2.3

Comparison of the color characteristics of the bioscoured fabrics with those of the scoured fabrics from Testfabric Inc. NJ., revealed that the differences in color were more significant and visually observable (Table III). We assume that the cellulase pretreatment enhanced the accessibility of desized cotton fabrics towards scouring chemicals, thereby increasing the rate and cleaning efficiency of the process. Accelerated removal of naturally coloured ingredients of greige cotton during alkaline scouring of the cellulase pretreated substrate is indicated by the respective CIELAB colour characteristics. The whiteness of the cellulase treated and thereafter scoured fabric is good enough - without further oxidative bleaching - for subsequent dyeing in medium or high color strength.

Cellulase pretreated (1g/l, 4 h) and control fabrics after alkaline scouring were further bleached with hydrogen-peroxide at different concentrations. Cellulase pretreated cotton fabrics showed higher whiteness at all concentrations of hydrogen-peroxide (Figure 5). Consequently, the cellulase pretreated fabrics requires less hydrogen-peroxide to attain a particular degree of whiteness.

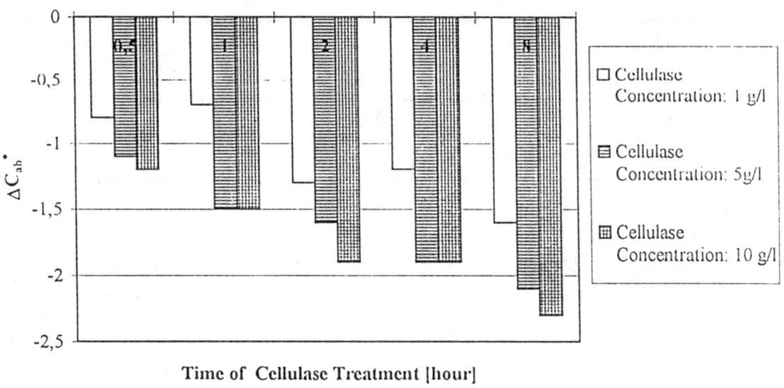

Figure 4. Effect of cellulase concentration and time of treatment on chroma-difference values (ΔC_{ab}^*) of cotton fabrics measured after consecutive enzymatic and scouring processes in comparison with desized and scoured cotton fabric without cellulase pretreatment

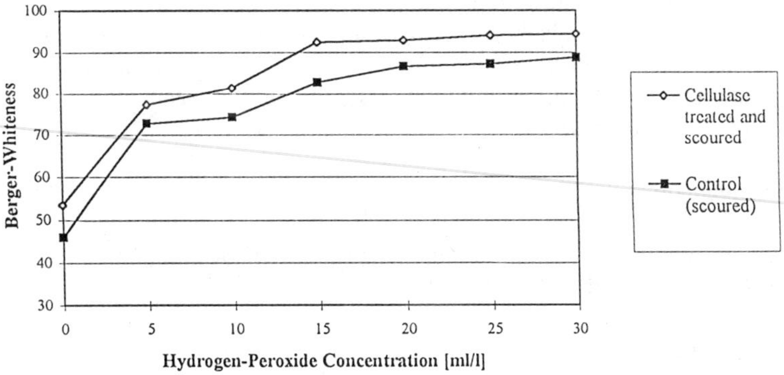

Figure 5. Berger-whiteness values of bioscoured and control fabrics after bleaching process with hydrogen-peroxide of different concentrations

Table III. CIELAB color- (ΔE_{ab}^*) and chroma-differences (ΔC_{ab}^*) of cotton fabrics measured after consecutive cellulase treatment and alkaline scouring process at different enzyme concentrations and times of treatment in comparison with Testfabric scoured cotton fabric

Cellulase [g/l]	*Parameter*	*Time of Cellulase Treatment [hour]*				
		0.5	*1*	*2*	*4*	*8*
1	ΔE_{ab}^*	2.8	2.9	3.4	3.6	3.9
	ΔC_{ab}^*	-2.8	-2.7	-3.2	-3.2	-3.5
5	ΔE_{ab}^*	3.2	3.8	3.8	4.4	4.5
	ΔC_{ab}^*	-3.1	-3.5	-3.6	-3.9	-4.1
10	ΔE_{ab}^*	3.3	4.0	4.3	4.3	4.9
	ΔC_{ab}^*	-3.1	-3.5	-3.9	-3.9	-4.3

Strength characteristics, degree of depolymerization and weight loss of the fabrics were within the limits accepted during conventional scouring process when enzyme assisted scouring of cotton was used in optimized conditions.

Conclusions

The removal of seed-coat fragments and other natural impurities of greige cotton may be started already during the cellulase treatment prior to the alkaline scouring process. Enzyme treatment increases the efficiency of chemicals used in subsequent traditional scouring. Cellulase action also increases the whiteness of fabrics considerably, especially in combined (cellulase + scouring) processes. This effect seems to be similar to that observed in kraft pulp biobleaching with xylanase or other enzymes in paper industry. These two beneficial effects of commercial (crude) cellulase enzymes have been demonstrated in increased whiteness, brightness and improved color characteristics of cotton textile.

Acknowledgement

This research was supported by the Hungarian National Science Foundation (Grant No. OTKA F 017363 and OTKA T 017201)

Literature Cited

1. Klahorst, S.; Kumar, A.; Mullins, M. M. *Textile Chem. Color.* **1994**, *26*, 13-18.
2. Screws, G. A. Jr.; Cerdoni, D. M. *AATCC Book of Papers,* Atlanta, G.A., 1992, pp 250-252.
3. Clarkson, K.; Collier, K.; Larenas, E.; Weiss, G. *AATCC Book of Papers,* Charlotte, N.J., 1994, pp 319-323.
4. Lange, N. K., In *Proceedings of the Second TRICEL Symposium on Trichoderma reesei Cellulases and Other Hydrolyses*, Espoo, Finland 1993. P. Suominen, and T. Reinikainen, Eds.; Foundation for Biotechnical and Industrial Fermentation Research Helsinki 1993, *8*, pp 263-272.

5. Kirk, T. K.; Jeffries, T. W. In *Enzymes for Pulp Paper Processing*, T. W. Jeffries, L. Viikari, Eds.; ACS Symposium Series 655, Am. Chem. Soc., Washington, 1996. pp. 2-14.
6. Viikari, L.; Suurnakki, A.; Buchert, J. **Ibid**. pp.15-24.
7. Ghose, T. K. *Pure and Appl. Chem*. **1987**, *59*, 257-268.
8. Kubicek, C. P. *Arch. Microbiol*. **1982,** *132*, 349-354.
9. Bailey, M. J.; Nevalainen, K. M. H. *Enzyme Microb. Technol*. **1981**, *3*, 153-157.
10. Bailey, M. J.; Biely, P.; and Poutanen, K. *J. Biotechnol*. **1992**, *23*, 257-270.
11. Völz, H. G. *Industrial Color Testing Fundamentals and Techniques,* VHC Weinheim-New York-Basel-Cambridge-Tokyo 1995, pp 28-37.

Chapter 18

Pectin-Degrading Enzymes for Scouring Cotton

M. Michelle Hartzell[1] and You-Lo Hsieh[2]

Division of Textiles and Clothing, University of California, Davis, CA 95616

In our previous study, a pectinase was found to improve the surface wetting properties of greige cotton fabrics following a water pretreatment at 100°C. This study further evaluated seven pectin-degrading enzymes, i.e., four pectinases, two pectinesterases, and a pectin lyase, for scouring raw cotton fabrics. Three of the pectinases significantly improved the wettability of cotton fabrics following a 100°C water pretreatment to the same extent as alkaline scouring. The other pectinase, pectinesterases and pectin lyase had no beneficial effects on improving the wettability of raw cotton fabrics. Reaction conditions for the three pectinase treatments were optimized in respect to temperature, concentration, pH, and time. The pectinase treated fabrics did not exhibit additional shrinkage, color change, nor significant strength loss from the fabrics pretreated in water at 100°C.

The majority of the non-cellulosic components in cotton are located on the fiber surfaces (the cuticles) (*1-3*). The cuticles contain approximately 0.6% waxes, 0.9% pectins, 1.3% proteins, 2.0% non-cellulosic polysaccharides, ash and other miscellaneous compounds, all of which protect the cells from potential environmental and pathogenic damage during cell growth and development (*4*). The surface waxes facilitate yarn spinning and fabric weaving by acting as a lubricant (*5*).

Conventionally, these hydrophobic non-cellulosic compounds are removed via an alkaline scouring process to facilitate uniform dyeing and finishing. Although hot aqueous alkaline scouring is highly effective for removing the non-cellulosics from raw cotton, the process conditions are limited to high pH and high temperature. We have surveyed several enzymes on their ability to improve the wetting properties of

[1]Current address: Department of Human Environments, Utah State University, Logan, UT 843322–2910.
[2]Corresponding author.

cotton under aqueous conditions (*6*). Enzymes, in general, are active under a much broader range of pH and at lower temperatures. Scouring with enzymes thus offers greater flexibility for scouring cotton. For instance, enzyme scouring enables cleaning cotton blends containing fibers which may not be suited for scouring under alkaline conditions.

A variety of enzymes are capable of reacting with specific components on cotton fiber surfaces. Cellulases, pectinases, proteases, and lipases have been studied and compared to sodium hydroxide scouring (*7-10*). In one study, both the cellulase and pectinase treatments generated more weight loss than either the protease and lipase treatments; cellulase and pectinase were assumed to remove a greater amount of non-cellulosic material with the pectinase hydrolyzing the pectin and the cellulase dislodging other non-cellulosic compounds by hydrolyzing the supporting cellulose in the primary cell wall (*7-8*). The fabric wettability of the cellulase treated cotton improved, but remained less wettable than the bleached fabric. Another study found that fabric wettability improved when the cotton fabrics were extracted with chloroform prior to cellulase and pectinase treatments (*9-10*). Cellulase reduced cotton yarn strength while pectinase alone had little effect on strength. The addition of cellulase to the pectinase treatment significantly decreased strength.

We demonstrated that a pectinase following a water pretreatment at 100°C reduced the water contact angle within the range of commercially scoured fabrics (*7*). The pectinase treated cotton fabrics experienced some strength loss due to the small percentage of cellulase present in the pectinase formulation. Cellulase was the only enzyme to improve fabric wettability when used alone, however, more significant strength losses were observed with the cellulase treatments. The pectinase, protease and lipase studied produced only limited effects on the fabric wettability. The 100°C water pretreatment is assumed to alter the wax composition and/or organization on the fiber surface to allow greater pectinase access to the pectin (*5*).

Pectin is composed mainly of a polygalacturonic acid backbone (85% methylated) interspersed with rhamnose and side chains of arabinose and galactose (*2*). Pectinases hydrolyze the polygalacturonic acid backbone into the individual D-galacturonic acid monomers (*11*). Pectinesterases hydrolyze the ester groups from the methylated acid groups in the polygalacturonic acid backbone, and pectin lyase cleaves the -C-C- and -C-O- linkages. Among the many pectin-degrading enzymes identified, nine different pectinases have been discovered thus far from many sources with varying substrate specificity, activity and protein structure (*12, 13*). One fungal source, *Aspergillus niger*, alone produces two separate pectinases.

This work aimed to further evaluate the use of pectin-degrading enzymes as scouring agents for raw cotton. Seven commercially available pectin-degrading enzymes were selected. As with our earlier study (*7*), completely aqueous conditions were employed for all enzyme reactions. The length of the pretreatment time in water at 100°C and the conditions of the enzyme reactions were optimized for greater pectinase effectiveness at reduced temperatures and close to neutral pH. The effects of the treatments on fabric wetting properties, fabric and yarn properties, yarn strength, and enzyme activity were measured.

Experimental

Materials. The substrate used in this study was a plain weave, one-hundred percent cotton fabric in the grey state (Nisshinbo California Incorporated). Fabric samples were either raveled to a dimension of 10 cm by 14 cm weighing approximately 1.5 grams, or 5 cm by 14 cm weighing approximately 0.75g. Very little starch sizing was present on the fabric as indicated by an iodine staining test, thus no attempts were made to desize the fabrics to avoid preliminary disruptions to the organization of the cotton fiber surface. A 0.33:1 (L/g) liquor:fabric ratio was used for enzyme, buffer, and water solutions.

Seven commercially available enzymes, four pectinases, two pectinesterases, and a pectin lyase, were used in this study (ICN, Costa Mesa, CA and Sigma, St. Louis, MO) (Table I). All chemicals were certified ACS grade except for the certified grade hexadecane (Fisher Scientific, Pittsburgh, PA), and the D-galacturonic acid (ICN), apple pectin (ICN), citrus pectin (ICN), and carbazole (Sigma). A Millipore, Mill-Q Water System was used for water purification. Water (72.6 dynes/cm) and hexadecane (26.7 dynes/cm) were used as the wetting liquids for this study.

Table I: Enzyme Sources and Stable Conditions

Enzyme (ID)	Source	E.C.#	Conditions: Temp, °C	pH	Activity u/mg pro
pectinase (AN1)	*Aspergillus niger*	3.2.1.15	40	4.0-7.0	778
pectinase (AN2)	*Aspergillus niger*	3.2.1.15	24-37	4.0-5.0	365
pectinase (AN3)	*Aspergillus niger*	3.2.1.15	40-70	4.5-5.5	11.8
pectinase (R1)	Rhizopus species	3.2.1.15	25	4	350
pectin lyase (PL1)	*Aspergillus japonicus*	4.2.2.10	25	5.5	100,000
pectinesterase (PE1)	orange peel	3.1.1.11	45-50	4.0-8.5	100
pectinesterase (PE2)	orange peel	3.1.1.11	50-60	4.5-5.5	62 solid

Methods. The greige cotton fabrics were subjected to a water pretreatment at 100°C prior to enzyme treatment. The fabrics were immersed in 100°C water for varying lengths and replications. Three minutes of centrifugation followed the final immersion before drying at 65% relative humidity and 70°F until constant weight was obtained *(7)*.

All enzyme treatments were performed in a phosphate buffer solution, but varied in time, temperature, concentration and/or pH. Treatments affecting the fabric wettability were duplicated. The initial enzyme concentration used was approximately 3650 units per gram cotton fabric. After achieving constant temperature, the fabric was added to the enzyme solution for the desired time duration. The enzyme treatments were performed at constant temperatures controlled by a water-bath (Dubroff Metabolic Shaking Incubator, Precision) set to 50 revolutions per minute. Enzyme activity was ceased by immersion into a pH 8.5 phosphate buffer, then rinsed in water. The fabric was centrifuged for three minutes (International Clinical Centrifuge), then dried at 65% relative humidity and 70°C for four days or until constant weight was obtained *(7)*.

Water contact angles (WCAs) were calculated from the wetting force ($\mathbf{F_w}$) measured on a tensiometer apparatus described earlier (*6, 14*). This method decouples the wetting force ($\mathbf{F_w}$) from the absorbed liquid:

$$\mathbf{F_w = p\gamma_{LV} \cos\theta} \qquad (1)$$

(where γ_{LV} is the surface tension of the wetting liquid, **p** is the perimeter of the fabric sample, and θ is the water contact angle). Assuming a zero contact angle, the perimeter of the sample was calculated from wetting force in hexadecane ($\mathbf{F}_{hexa}$) obtained in the second measurement:

$$\mathbf{p} = \frac{\mathbf{F}_{hexa}}{\gamma_{LV}} \qquad (2)$$

With **p** known, the water contact angle may be determined from the wetting force in water ($\mathbf{F}_w$) using equation 1.

Vertical liquid retention capacity ($\mathbf{C}_v$, µl/mg) and water retention ($\mathbf{C}_m$, µl/mg) values were derived from the weight of the total liquid retained ($\mathbf{B}_{sp}$) in hexadecane and water, respectively. When deriving $\mathbf{C}_v$ or $\mathbf{C}_m$, the liquid and water retention values were converted to volume and normalized by the weight of the specimen ($\mathbf{W}_s$):

$$\mathbf{C} = \left[\frac{\mathbf{B}_{SP}}{\mathbf{W}_S}\right] / \rho \qquad (3)$$

A minimum of three measurements were taken for both water contact angles and water retention.

Yarn breaking strength (ASTM method 2256), fabric count and fabric thickness (ASTM 1910), weight loss and CIE L*a*b* spectrophotometer measurements were used for fabric characterization. Test methods and procedures were described earlier *(7)*.

The activity (µg pectin hydrolyzed into galacturonic acid monomers per minute) of the pectinases at their optimum conditions were derived by using the galacturonic acid assay modified by Taylor and Buchanan-Smith *(15)*. This assay utilizes carbazole in acidic conditions to measure the quantity of D-galacturonic acid liberated by pectinase hydrolysis. The three replicates of pectinase treated pectin was further hydrolyzed by the presence of concentrated sulfuric acid, and incubated with a 0.1% (w/v) carbazole reagent for four hours at 40°C. The carbazole reacts with the acid group of the D-galacturonic acid to form a pink color measured via visible absorbance (Figure 1). The concentration of D-galacturonic acid was determined by the absorbance at 525nm (Hitachi U-2000 Spectrophotometer).

Results and Discussion

Initial Survey of Enzymes. The effects of the seven pectin-degrading enzymes were surveyed using the mid-range temperatures and pHs within those specified by the manufacturers (Table II). The fabrics were immersed in three fresh boiling (100°C) water baths, each for two-minutes (3x2m). To facilitate the detection of the enzymatic effects, relatively high enzyme concentrations were employed. The same enzyme activity level per mass of cotton, i.e., 3633 to 3687 u/g, was maintained to differentiate the effectiveness of these enzymes. The untreated cotton fabric has a WCA of 93.9° (±3.3). Three pectinases, i.e., AN1, AN2, and AN3, substantially reduce the water wetting contact angles of the cotton fabrics into the range of 45° to 60° (Figure 2). Their effects on cotton fabric wettability are similar to commercial scouring *(6)*. The two pectinesterases (PE1 and PE2) are much less effective on

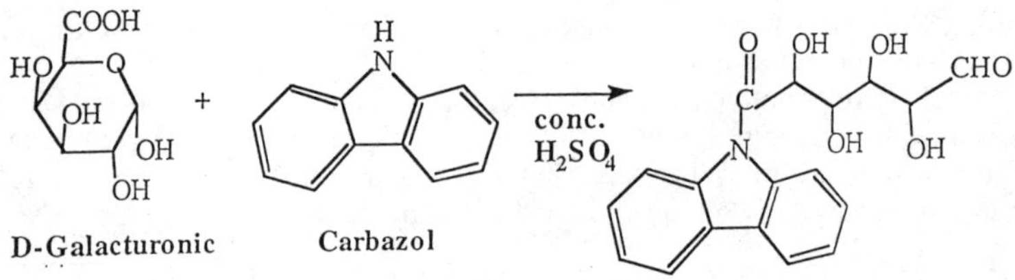

Figure 1. Reaction of Carbazole reagent with D-galacturonic acid detected in pectinase assay.

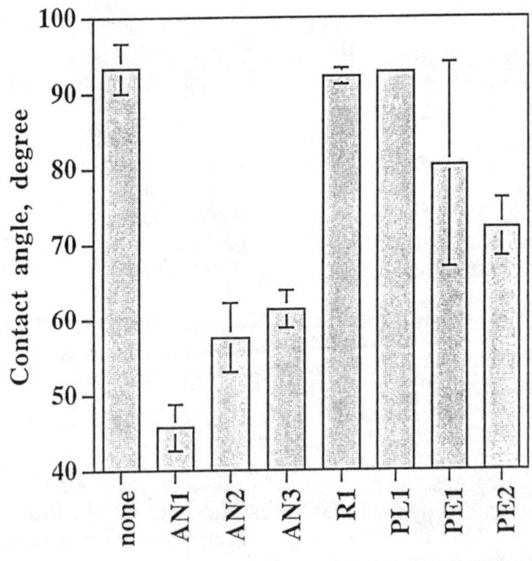

Figure 2. Water contact angles of pectin-degrading enzyme treated fabrics (3x2m 100°C water pretreatment).

improving the water wettability of cotton fabrics. The pectin lyase (PL1) and the Rhizopus pectinase (R1) show no effect (Figure 2).

Table II: Survey Conditions

Enzyme ID	Survey Conditions: Temp., °C	pH	Concentration	Activity u/g cotton
AN1	40	5.0	14.2 g/L	3684
AN2	30	5.0	30.0 ml/L	3650
AN3	50	5.0	30.0 ml/L	3650
R1	25	4.0	31.6 mg/L	3687
PL1	25	5.0	110.5 mg/L	3683
PE1	47	7.0	10.0 ml/L	3633
PE2	50	5.0	236.0 mg/L	3658

Water Pretreatment

We have shown that three consecutive two-minute rinses in water at 100°C (3x2m) improve the effectiveness of pectinase and cellulase on cotton fabric (*7*). The 3x2m pretreatment in water at 100°C reduces the WCA to 77.6° (±16.7) and increases water retention from 0.15 μl/mg (±0.10) to 1.20μl/mg (±1.08) (Figure 3). Both fabric thickness and lightness increase.

The effects on fabric wettability, shrinkage, and lightness are lessened when the exposure to water is reduced to two minutes (Figure 3a; Table III). Shorter pretreatment times of 30 and 10 seconds, however, improve fabric wettability and water retention more so than the fabrics pretreated for two minutes (Figure 3a and 3b). Both shrinkage and lightness of these fabrics are comparable to those fabrics treated either once or three times for two minutes (Table III). Therefore, the duration of the pretreatment in 100°C water was further reduced to two seconds. Both the water contact angle and water retention values of the fabric treated for 2 seconds surpass those treated for 30 and 10 seconds (Figure 3a and 3b). The shorter pretreatment times of 2, 10, and 30 seconds seemingly allow the waxes to melt, exposing more hydrophilic components.

However, subsequent pectinase treatments on the two-second (2s) pretreated fabrics do not produce water wetting properties equal to those on the 3x2m pretreated fabrics (Figure 4). It is likely that waxes melt and bead up from the short 2s exposure to water at 100°C, but do not disperse into the water. Prolonged exposure (one hour) to the enzyme-buffer solution at 40°C may cause softening and spreading of the waxes over the fiber surface preventing enzyme access to the pectin.

The 2s pretreatments in 100°C water were repeated twice (2x2s) and four-times (4x2s). Although higher repetitions of the 2s pretreatment cause the fabrics to be less wettable and absorbent (Figure 3), the subsequent pectinase reactions on these cotton fabrics produce similar wetting properties as the scoured fabrics (Figure 4) (*6*). Therefore, 2x2s in 100°C water is the shortest pretreatment conditions for pectinase AN1 to take effect.

The cotton fabrics lose 4.3% to 5.8% weight following all water pretreatments at 100°C, irrespective of the lengths of time or the number of repetitions. The raw cotton fabric has a thickness of 320 μm (Table III). Exposures to water at 100°C

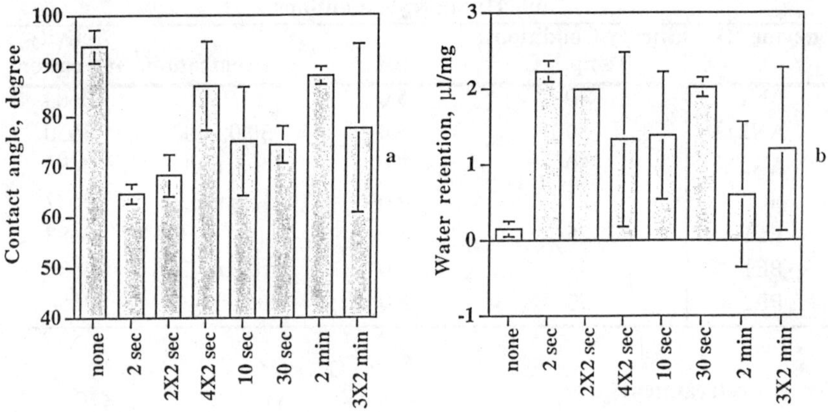

Figure 3. Water contact angles and retention of cotton fabric after pretreatments in water at 100°C.

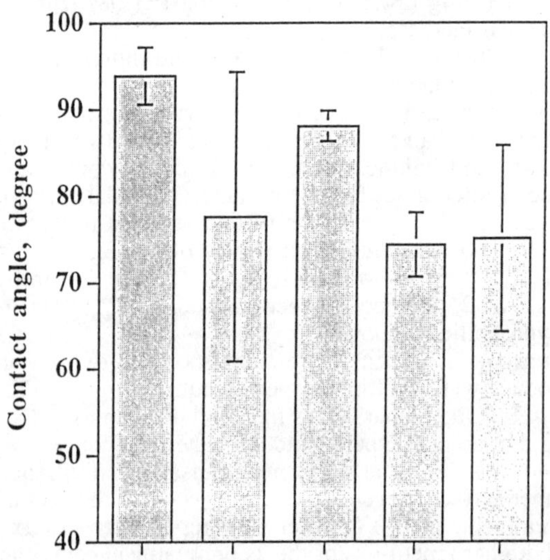

Figure 4. Water contact angles of *Aspergillus niger* pectinase (AN1, 40°C, pH 5.0, 14.2 g/L) treated fabric after varying pretreatment times.

substantially increase the fabric thickness to between 400 and 495 μm. The increases in fabric thickness are similar among fabrics treated for 10 seconds to 2 minutes, but the highest increase in fabric shrinkage occurs with the 3x2m pretreated fabrics. The short 2-second exposure substantially increases fabric thickness, exceeding those treated for 10s to 2m. A high standard deviation accompanies this thickness value suggesting non-uniform yarn relaxation during the short boiling period. Fabric shrinkage from the 2x2s and 4x2s pretreatments is less than the 2-second pretreatment.

Table III: Characteristics^ of Cotton Fabrics Treated in Water at 100°C

Time	Weight Loss (%)	Thickness (μm)	Lightness L*
none	0.0	320 (9)	85.1 (0.1)
3 X 2 min.	-5.5	495 (28)	86.5 (0.8)
2 min	-5.8	405 (13)	85.5 (0.2)
30 sec	-5.6	411 (7)	85.4 (0.1)
10 sec	-4.9	400 (12)	84.9 (0.1)
2 sec	-4.5	441 (21)	84.6 (0.2)
2 X 2sec	-5.3	425 (4)	85.1 (0.3)
4 X 2 sec	-4.3	426 (4)	85.5 (0.1)

^ number in () denotes the standard deviation

The raw cotton fabric has a lightness L* value of 85.1. The 2m, 3x2m, and 4x2s pretreatments at 100°C produce slightly lighter fabric colors (Table III). The slight improvement in fabric lightness indicates some removal of the non-cellulosic compounds during each successive immersion.

Optimization of Enzyme Reactions

Generally, the advantages of enzyme reaction conditions over alkaline scouring include the lower reaction temperatures and a wider range of pH. For these pectin-degrading enzymes, optimizing their temperature and pH as well as concentration and time may offer additional advantages. The 2x2s pretreatment in water at 100°C was used for all the following reactions because it was the shortest pretreatment for the effective reaction of pectinase AN1. This pretreatment causes the cotton fabric to lose 5.3% weight and increase thickness (425 μm) while retaining the fabric lightness (L* = 85.1). All fabric characteristics except for weight loss of the enzyme treated fabrics are compared with those pretreated 2x2s in water at 100°C.

Temperature. Pectinases are generally active and stable between 40°C and 70°C (*11*). Pectinases AN2 and AN3 improve the wettability of cotton substantially at 30°C and 40°C (Figure 5a). Pectinase AN2 has no effect on cotton fabric wettability at 25°C, 55°C or 70°C. Pectinase AN3 has very little effect on fabric wettability at 55°C, and fabric wettability remains completely unchanged when incubated for one hour at 70°C. The effects of pectinase AN1 are similar to AN2, except that its effects on wettability are less at and below 35°C (Figure 5a). The improved water wettability

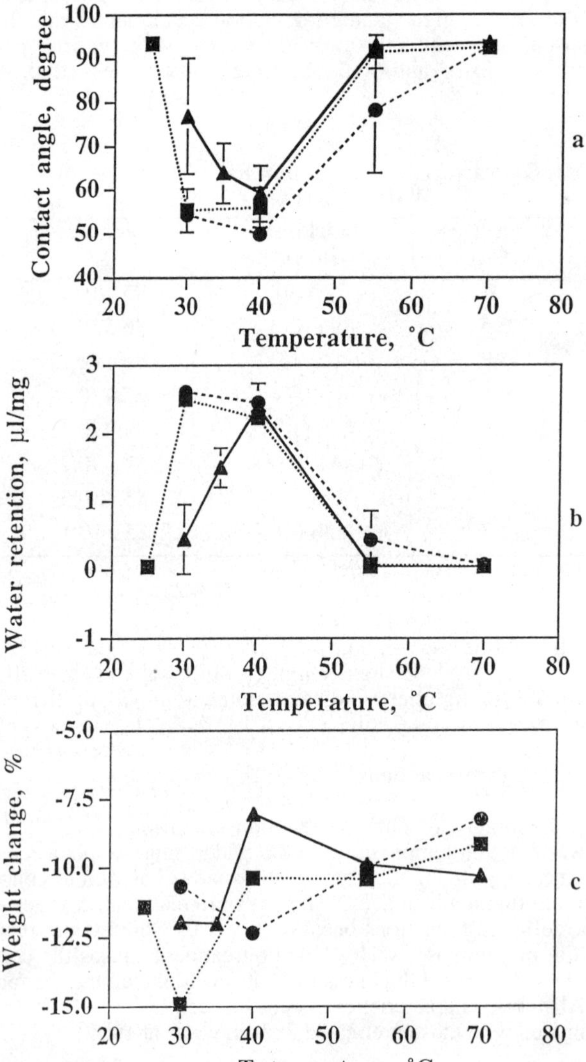

Figure 5. Temperature effects of *Aspergillus niger* pectinases on fabrics (2x2s 100°C water pretreatment):

▲ AN1 (14.2 g/L, pH 5.0)
■ AN2 (30 ml/L, pH 5.0)
● AN3 (30 ml/L, pH 5.0)

with AN1, AN2 and AN3 is coupled with increased water retention of the cotton fabrics (Figure 5b).

Compared to the 2x2s 100°C water pretreated fabric, no changes in fabric thickness nor lightness are detected for these enzyme treated fabrics at varying temperatures. Weight loss remains greater for fabrics treated at 30°C and 40°C than treatments at higher temperatures (Figure 5c). Nitrogen contents of the fabrics treated with pectinases AN1, AN2, and AN3 are much lower than that of the raw cotton (Table IV). The very low amounts of nitrogen indicate very low amounts of proteinacious materials on the fibers. These materials are likely residual cotton cell wall proteins. At such low levels, absorption of enzyme proteins from the pectinase reactions appears unlikely.

Table IV: Nitrogen Contents of Pectinase Treated Fabrics

Enzyme	Temperature (°C)	Concentration	Nitrogen Content (%)
none	na	na	0.41
AN1	40	14.2 g/L	0.10^
AN2	30	30.0 ml/L	0.14
AN3	40	30.0 ml/L	0.18

^ possibly none

Among the pectinases studied, three are from *Aspergillus niger* which is known to produce two forms of pectinases (*13*). Because the effects of pectinases AN2 and AN3 on cotton are similar as a function of temperature, it is likely that they are the same form of *Aspergillus niger* pectinase from two separate commerical sources. Therefore, either AN2 or AN3, but not both, were evaluated to represent this form of *Aspergillus niger* pectinase. The different effects of pectinase AN1 on cotton fabrics suggest that AN1 is the other *Aspergillus niger* pectinase.

pH. Information provided by the manufacturers shows that the stable and active pH ranges for pectinases AN1, AN2, and AN3 are 4.0-7.0, 4.0-5.0, and 4.5-5.5, respectively. In addition to the pH 5.0 used in the initial survey, the effects of pectinases AN1 and AN2 on cotton fabric wettability and retention were also examined at higher pH of 6.0 and 7.0 (Figure 6a and 6b). No changes in fabric wetting are observed for either pectinases AN1 and AN2 at these pH values. Pectinases AN1 and AN2 were also used without the buffer. Pectinase AN1 in water is ineffective whereas pectinase AN2 in water reduces the cotton fabric wettability to the same level as in the pH 5.0 buffer. The pH of pectinase AN1 in water remains to be 7.0 whereas AN2 in water reduces the pH to 5.0, indicating inclusion of buffer in the AN2 formulation. Both AN1 and AN2 require pH at 5.0 to be effective in improving cotton fabric wettability. Similar behavior is expected from AN3 as the two enzymes are assumed to be the same pectinase enzyme.

The effects of these two pectinases on fabric thickness and shrinkage are less at pH 6.0 and 7.0 and are consistent with the effects on water wettability. There is no change in lightness nor weight loss (Figure 6c) for these enzyme treated fabrics at varying pHs.

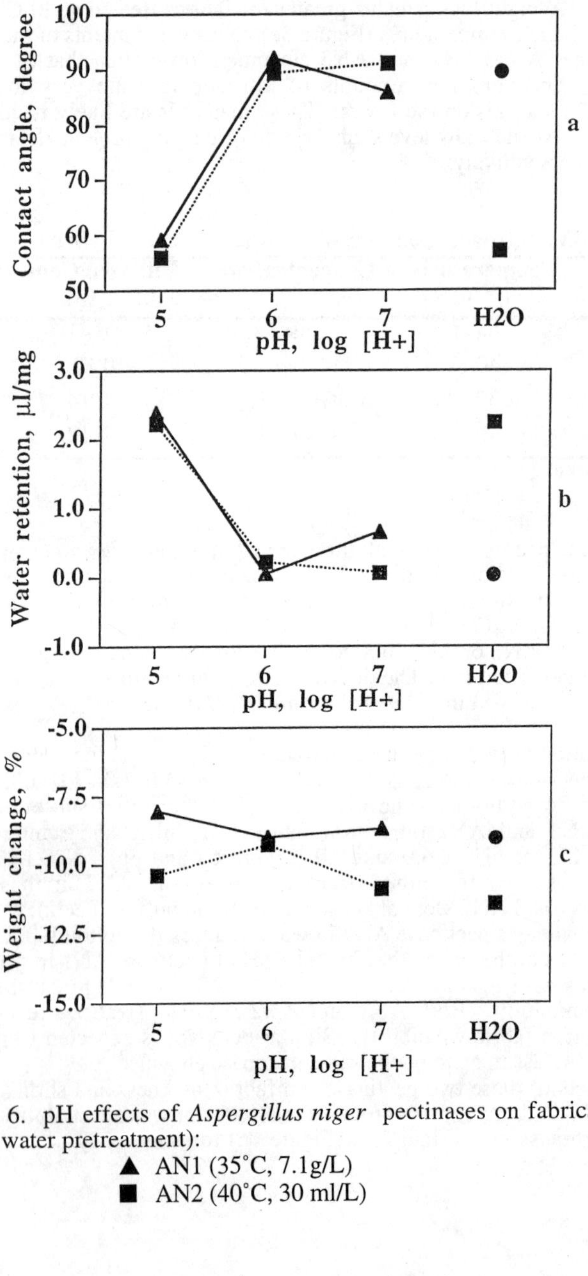

Figure 6. pH effects of *Aspergillus niger* pectinases on fabrics (2x2s 100°C water pretreatment):

▲ AN1 (35°C, 7.1g/L)

■ AN2 (40°C, 30 ml/L)

Concentration. AN1 produces the most improvement in cotton fabric wettability and water retention when used at 14.2 g/L (Figure 7a and 7b). A 50% reduction in concentration lowers the fabric wetting properties, but wettability is still within the range of the commercially scoured fabrics (*6*). The effects are nearly identical for AN3. Reactions at further reduced concentrations do not improve fabric wettability.

Higher weight loss is observed with increasing AN3 concentrations, but weight loss of the AN1 treated fabrics remains mostly constant with changing concentrations (Figure 7c). Neither pectinase AN1 nor AN3 alters the overall thickness and lightness of the fabrics at varying concentrations. AN2 is expected to behave in a similar manner as AN3.

Time. The effects of pectinase AN1 reaction time were evaluated on fabrics pretreated under two conditions, i.e. 3x2m and 2x2s. In both cases, water wettability and retention improve with increasing reaction time (Figure 8a and 8b). The overall effects are greater on the fabrics pretreated for 3x2m, particularly for water retention. Earlier works noted that longer treatment times were necessary for pectinases to adequately react with the cotton fabric (*7, 10*). The same trend is observed here.

Although longer reaction times are needed for enhanced fabric wettability, there is no change in fabric thickness nor lightness for either set of fabrics. Weight loss is higher for the one-hour treatments compared to the shorter reactions. Less weight loss is observed on fabrics pretreated 2x2s in water at 100°C.

Optimal Conditions. AN2 and AN3, which have similar effects on cotton fabrics despite different formulations and sources, are thought to be one *Aspergillus niger* pectinase and AN1 the other. The optimum reaction temperature for AN2 and AN3 is 40°C. AN2 produces optimal effects on fabric wetting at pH 5.0 while the optimal concentration for AN3 is 30 ml/L. Without the buffer, pectinases AN2 in water effectively improves cotton fabric wettability. Pectinase AN1 shows the greatest effect on improving cotton fabric wettability at 14.2 g/L, 40°C, and pH 5.0. Nitrogen content of the raw cotton fabric is 0.41%. Cotton fabrics treated with pectinases AN1, AN2, and AN3 contain less than 0.18% nitrogen by weight, indicating the loss of proteinacious substances. Protein absorption at these extremely low nitrogen levels is considered unlikely.

Yarn Strength

Earlier studies reported high strength losses with the use of cellulase-containing treatments (*7-10*). Only the treatment with AN1 (40°C, pH 5.0, 14.2 g/L) generates a significant loss in strength at the 0.1% significance level (Table V). Fabrics treated with either AN2 or AN3 have shown to retain full strength. This difference in yarn strength is likely due to the higher activity of AN1 compared to AN2 and AN3. All treatments following the 3x2m 100°C water pretreatment also result in greater cotton strength loss than those pretreated for shorter lengths of time.

Pectinase Activity

Pectinase AN1 demonstrates greater activity than pectinases AN2 and AN3 in a polygalacturonic acid assay (*15*) (Figure 9). The activities of AN1 at 35°C and 40°C have no statistical preference over methylated citrus or non-methylated apple pectin. This is also true for pectinases AN2 and AN3.

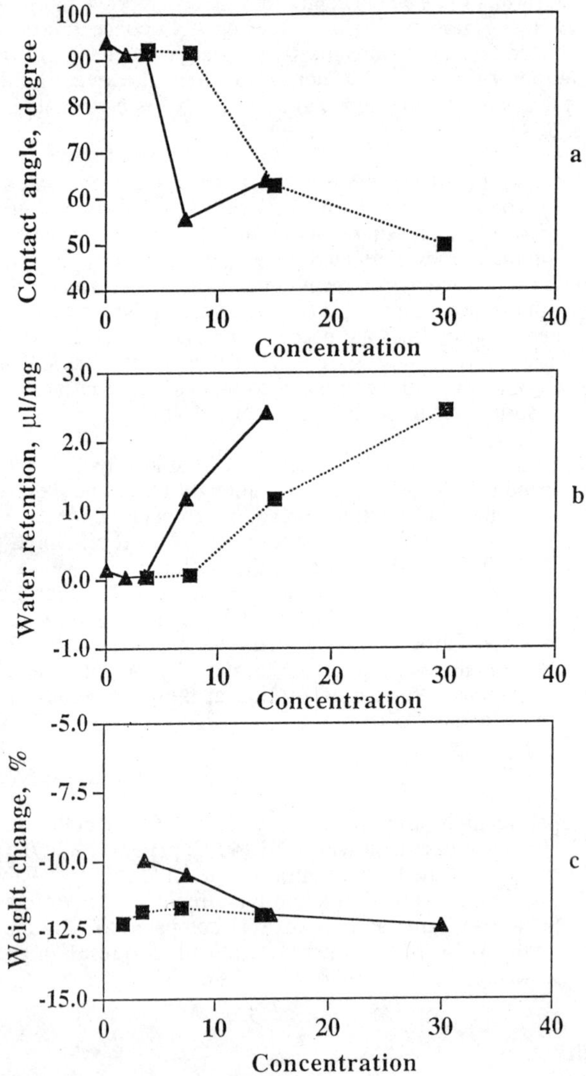

Figure 7. Concentration effects of *Aspergillus niger* pectinases on fabrics (2x2s 100°C water pretreatment):
▲ AN1 (35°C, pH 5.0)
■ AN3 (40°C, pH 5.0)

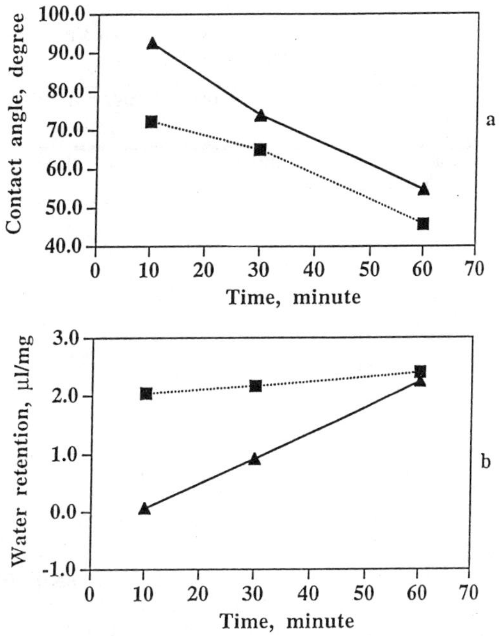

Figure 8. Reaction time effects on *Aspergillus niger* pectinase AN1 on fabrics (2x2s 100°C water pretreatment):

■ 3x2 min. in H_2O at 100°C

▲ 2x2 sec. in H_2O at 100°C

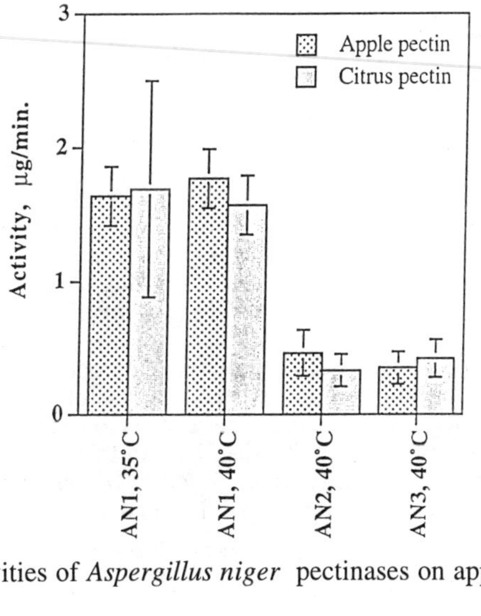

Figure 9. Activities of *Aspergillus niger* pectinases on apple and citrus pectins.

Table V: Yarn Properties of Fabrics Treated with Pectinases Under Optimal Reaction Conditions^

Pectinase	Temperature (°C)	Concentration	Linear density (mtex)	Tenacity (mN/tex)
none	na	na	20.5 (1.1)	191.1 (20.7)
AN1	35	14.2 g/L	16.4 (1.2)	177.7 (28.2)
AN1	40	14.2 g/L	19.0 (1.1)	155.8 (17.0)*
AN2	30	30.0 ml/L	16.2 (2.2)	196.6 (22.0)
AN2	40	30.0 ml/L	16.8 (1.4)	182.6 (21.9)
AN3	40	30.0 ml/L	16.7 (1.7)	177.3 (25.2)

^ All reactions at pH 5.0 for one-hour
* 0.1% significance by t-test

Summary

Our previous work has shown that a pectinase treatment under completely aqueous conditions greatly improves the water wetting and retention properties of raw cotton fabrics (*7*). This paper reports the effectiveness of seven pectin degrading enzymes, i.e., four pectinases, two pectinesterases, and a pectin lyase, on improving the water wetting and retention properties of cotton fabrics. To survey the effectiveness of these pectin-degrading enzymes, the cotton fabrics were pretreated in 100°C water (3x2m) as before. Of the seven enzymes surveyed, only three pectinases produce substantial improvements in water wetting and retention properties of the cotton fabrics. AN1 produces the greatest improvement whereas pectinases AN2 and AN3 behave similarly. The two pectinesterases, the fourth pectinase, and the pectin lyase have no effect on improving cotton fabric wettability.

Further optimization of these enzyme reactions include shortening the pretreatment time and moderating the pH and concentrations. The duration of the 100°C water pretreatment has been further reduced from three two-minute (3x2m) immersions to two immersions for 2 seconds (2x2s) while achieving similar effects. Repeating the 2-second treatment twice in 100°C water adequately alters the cotton fiber surface to allow pectinase access to the pectin. This 2x2s pretreatment in 100°C water was adopted as the fabric pretreatment for the optimization of the pectinase reactions. The three pectinases were optimized for efficiency and activity in terms of temperature, pH, concentration, and reaction time. All three pectinases improve fabric wettability when the treatments are performed at pH 5.0 and 40°C with AN1 retaining activity at 35°C, and AN2 and AN3 retaining activity at 30°C. Lowering or increasing the reaction temperatures beyond those mentioned renders the pectinases ineffective.

The pectinases were only effective in improving cotton fabric wettability at pH 5.0. When used in non-buffered water, the AN2 solution maintains a pH of 5.0 indicating buffers in the formulation. The activity of pectinase used per gram of cotton fabric was approximately 3650 units. Half of this concentration also improves fabric wettability, but to a slightly lower extent. One-hour reaction times appear necessary to achieve improved fabric wetting and retention properties similar to alkaline scouring.

Shrinkage occurs during the 100°C water pretreatment. Thickness increases from 320 μm (±9) to 441 μm (±21) for the 2x2s pretreatment. Lightness and weight loss are slightly increased by the water pretreatments. Subsequent treatment in pectinase does not result in further changes in thickness nor lightness. Further weight loss is additive with the pectinase treatments. The more active AN1 generates greater weight loss than AN2 or AN3. Yarn strengths from fabrics treated with pectinases AN2 and AN3 are not altered by the enzyme treatments. A significant loss in strength to the 0.1% level is observed only with pectinase AN1 under the optimized condition. The greater effects of pectinase AN1 on cotton strength is consistent with its higher activity as determined by the pectin assay.

Pectinase treatments following a 100°C water pretreatment improves the water wetting and water retention properties of cotton fabric equal to those of industrially scoured fabrics *(2)*. The brief duration of the water pretreatment and the lower temperatures and pH of the pectinase reactions afford opportunities to substantially reduce energy consumption and waste water treatment, and offer flexibility in conditions for scouring cotton goods. Furthermore, unlike the enzymatic processes reported by others, the combined water pretreatment and pectinase reactions maintain the retention of full strength and cause considerably less dimensional changes in cotton fabrics.

Acknowledgment

Support from Cotton Incorporated (#96-358) is greatly appreciated

Literature Cited

1. Basra, A. S.; Malik, C. P. *Intl. Rev. Cytology*, **1984**, *89*, 65-111.
2. Heredia, A.; Guillen, R.; Jimenez, A.; Fernandez-Bolanos, J. *Rev. Esp.Ciencia y Tech.de Alimentos*, **1993,** *33*, 113-131.
3. Tripp, V. W.; Moore, A. T.; Rollins, M. L. *Text. Res. J.,* **1951,** *21*, 886-894
4. Juniper, B. E.; Jeffree, C. E., *Plant Surfaces.* Edward Arnold Limited: London, UK, 1983: pp. 1-2, 15-17..
5. Freytag, R.; Donze, J. J. In *Alkali Treatment of Cellulose Fibers*; Lewin, M; Sello, S., Eds.; *Handbook of Fiber Science and Technology: Volume 1, Chemical Processing of Fibers and Fabrics.* Marcel Dekker: New York, NY, 1983, Part A; pp 111-116.
6. Hsieh, Y. -L.; Yu, B. L., *Text. Res. J* . **1992,** *62*, 677-685.
7. Hartzell, M. M.; and Hsieh, Y. L. *Text. Res. J* ., in press.
8. Rößner, Ulrike., *Meilliand Textilberichte* **1993,** *74*, 144-148.
9. Achwal, W. B. *Coulorage*, **1992**, *7,* 35.
10. Bach, E.; Schollmeyer, E. *Excerpts from the Dissertation of Elke Bach*, 1993.
11. *Enzyme Handbook;* Schomburg, D.; Salzmann, M., Eds.; Springer-Verlag: NY, 1996.
12. Schroter, K. H.; Arkema, A.; Kester, H. C. M.; Visser, J.; Dijkstra, B. W. *J. Mol. Biol.*, **1994**, 243, 351-352.
13. Stratilova, E.; Markovic, O.; Skrovinova, D.; Rexova-Benkova, L.; Jornvall, H.; *J. of Protein Chem.,* **1993**, 12, 15-22.
14. Hsieh, Y.-L. *Text. Res. J .,* **1995**, *65,* 299-307.
15. Taylor, K. A.; Buchanan-Smith, J. G. *Anal. Biochem.,* **1992,** *201*, 190-196.

Chapter 19

The Effect of Cellulases on the Biodegradation and Morphology of Naturally Colored Cotton Fibers

Barbara R. Evans[1], Ida Lee[1], Jonathan Woodward[1], and Sally V. Fox[2]

[1]Chemical Technology Division, Oak Ridge National Laboratory, Oak Ridge, TN 37831–6194

[2]Natural Cotton Colours, Inc., 67 Tegner Street, Wickenburg, AZ 85390

The effect of cellulases on the biodegradation and structure of natural colored cotton (FoxFibre) was examined. Cottons containing brown pigments, the varieties Buffalo (mocha brown) and Coyote (reddish brown), were quite resistant to hydrolysis by *Trichoderma reesei* cellulase (Celluclast) and purified cellobiohydrolase I (CBH I) and endoglucanase II (EG II), while white and Palo Verde (sage green) varieties derived from the same germplasm stock were more susceptible to hydrolysis. No differences in the binding isotherms of CBH I and EG II to the different FoxFibre cottons were observed. Atomic force microscopy of the colored cotton fibers treated with CBH I detected peeling of the outer surface. The surface of cotton fibers treated with EG II appeared to be smoothed or polished. After alkaline extraction of the pigments, the brown cottons were rapidly hydrolyzed by Celluclast, CBH I, or EG II. Extracts of the colored cotton contained yellow pigments with absorption peaks at 270 nm. The extracts were found to be strong inhibitors of cellulase activity, but had no effect on β-glucosidase. Incubation of the extracts with β-glucosidase decreased the inhibition of cellulases. Analysis with nmr, IR, and hplc indicated that the extracted pigments were associated with oligosaccharides.

Microorganisms carry out the biodegradation of cellulose using cellulase mixtures that consist of multiple enzymes with complementary functions. A typical and widely studied example is the cellulase complex secreted by the fungus *Trichoderma reesei*, which is composed of exoglucanases (cellobiohydrolases, CBH I and CBH II), endoglucanases (EG I, II, and III), and β-glucosidases (*1*). The complementary functions of these cellulase components have been examined. For example, the major cellulase component, CBH I, binds to and disperses crystalline cellulose, while the endoglucanase EG II rapidly hydrolyzes amorphous cellulose (*2*). Based on hydrophobic cluster analysis of amino acid sequences, CBH I has been assigned to cellulase family C, and EG II is assigned to the cellulase A family. Like other fungal and bacterial cellulases, CBH I and EG II share a basic structure, being composed of a cellulose binding domain (CBD) and a catalytic domain joined by a glycosylated linker region (*3,4*). The respective roles of these two domains in the process of cellulose hydrolysis and particularly in the swelling and dispersion of cellulose microfibrils is still being defined. Basically, the question remains whether binding alone is sufficient to result in swelling and dispersion of crystalline cellulose, or if catalysis is coupled to the dispersion process. Reported studies relevant to this problem include the examination of the properties of domains isolated by partial proteolysis of whole enzymes (*5,6*), characterization of recombinant CBDs expressed in non-cellulolytic bacteria (*7*), and site-specific mutagenesis of active site amino acids of CBH I (*8*). Another strategy is the catalytic inactivation of the cellulase active site. In our previous studies, CBH I that had been catalytically inactivated by incubation with water-soluble carbodiimide or hexochloropalladate was able to bind to cotton linters but could not disperse them or carry out hydrolysis (*9,10*). Examination of cotton linters treated with hexachloropalladate-inactivated CBH I with atomic force microscopy at lower magnification (2 × 2 µm) showed no dispersion of the fibers, while at higher magnification (300 × 300 µm), small holes could be detected on the fiber surface (*11*). Naturally colored cotton varieties have been cultivated for millennia by the native peoples of the Americas. As many as six colors--pink, yellow, lavender, brown, green, and red--of *Gossypium barbadense* were grown in Peru. These cotton varieties have also been reported to be insect- and drought-resistant, but the fibers are often short-staple or brittle, properties that are not suited to machine-spinning and weaving (*12*). As consumers have become more concerned about the environmental and health effects of cotton fabric production, there has been a renewal of interest in the cultivation and commercialization of naturally colored cotton. Starting with a parent strain obtained from the United States Department of Agriculture, naturally colored cotton varieties of *Gossypium hirsutum* have been developed and improved for commercial textile production by Naturally Coloured Cottons, Inc., using selective breeding (*13*). Three varieties of these cottons have been registered and are currently being marketed: Coyote (reddish brown), Palo Verde (sage green), and Buffalo (mocha brown). It is currently believed that the brown pigments are composed of both water-soluble and insoluble tannins that are deposited in the lumen of the fiber. The green pigment is thought to be a suberin, a plant compound composed of aromatic and aliphatic polymers, that is deposited between all 30-45 layers of cellulose in the fiber (*14,15*).

It appeared to us that important insights into the mechanism of cellulase binding and hydrolysis could be obtained by the investigation of the effects of cellulases on these naturally colored cottons. To our knowledge, no studies of cellulase action on naturally colored cottons have been reported previously. In this study, we report the results of our studies on the action of cellulases on the naturally colored Foxfibre cottons.

Materials and Methods

Development of colored cotton varieties. FoxFibre, Coyote, Palo Verde, and Buffalo are registered trademarks of Natural Cotton Colours, Inc., Wickenburg, Arizona.

Cotton germplasm was obtained from Dr. Gus Heyer at the USDA Research Station, Shaftner, California. Selective breeding of this germplasm was used to develop improved strains that were chosen for plant type, fiber color intensity, fiber color uniformity, fiber length improvement, and fiber strength improvement. Initial machine spinning tests were conducted in 1988 at the International Textile Research Center at Texas Technical University, Lubbock, Texas. A process of single plant selections over the course of six generations produced Coyote (*16*), the first variety of brown-colored cotton that was suitable for machine spinning and knitting. A second brown cotton variety, Buffalo (*17*), was developed by crossing an ancestor of Coyote with Pima S5. Single plant selections were conducted over many generations to improve the color intensity, and fiber length, strength and uniformity of Buffalo compared to the parent strain Coyote. Both of these strains, Coyote and Buffalo, express an innate fire retardancy, a characteristic that can be selected for. The green cotton variety known as Palo Verde (*18*) was developed by crossing a variety, "Green", that had also been isolated from the original germplasm obtained from USDA, with an F1 of Pima S5 and an ancestor of Coyote. Selections were made of individual plants over many generations using as the selection criteria yield improvement without deterioration of fiber color intensity or fiber suitability for spinning.

Purification of enzymes. The recombinant thermophilic cellulases Clonezyme CEL-001-01 and CEL-001-02 (Recombinant Biocatalysis, Inc., La Jolla, California) were used without further purification.

The β-glucosidase from *Aspergillus niger* was prepared by filtering Novozym 188 (a generous gift from Novo Nordisk Bioindustrials) on Biogel P-60.

Cellobiohydrolase I (CBH I) from *Trichoderma reesei* was purified from Celluclast (a generous gift from Novo Nordisk Bioindustrials) using a modification of the method described previously (*19*). After the first chromatofocusing chromatography, the CBH I was reinjected on the Pharmacia Mono-P column and eluted as before.

Endoglucanase II (EG II) was purified from the fraction of Celluclast that elutes from DEAE Sepharose in 50 mM sodium acetate, pH 5. This ion-exchange fraction was applied to a Pharmacia Mono-P chromatofocusing column equilibrated in 25 mM Bis-Tris HCl, pH 7.1. Proteins were eluted with a linear gradient of Polybuffer 74, 1:10 diluted, pH 4.0. The EG II pool was further purified by ion-exchange chromatography on a Pharmacia Mono-Q column equilibrated in 20 mM Tris-HCl, pH 8, with a linear gradient from 0 to 1.0 M sodium chloride. This EG II preparation appeared homogeneous on SDS-PAGE and isoelectrofocusing polyacrylamide gels (data not shown).

Enzyme assays. Commercial cotton linters (Buckeye, Memphis, Tennessee) were a generous gift from Tennessee Eastman, Kingsport, Tennessee. These linters were 99.7% α-cellulose, with an average chain length of greater than 2,600 glucose units.

Binding assays of purified CBH I and EG II were carried out at 23°C in 50 mM sodium acetate, pH 5.0, with 10 mg/ml cotton as described (*20*), but with the following modifications. Due to the elution of compounds adsorbing in the UV-region by the buffer during the incubations, the absorbance of a control of cotton in buffer had to be subtracted from the absorbance at 280 nm of the supernatants from the binding reactions before it was used to determine the amount of protein bound compared to no-cotton enzyme controls. The molar adsorption coefficients reported in the literature for CBH I (*5*) of 73,000 M^{-1} cm^{-1} and for EG II of 77,000 M^{-1} cm^{-1} (*21*) were used to calculate the concentrations of free and bound enzyme.

Hydrolysis of cotton samples by *Trichoderma* cellulases was carried out at 45°C in 50 mM sodium acetate or citrate buffer, pH 5.0. Hydrolysis with the thermophilic cellulases CEL-001-01 and CEL-001-02 was carried out at 70°C in 50 mM potassium phosphate buffer, pH 7.0. Assays of β-glucosidase activity were carried out in a reaction mixture containing 10 mM cellobiose in 50 mM citrate, pH 5.0. Hydrolysis products were analyzed with the dinitrosalicylic acid reducing sugar reagent (*22*), the Sigma hexokinase glucose reagent, and by hplc on a Waters Sugar-Pak column.

Atomic force microscopy. For each sample, Buffalo cotton (2 mg) was incubated in 2 ml of 50 mM sodium acetate, pH 5.0, for 24 h at 45°C. The samples were incubated with buffer only, with 0.2 mg/ml CBH I, or with 0.2 mg/ml EG II. All cotton samples were washed with double-distilled water and dried under low vacuum in a vacuum desiccator. Atomic force microscopy was performed as described (*11*), except that an amplitude of 130 mV was used.

Extraction and analysis of chromophores from colored cotton. For UV-visible spectra and inhibition studies of the cotton pigments, samples of cotton (10 mg) were extracted with 2 ml of 5 M sodium hydroxide for 3 days at 23°C. The alkaline extracts were neutralized by the addition of hydrochloric acid to a final pH of 5-6, then diluted to a final total volume of 8 ml with distilled water. UV-visible scans were carried out with a Perkin-Elmer Lambda Array spectrophotometer on neutralized extracts (20% final volume) in 80% ethanol.

For chemical analysis of the extracted pigments, cotton samples were extracted with 0.5 N sodium hydroxide. The extracts were neutralized with hydrochloric acid, evaporated, and redissolved in chloroform. Analysis by Fourier transform infrared spectroscopy (FTIR) and ^{1}H-nuclear magnetic resonance (nmr) was carried out by the Chemical and Analytical Division at Oak Ridge National Laboratory.

Results and Discussion

Resistance to cellulase digestion. Production of reducing sugar and glucose from cellulase digests of cotton linters and the FoxFibre varieties were compared. All of the FoxFibre cottons were much more resistant to cellulase attack than commercial cotton linters. The brown-colored cotton varieties Coyote and Buffalo were found to be very resistant to cellulase digestion, even at loadings as high as 4% Celluclast. The green-colored cotton variety Palo Verde was found to be slightly more susceptible to cellulase hydrolysis than the unpigmented white FoxFibre cotton (Figure 1). Examination of the cotton fibers after 72 h cellulase digestion confirmed that white cotton and Palo Verde cotton were more susceptible to cellulase attack than the Coyote and Buffalo cottons. Analysis of the reaction products obtained with Celluclast by hplc identified glucose as the major product in all of the digests, with approximately 10% residual cellobiose. Conversion of cellobiose to glucose was expected as the β-glucosidase activity of the Celluclast preparation used for the digests was 0.189 μmol glucose mg^{-1} min^{-1}.

The action of the purified *T. reesei* cellulase components, CBH I and EG II, was examined. After 24 h incubation, purified EG II could only produce very low amounts of reducing sugar from the FoxFibre cottons; purified CBH I was more effective (Table I). Preincubation of FoxFibre Coyote, Palo Verde, and Buffalo (10 mg/ml cotton) with *A. niger* β-glucosidase (0.04 mg/ml) for 16 h at 45°C before addition of CBH I (final concentration 0.2 mg/ml) did not increase the rate of hydrolysis of the cotton by CBH I (data not shown).

The experiments with the *T. reesei* cellulases were all carried out at pH 5.0 and 45°C. In order to find out if temperature or pH might be implicated in the resistance of the colored cottons to digestion with cellulases, the cottons were incubated with the thermophilic cellulases CEL-001-01 and CEL-001-02 at pH 7.0 and 70°C. All of the FoxFibre cottons were found to be very resistant to attack by these thermophilic cellulases (Table I).

Table I. Yields of reducing sugar from digests of cotton varieties with various cellulases.

mM Reducing Sugar Produced after 24 h Incubation					
Enzyme	Linters	White	Coyote	Palo Verde	Buffalo
Celluclast	5.341	1.869	0.1941	1.581	0.2785
CBH I	2.054	0.2887	0.1562	0.2168	0.2674
EG II	0.999	0.01719	0.01990	0.01565	0.004323
CEL-001-01	0.3718	0.008418	0.03088	0.01473	0.02162
CEL-001-02	0.5896	0.01543	0.006558	0.01407	-0.0124

Comparison of binding isotherms of CBH I show that binding capacity for cotton linters, white FoxFibre cotton, and Coyote cotton is similar. However, binding of CBH I to the FoxFibre cottons takes much longer to reach equilibrium than binding to Buckeye cotton linters (Figure 2). Binding isotherms of CBH I are similar for the different Foxfibre cottons, implying that the resistance to hydrolysis exhibited by Coyote and Buffalo, as compared to white and Palo Verde, is not the result of inhibition of binding. Although able to bind quickly and efficiently to Buckeye cotton linters, EG II bound to all of the FoxFibre cottons much less effectively than CBH I did (data not shown).

The effects of treatment of the Buffalo cotton with CBH I or EG II for 24 h at 45°C were examined with atomic force microscopy (AFM). On AFM images of the same magnification, the macrofibrils of the Buffalo fibers appeared to be twice as big as the macrofibrils of the commercial cotton linters in the previously published AFM images (*11*). Examination of the AFM scans found that, compared to a control of Buffalo cotton incubated with buffer only (Figure 3), Buffalo cotton treated with CBH I exhibited peeling of the outer fiber surface (Figure 4). Compared to previously published AFM scans of the action of the CBH I on Buckeye cotton linters that show considerable dispersion and disruption of the fibers (*11*), the Buffalo cotton appeared to show a minimal effect that was confined to the outer surface of the fiber. The EG II-treated sample (Figure 5) appeared smoothed and polished compared to the untreated control (Figure 3); no disruption of the surface could be detected.

The effect of alkaline extraction on the hydrolysis of the colored cotton was investigated. Cotton samples from which the pigments had been extracted with 5 M sodium hydroxide were washed with distilled water and dried overnight in a vacuum oven at 60°C. As treatment with strong alkali swells cotton fibers, the susceptibility of the brown cottons to cellulase hydrolysis after extraction was compared to that of the samples of FoxFibre white and Palo Verde cottons that had been extracted with 5 M sodium hydroxide. After alkaline extraction, Palo Verde and Buffalo cottons were equally susceptible to cellulase attack and were rapidly hydrolyzed by Celluclast (Figure 6). Similar results were obtained for Coyote and white FoxFibre cottons after alkaline extraction (data not shown).

Characterization of chromophores from colored cotton. Upon visual examination with the light microscope at 150 × magnification (transillumination), it was observed that the brown pigment in Coyote and Buffalo was not uniformly distributed throughout the fibers, but was darker in some fibers and looked like streaks in the middle of the fibers. More colored fibers could be detected in the Buffalo than in the Coyote. No coloration was detectable in the Palo Verde at 150 ×, despite its green color without magnification.

Preliminary experiments were carried out to examine the effect of different solvents on the naturally colored cotton samples. Incubation of cotton samples (10 mg/ml) for up to 4 days in ethanol, glacial acetic acid, or 5 M sodium chloride did not appear to have any effect on the coloration. Sodium hydroxide at 1 or 5 M was found to extract the chromophores from the cotton. In UV-visible absorption scans

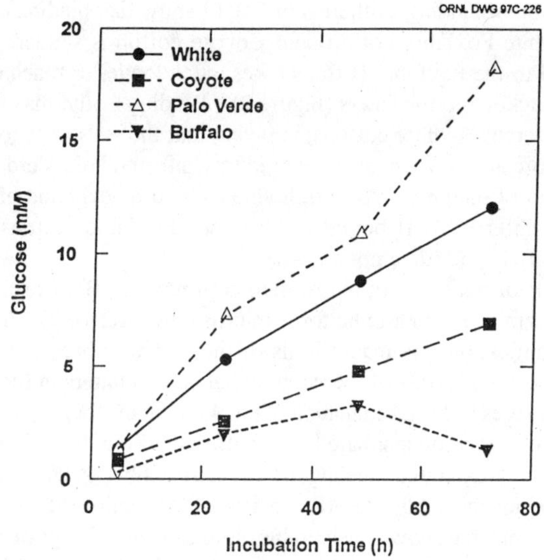

Figure 1. Hydrolysis of FoxFibre cotton varieties (10 mg/ml) by Celluclast (0.2 mg/ml) at pH 5.0 and 45°C.

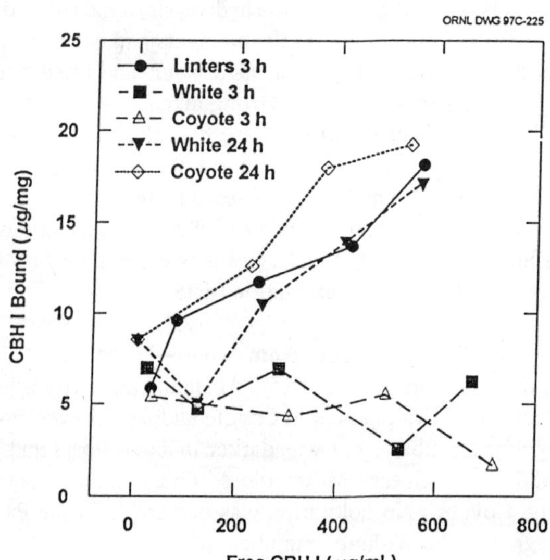

Figure 2. Binding of CBH I to Buckeye cotton linters after 3 h, and to FoxFibre white and Coyote cottons after 3 h and after 24 h incubation at 23°C and pH 5.0, with 10 mg/ml cotton.

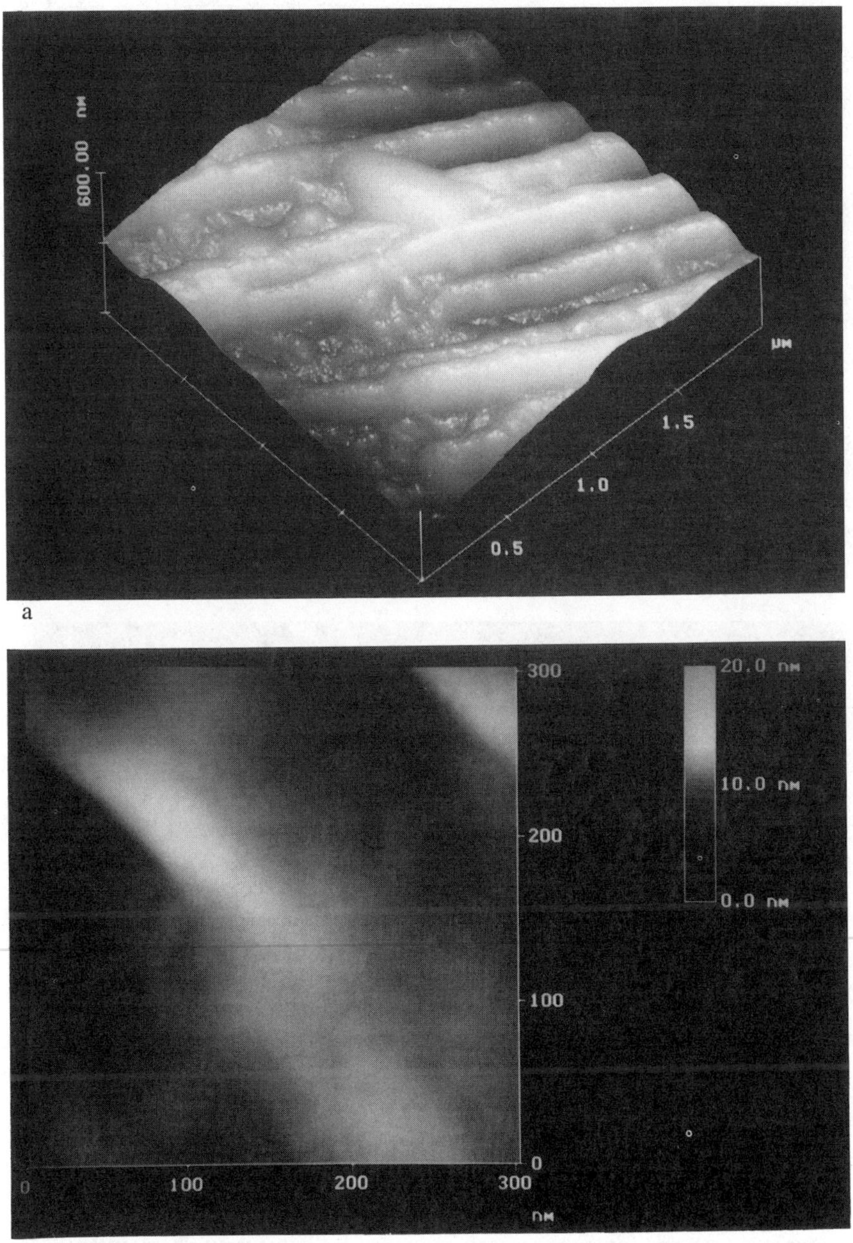

Figure 3. Atomic force microscope images of Buffalo cotton fibers in (a) 2 × 2-µm and (b) 300 × 300-nm scans.

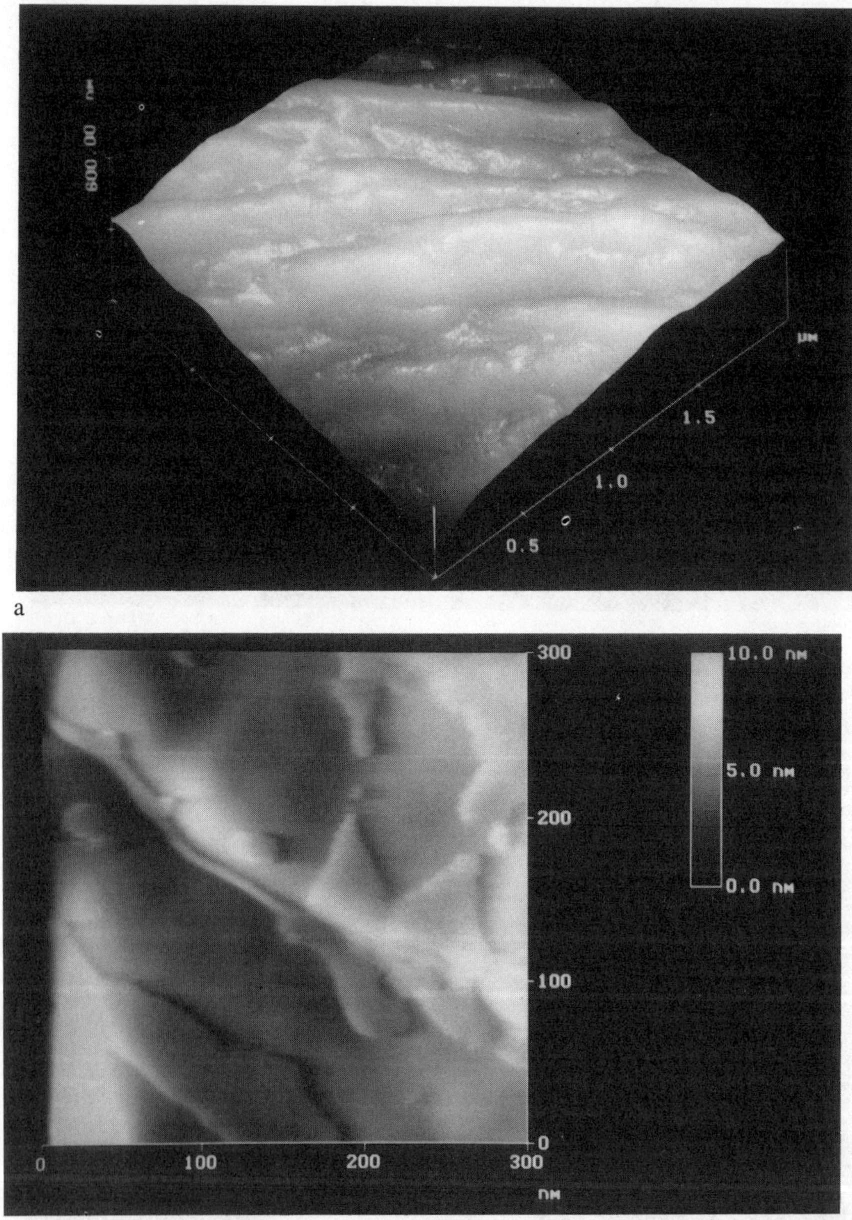

a

b

Figure 4. Atomic force microscope images of Buffalo cotton fibers treated with CBH I for 24 h in (a) 2 × 2-μm and (b) 300 × 300-nm scans.

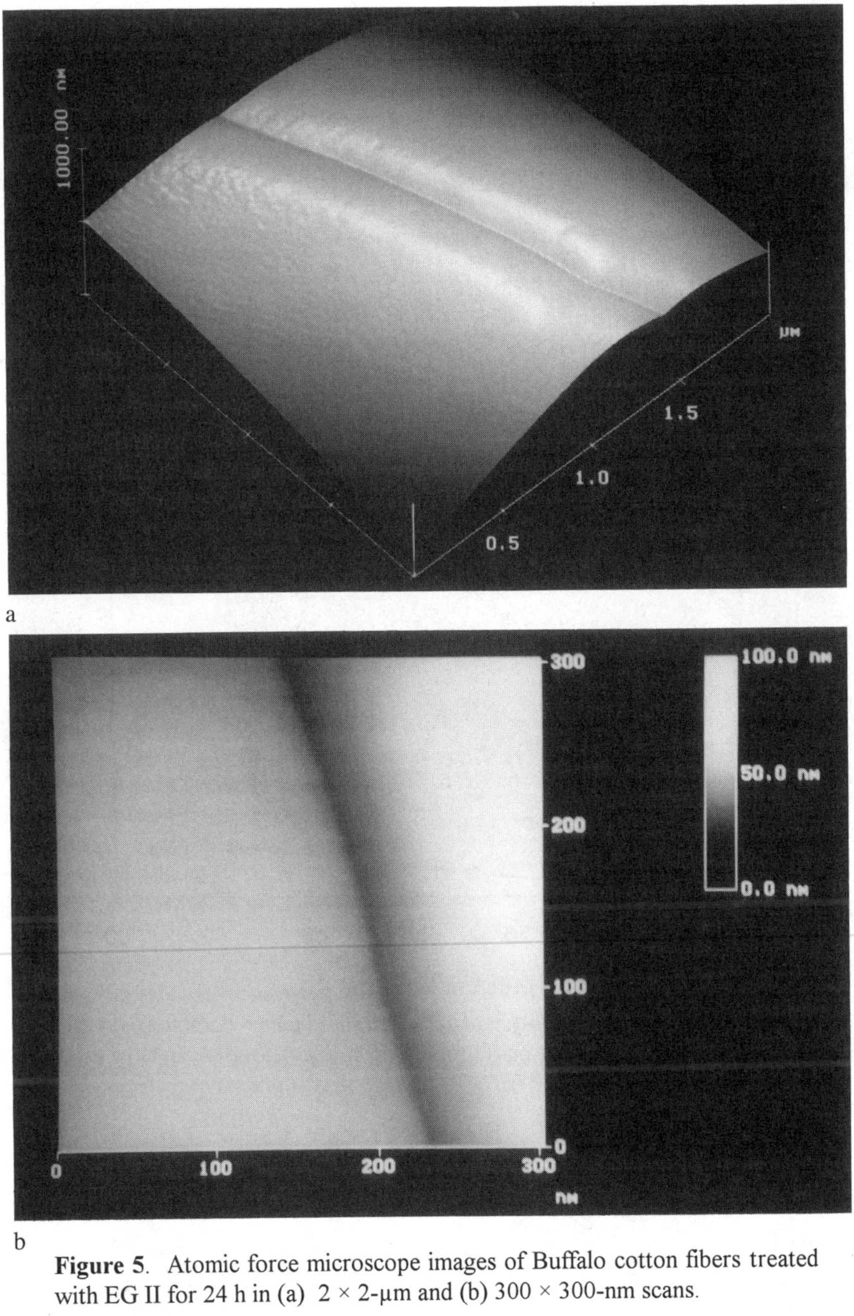

Figure 5. Atomic force microscope images of Buffalo cotton fibers treated with EG II for 24 h in (a) 2 × 2-µm and (b) 300 × 300-nm scans.

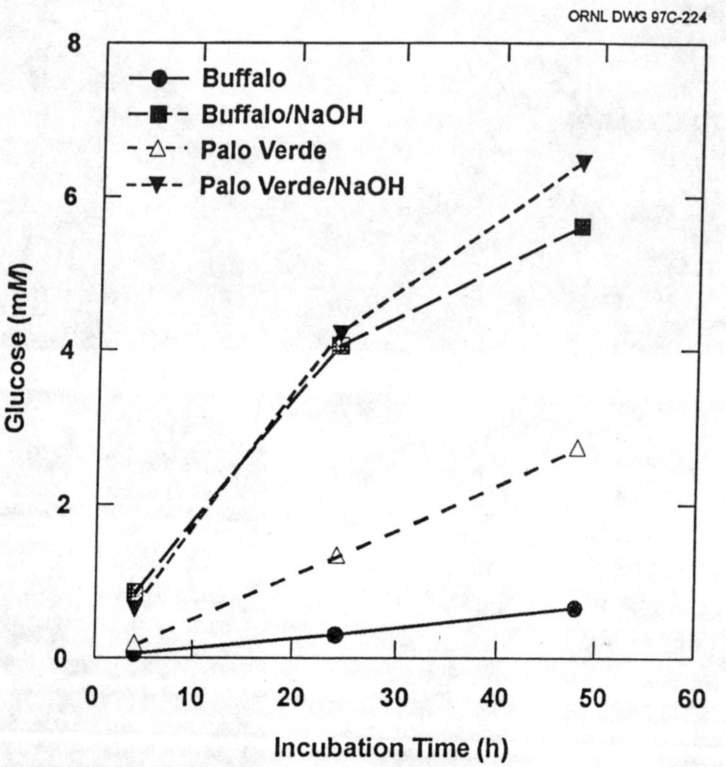

Figure 6. Comparison of hydrolysis of untreated and alkaline-extracted samples of Buffalo and Palo Verde cottons (1 mg/ml) with Celluclast (0.2 mg/ml) at pH 5.0 and 45°C.

of the pH-adjusted extracts in 90% ethanol, Coyote, Palo Verde, and Buffalo extracts had a large absorption peak at 195-205 nm, with a prominent shoulder at 260-270 nm corresponding to the expected absorption of aromatic ring structures that was not present in the white cotton extract (Figure 7). When white and Coyote extracts were analyzed on the Sugar-Pak HPLC column, only one component was detected by the refractometer, which eluted at the same position as cellotriose. Digestion of the extracts with β-glucosidase (*A. niger*) for 4 h approximately doubled the amount of glucose in the extracts from the colored varieties, but not in the extract from the white FoxFibre control (Figure 8).

The pigment extracted with sodium hydroxide from Coyote cotton was subjected to instrumental analysis. The infrared spectrum had significant hydroxyl and C-O bands, as well as CH3 and CH2 stretching bands, and, when searched against a Stadtler infrared standard library, closely matched calcium alginate (Figure 9). The nmr data identified a poly-alcohol-like compound (saccharide or sugar), that would be similar to an alginate-type compound (Figure 10). Analysis of alkaline extract from FoxFibre white cotton identified the same compounds found in the Coyote extract, but at lower concentration.

The pH-adjusted extracts were tested for inhibition of cellulases and β-glucosidase. In the case of the cellulases, inhibition of hydrolysis of the white cotton was observed when 1/5 volume of extract from the Coyote, Palo Verde, or Buffalo cotton was added to the reaction mixture; addition of extract from white cotton did not cause inhibition (Table II). The extract from Buffalo cotton was found to inhibit the cellulase activity of Celluclast on white FoxFibre cotton if added at the start of the hydrolysis reaction, but did not inhibit hydrolysis if added after the reaction mixture had been incubated with cellulase for 1 to 8 h (Figure 11). When the extracts from the FoxFibre cottons were incubated with *A. niger* β-glucosidase, no inhibition of the cellobiase activity could be detected (Figure 12). After incubation with β-glucosidase, the extracted pigments lost their inhibitory activity on the cellulases (data not shown).

Table II. Inhibition of Celluclast Hydrolysis of FoxFibre White Cotton by Cotton Extracts

Cotton Extract	% Control Activity
White	97.1
Coyote	26.9
Palo Verde	13.6
Buffalo	4.69

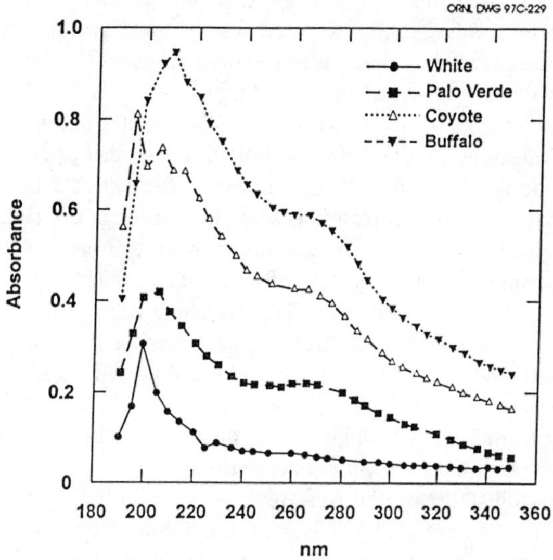

Figure 7. Absorption spectra of extracts from FoxFibre cottons in 80% ethanol.

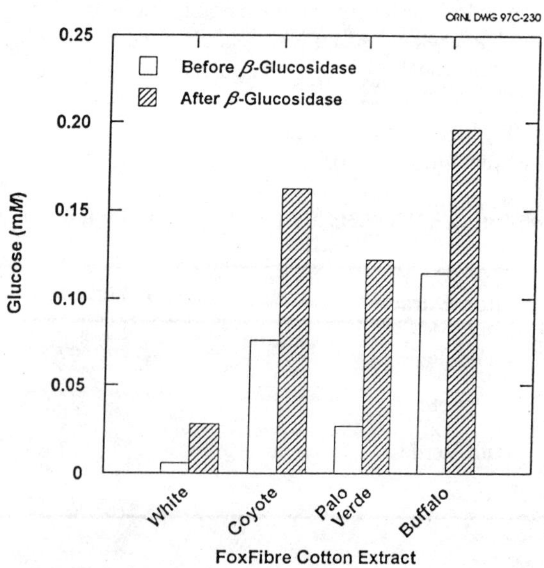

Figure 8. Production of glucose from cotton extracts by 4 h digestion with *A. niger* β-glucosidase (0.04 mg/ml) at pH 5.0 and 45°C.

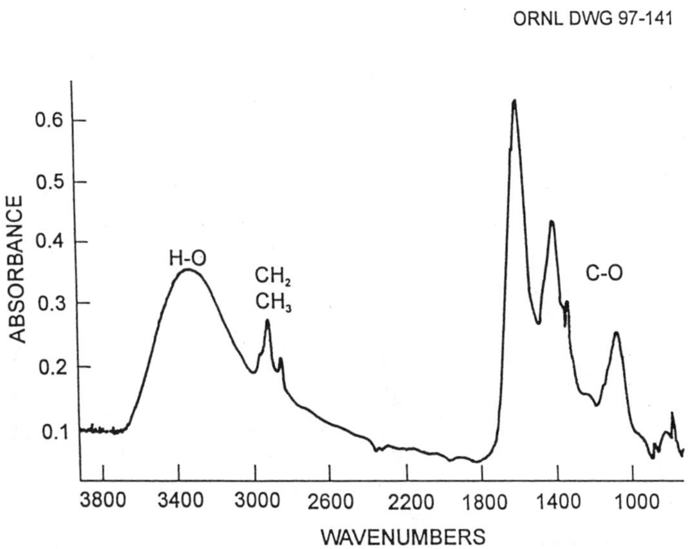

Figure 9. FTIR spectrum of pigment extracted from Coyote cotton with 0.5

ORNL DWG 97-142

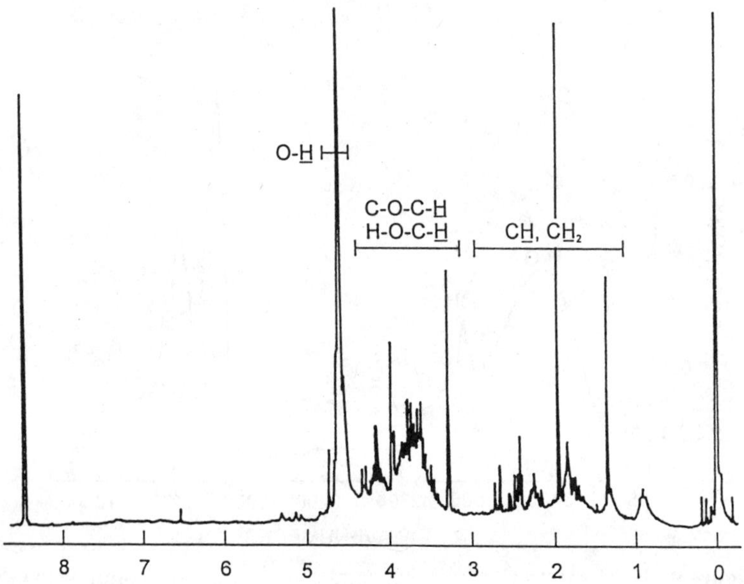

Figure 10. [1]H-nmr spectrum of pigment extracted from Coyote cotton with 0.5 M sodium hydroxide.

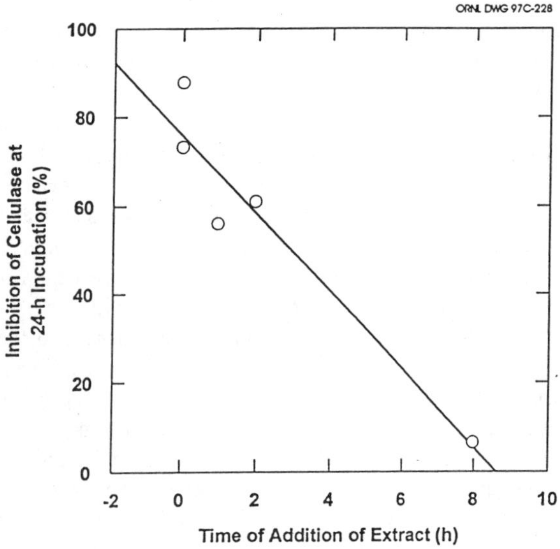

Figure 11. Effect of time of addition of Buffalo extract (1/10 reaction volume) to digests of FoxFibre white cotton by Celluclast (0.2 mg/ml).

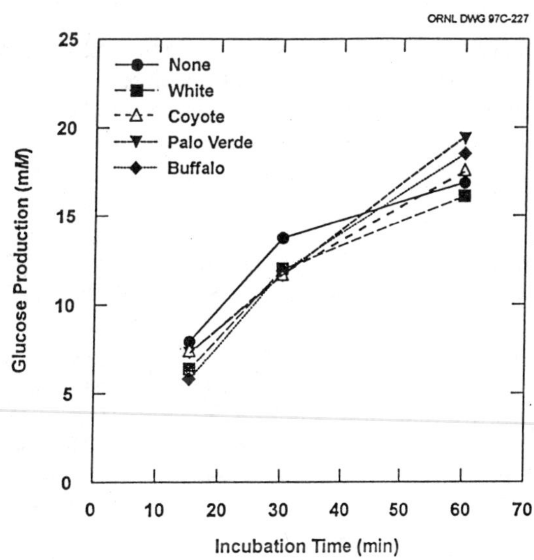

Figure 12. Effect of cotton extracts on activity of *A. niger* β-glucosidase (0.04 mg/ml) on cellobiose (10 mM) at pH 5.0 and 45°C.

Conclusions

Two naturally colored cottons (FoxFibre varieties Coyote and Buffalo) contain pigments that inhibit cellulase attack of the cotton fibers. Cellulases are able to bind to but not hydrolyze and disrupt these colored cotton varieties. Inhibition appears to occur as the cellulases begin to hydrolyze the cotton after binding to the surface of the fibers. Reduction of the amount of pigment incorporated into the cotton fibers, whether by strain selection or by alkaline extraction of brown cotton, renders the cotton susceptible to cellulase hydrolysis. The pigments appear to consist of aromatic moieties associated with or attached to short glucose oligomers that can be extracted with sodium hydroxide. This sort of structure would be consistent with the reported structure of another yellow cotton pigment, quercimeritrin (3,3',4',5,7-pentahydroxyflavone-7-D-glycoside), that is found in the flowers of *Gossipium herbaceum* (*23*). The extracted pigments are susceptible to hydrolysis by β-glucosidase. As treatment of the brown cotton fibers with β-glucosidase does not aid cellulase digestion, it appears that the pigments in the fibers are not accessible to β-glucosidase. The results of these studies on the action of cellulases on colored cottons support a cellulase mechanism in which hydrolysis and cellulose disruption are coupled. Further studies would be useful to determine the effect of the extracts on cellulase activity towards other substrates, as well as their effects on other enzymes.

Acknowledgments

Research was supported by Office of Basic Energy Sciences, U. S. Department of Energy under contract DE-AC05-96OR22464. Analyses using nmr and IR were carried out by Joseph Oswald and David Williams of the ORNL Analytical Chemistry Division.

Literature Cited

1. Goyal, A; Ghosh, B.; and Eveleigh, D. *Biores. Technol.* **1991**, *36*, 37-50.
2. Nidetzky, B.; Hayn, M.; Macarron, R.; Steiner, W. *Biotech. Lett.* **1993**, *15*, 71-76.
3. Claeyssens, M.; Henrissat, B. *Protein Sci.* **1992**, *1*, 1293-1297.
4. Esterbauer, H.; Hayn, M.; Abuja, P. M.; Claeyssens, M. in *Enzymes in Biomass Conversion* (eds. Leatham, G. F.; Himmel, M. E.), ACS Symposium Series, ACS, Washington, D. C., 1991, No. 460, 301-312.
5. Tomme, P.; Van Tilbeurgh, H.; Pettersson, G.; Van Damme, J.; Vandekerchove, J.; Knowles, J.; Teeri, T.; Claeyssens, M. *Eur. J. Biochem.* **1988,** *170*, 575-581.
6. Din, N.; Gilkes, N. R.; Tekant, B.; Miller, R. C. Jr.; Warren, R. A. J.; Kilburn, D. G. *Bio/Technol.* **1991**, *9*, 1096-1099.

7. Linder, M.; Teeri, T. T. *Proc. Natl. Acad. Sci. USA* **1996**, *93*, 12251-12255.
8. Ståhlberg, J.; Divne, C.; Koivula, A.; Piens, K.; Claeyssens, M.; Teeri, T. T.; Jones, T. A. *J. Mol. Biol.* **1996**, *264*, 337-349.
9. Donner, T. R.; Evans, B. R.; Affholter, K. A.; Woodward, J. in *Bioconversion of Fuels* (Eds. Himmel, M. E.; Baker, J. O.; Overend, R. P.) ACS Symposium Series, ACS, Washington, D.C., 1994, No. 566, pp. 75-83.
10. Shultz, M. D.; Lassig, J. P.; Gooch, M. G.; Evans, B. R.; Woodward, J. *Biochem. Biophys. Res. Comm.* **1995,** *209,* 1046-1052.
11. Lee, I.; Evans, B.R.; Lane, L.M.; Woodward, J. *Bioresource Technol.* **1996**, *58*, 163-169.
12. Schuster, A. M. H.; Walker, H. B.; Ferreyra, R. *Archeology* **1995**, *July/August*, 40-45.
13. Thompson, J.; Wolinsky, C. *National Geographic* **1994**, *185*, 60-87.
14. Yatsu, L. Y.; Espelie, K. E.; Kolattukudy, P. E. *Plant Physiol.* **1983**, *73*, 521-524.
15. Ryser, U.; Meyer, H.; Holloway, P. J. *Protoplasma* **1983**, *117*, 196-205.
16. Fox, S. V., PVP # 8900169, USDA Plant Variety Protection Office, **1990**.
17. Fox, S. V., PVP # 9400039, USDA Plant Variety Protection Office, **1996**.
18. Fox, S. V., PVP # 9400040, USDA Plant Variety Protection Office, **1995**.
19. Offord, D.A.; Lee, N. E.; Woodward, J. *Appl. Biochem. Biotechnol.* **1991**, *28/29*, 377-386.
20. Woodward, J.; Affholter, K. A.; Noles, K. K.; Troy, N. T.; Gaslightwala, S. F. *Enzyme Microbiol Technol.* **1992**, *14*, 625-630.
21. Saloheimo, M; Lehtovaara, P; Penttilä, M.; Teeri, T. T.; Ståhlberg, J.; Johansson, G.; Pettersson, G.; M. Claeyssens, M.; Tomme, P.; and Knowles, J. K. C. *Gene* **1988**, *63*, 11-21.
22. Miller, G. L. *Anal. Chem.* **1959**, *31*, 426-428.
23. Merck Index, 1989, Merck & Co., Inc., Rahway, NJ.

Chapter 20

Effect of Enzyme Treatment on Chemically Pretreated Linen

K. K. Wong, X. M. Tao, C. W. M. Yuen, and K. W. Yeung

Institute of Textiles and Clothing, The Hong Kong Polytechnic University, Hong Kong

Due to environmental awareness in recent years, enzymes which are bio-degradable and environmentally friendly have been increasingly used for the modification of cellulosic fibres. The aim of this paper is to study enzyme treatment on linen pre-treated with caustic soda. The factors concerned were the mercerisation tension, concentration of cellulase, and incubation time. Changes in crystallinity of fibres, moisture regain, fabric weight, fabric surface thickness, tensile strength, stiffness and fibre surface appearance of material were observed and studied. It was confirmed by the experiments that the fibre structure, accessibility and physical properties of linen pre-treated with caustic soda could further strengthen the effectiveness of enzyme treatment by exhibiting high reduction of fabric stiffness, minimising the loss of tensile strength and shortening the original treatment time. Investigation by scanning electron microscopy revealed that pretreatment could further enhance the removal of surface fibrils from the fibre surface by enzyme treatment.

In recent years, the textile industry has greatly benefited from the use of enzymes in yarn, fabric and garment treatments. The concept of cellulase treatment for the cellulosic fibre was first developed in Japan for the wet treatment of cotton fabric *(1)*. It was referred to as bio-polishing to remove the small fibre ends from the yarn surface to create a smooth fabric appearance and introduce a degree of softness without using traditional chemical treatment *(2)*. The cellulase treatment made 'stonewash' effect on denim fabric a reality, eliminating the use of pumice stones and associated problems such as irregularly lightened areas by excessive mechanical abrasion on the fabric surface. The mechanism of cellulase treatment was identified as the specific catalytic action on the beta-1,4-glycoside bond of the cellulose molecule *(3, 4)*. However, synergism in cellulase systems and the mechanism of enzymatic hydrolysis of cellulosic materials are complicated and not well understood

(5-8). In addition, significant strength reduction and the slow reaction rate of the enzymatic reaction limit its industrial application.

In the course of enzymatic hydrolysis of cellulosic materials, structural features were found to be the most important factor which governed the yield and rate of reaction *(9)*. Key structural features such as crystallinity and accessible surface area determine the susceptibility of cellulosic materials to enzymatic degradation. Studies on other cellulosic fibres such as linen, ramie and regenerated cellulose have been carried out *(10, 11)*. In order to enhance the reactivity of the cellulose to the catalytic reaction, physical and chemical pretreatments were employed *(8, 9, 12-14)*. Mercerisation was used for cotton yarn to reduce fibre crystallinity *(13, 15)*. The application of sodium hydroxide (mercerisation) to flax yarns increased its accessibility and yarn strength while tension was applied during the treatment *(16)*. Flax fibres contain a certain amount of natural lignin which inhibits the efficient enzymatic hydrolysis of cellulose *(17)*. Thus the objective of our current work is to investigate the effect of mercerisation pretreatment on the enzymatic hydrolysis of linen material.

Experimental

Sample Preparation

Scoured and semi-bleached linen used in this experiment was supplied by a fabric manufacturer in China. Fabric weight was 154g/m^2 with a sett 21/20 per cm (54/50 per inch) and yarn count 118tex/118tex (14lea/14lea). Table I lists the specifications of fabric samples. A commercial cellulase, Cellulsoft L from Novo Nordisk, was used with a specified activity of 1500NCU/g. All other chemicals used were of reagent grade.

Three levels of mercerisation tension were applied to the linen material, i.e., slack mercerisation, 95% and 100% tension mercerisation. For slack mercerisation, the fabric samples were immersed in 20% sodium hydroxide solution for 10 minutes at 20$\pm$2°C. Neutralisation was carried out by rinsing the mercerised samples in distilled water for 10 minutes followed by the addition of 1% acetic acid and then rinsed again until the rinse water was neutral. A stainless steel needle frame was used for controlling the fabric tension. The tension mercerised fabric sample was removed from the frame after final rinsing and then allowed to dry in air without restraint. The samples were spin dried for 5 minutes to remove excessive water and then air dried under standard conditions (65$\pm$2% RH and 21$\pm$1°C).

Enzyme treatments were carried out separately at 1% and 2% of concentration based on the weight of the fabric sample. The treatment time ranged from one to four hours at the temperature of 50$\pm$ 2°C with the liquor ratio of 1:30 in a Launder-Ometer using a 1 litre pot. The machines ran at 42 rpm. to provide mechanical agitation during the enzyme treatment. For enzymatic hydrolysis, 1 gram of cellulase was dissolved at room temperature in 500 ml of sodium acetate buffer at pH4.5 to prepare a 0.2% cellulase stock solution. For each gram of conditioned fabric, 5 ml of the cellulase stock solution was added for 1% enzyme dosage and 10 ml was added for 2% enzyme dosage. Fabric samples were immersed in the buffer solution and the temperature was raised to 50°C within 10 minutes. After the fabric samples were treated separately from 1 hour to 4 hours, the cellulase was deactivated by washing thoroughly in hot water at 70°C for 10 minutes. Finally, the samples were spin dried for 5 minutes to remove excessive water, then air dried in standard conditions.

Material Characterisation

Moisture regain of the fabric sample was determined according to ASTM D2654-89a and five samples were used in total. X-ray crystallinity was measured by Philips PW3710 diffractometer using CuK_{α} radiation at an operating voltage of 40kV and current of 55mA from angles 5° to 30°. Pellets were prepared from 0.25g ground fibres *(18)* and the crystallinity ratio was calculated *(19)* according to the equation: Cr.R. = $1 - I_1/I_2$, where I_1 is the minimum intensity between 2θ = 18° and 19° for cellulose I and between 2θ = 13° and 15° for cellulose II, and I_2 is the maximum intensity of crystalline peak between 2θ = 22° and 23° for cellulose I and between 2θ = 18° and 22° for cellulose II.

The fabric weight loss was determined based on the conditioned weight at 65±2% RH and 21±1°C before and after the cellulase treatment. Fabric thickness was measured by the FAST- 1 (Compression Meter of FAST system) and 15 samples were used. The untreated control samples were washed in distilled water and air dried before measurement. Fabric thickness was measured at $2gf/cm^2$ and $100gf/cm^2$, and the surface thickness was calculated by the thickness difference between $2gf/cm^2$ and $100gf/cm^2$.

Fabric strength was measured by an Instron tensile tester model 4466 according to ASTM D5035-95 at standard testing atmospheric conditions. The cross-head speed was 304 mm/minute. The width of sample was 5.08 cm (2 inch) while the gauge length was 7.62 cm (3 inch) and the results obtained were the averages of five readings. Fabric bending properties were measured by KES-FB2 (Pure Bending Tester) from Kato Tech. Co. Ltd. The sensitivity was selected at 5x1 to provide the full scale momentum of 50gfcm. The width of fabric sample was 10 cm and the average of five readings was taken. The bending rigidity was calculated by the slope between the curvature K= 0.5 and $1.5cm^{-1}$, and the bending hysteresis was calculated at the curvature K= $0.5cm^{-1}$. Scanning electron microscopy (SEM) examination was carried out on the randomly selected fabric surfaces at different areas of the untreated and treated fabrics using a Lecia Cambridge Stereoscan 440 at 5kV. The paired comparison *t*-test was used to determine where significant differences occurred. A 95% confidence limit was used in the statistical analysis.

Results and Discussion

Accessibility and Crystallinity of Samples

The increased accessibility was probably due to the action of sodium hydroxide solution which swelled and broke the hydrogen and Van der Waals bonds, thus freeing the polymers to rearrange and move further apart for reorientation leading to an expanded fibre structure. By evaluating the changes in absorption moisture regain, the accessibility of the fibres can be determined since the quantity of moisture held by fibres is proportional to the non-crystalline, less ordered regions within the fibre. Moisture regain of the control and mercerised sample is shown in Table II.

Significant increase of the moisture regain value was obtained after the mercerisation treatment especially for the slack mercerised sample. This observation was consistent with the pervious report which showed that the fibre accessibility was increased to a greater extent under the slack conditions in flax yarn mercerisation *(16)*.

The X-ray crystallinity ratio of the cellulosic material is another measure to determine the degree of modification of the fibre accessibility. The crystallinity was calculated based on the ratio of the amorphous and crystalline region of the cellulosic material. The experimental results are shown in Table II. Significant reduction in the X-ray crystallinity was found for the slack mercerised sample when compared with the untreated sample. However, the magnitude of change was less prominent in the tension mercerised samples since the lateral movement of the polymer chains was restricted by the applied tension during the caustic swelling. As shown in Table II, there was no additional change in moisture regain and X-ray crystallinity brought about by the enzymatic treatments, indicating that there were no changes in either the accessible surface area or crystallinity. This phenomenon *(6, 8)* might be attributed to the mechanism that the large molecule could not penetrate the structure but only attacked the ends of accessible chains on the crystallite surface.

Fabric Weight Loss and Thickness

As the enzymatic hydrolysis can remove the surface fibrils from the fibre surface, fabric weight loss is thus considered as the primary index for the degree of reaction to occur. The results of fabric weight loss measurement are shown in Table III. In most cases, the weight loss would increase as the incubation time was prolonged, particularly a predominant weight loss was found at 2% enzyme dosage. When doubling the enzyme dosage from 1 to 2%, approximately twice the weight loss was obtained. Table IV showed that fabric thickness was increased after the mercerisation pretreatment. The highest weight loss after enzyme treatment was found in the unpre-treated sample followed the by the 95% and 100% tension mercerised samples. The slack mercerised sample had the lowest crystallinity ratio and its weight loss was very much lower than that of the unpre-treated sample. This may suggest that the macroscopic features such as fabric thickness and construction also can affect the rate of reaction especially at short treatment time *(2, 10)*. Thicker fabrics appear to have lower weight loss in the enzymatic treatments. It seems that the weight retention ability of the enzyme treatment can be improved by mercerisation because of the increment in fabric thickness.

The slack mercerised fabric was allowed to swell without restriction and had the highest fabric thickness followed by the 95% and 100% tension mercerised sample. As shown in Table IV, there was no significant change of fabric thickness measured at 100gf/cm^2 before and after the enzyme treatment among all the samples. However, a significant difference of fabric surface thickness was found. Fabric surface thickness was reduced very quickly after the first hour of enzyme treatment and the changes then started to slow down and approach an equilibrium. Even though the period of treatment was prolonged to four hours, there was no significant change in equilibrium value. In the case of reduction of surface thickness, the effect of concentration was less significant than the fabric weight loss. However, the

Table I. Fabric specifications.

Fabric type	Fabric count[a], warp/weft	Weight/area, g/m^2
Untreated	21/20	154
Slack mercerised	25/24	226
95% tension mercerised	24/21	180
100% tension mercerised	22/20	164

[a] Number of yarns per cm.

Table II. Moisture regain and x-ray crystallinity indices of untreated and after 4 hour treatment with 2% enzyme dosage of the linen materials

	Moisture regain, %		X-ray crystallinity index	
Fibre type	Without enzymatic treatment	4 h enzymatic treatment	Without enzymatic treatment	4 h enzymatic treatment
Untreated	6.42	6.21	0.81	0.82
Slack mercerised	7.80	7.86	0.77	0.78
95% tension mercerised	7.91	7.76	0.78	0.78
100% tension mercerised	7.57	7.31	0.79	0.80

Table III. Fabric weight loss under 1% or 2 % enzyme dosage for various periods of incubation time.

	Weight loss, %			
Fabric type	1 h	2 h	3 h	4 h
Linen, 1%	0.98	1.25	1.29	1.34
Linen, 2%	1.82	2.50	2.97	3.01
Slack mercerised, 1%	0.33	0.25	0.41	0.45
Slack mercerised, 2%	0.62	1.06	1.87	1.75
95% tension mercerised, 1%	0.92	1.01	1.28	1.21
95% tension mercerised, 2%	1.54	2.03	2.26	2.22
100% tension mercerised, 1%	0.84	0.84	0.85	0.99
100% tension mercerised, 2%	1.47	1.59	1.94	2.20

highest surface thickness loss could be found in the unpre-treated sample followed by the 95% and 100% tension mercerised samples. The observed variation of fabric surface thickness further supported the theory that macroscopic features such as fabric thickness could affect the rate of reaction especially at a short treatment time, and the denser fabric structure could also retard the access of the enzyme.

Fabric Strength

Data of breaking strength values of the untreated and enzyme treated linen are shown in Table V. When comparing the initial fabric strength of the control and the mercerised sample, it was found that the improvement of fabric strength was apparent. The increase of the fabric densities in the case of slack mercerised sample as well as the 95% tension mercerised sample confirmed that the strength loss actually occurred. According to pervious findings *(16)*, the slack mercerised flax fibres exhibited a decrease in crystallinity as a result of the unrestrained lateral swelling of the fibre. Because of the absence of primary wall to exert restraint on lateral swelling and consequent disorientation of the swollen fibres, strength of flax fibre would decrease after slack mercerisation. Nevertheless, the fibre strength would increase after tension mercerisation because the lateral movement of the polymer chains in flax was restricted, so strength losses did not occur.

The relative breaking load of the unpre-treated and mercerised linen are shown in Figure 1a and 1b. The breaking strength of the enzyme treated fabrics were reduced as the treatment time increased and significant reduction was found after one hour of incubation especially at 2% enzyme dosage. Tensile strength of the unpre-treated sample lost distinctly. For 1% enzyme dosage, strength retention ability of the mercerised samples for prolonged incubation was higher than the unpre-treated sample. In the case of cotton, the reactivity would increase after the mercerisation and made it more susceptible to chemical attack. However, the mercerisation could alter the molecular structure of cotton in such a way that it showed higher strength retention than the corresponding non-mercerised cotton when subjected to degradative treatment under identical conditions *(20)*. The higher retention of breaking strength of the mercerised samples was attributed to the removal of weak points present in the fibre structure so that the uniform structure was capable of distributing stress evenly along the fibres. However, strength retention ability was reduced as the enzyme dosage was increased to 2%. Significant reduction of breaking strength was also found in the unpre-treated sample. For the slack mercerised sample, strength reduction behaviour was similar to the unpre-treated sample which suggested that lower crystallinity could enhance the reactivity of the substrate for enzyme treatment at higher enzyme dosage and prolonged period of treatment time.

Fabric Bending Properties

Of the most commonly used fibres in apparel textile, flax is one of the fibres having the highest stiffness and lowest resilience. From the experimental results shown in Figure 2a, the increase of bending rigidity of the mercerised sample was triple under both 95% and 100% tension mercerisation, and the fabric had a stiff and hard feel because of its compact structure. As for the slack mercerised sample, the increment

Table IV. Fabric thickness of untreated and treated with 1% or 2% enzyme dosage for various periods of incubation time.

	Fabric thickness, mm														
	Untreated			1 h			2 h			3 h			4 h		
Fabric type	T2[a]	T100[b]	ST[c]	T2	T100	ST	T2	T100	ST	T2	T100	ST	T2	T100	ST
Linen, 1%	0.57	0.27	0.30	0.46	0.29	0.17	0.45	0.29	0.16	0.46	0.29	0.17	0.46	0.30	0.16
Linen, 2%				0.46	0.29	0.17	0.45	0.29	0.16	0.43	0.29	0.14	0.45	0.29	0.16
Slack mercerised, 1%	0.66	0.38	0.28	0.60	0.37	0.23	0.62	0.38	0.24	0.61	0.38	0.23	0.61	0.38	0.23
Slack mercerised, 2%				0.61	0.38	0.23	0.59	0.37	0.22	0.61	0.38	0.23	0.61	0.38	0.23
95% tension mercerised, 1%	0.60	0.36	0.24	0.50	0.33	0.17	0.52	0.35	0.17	0.53	0.35	0.18	0.50	0.34	0.16
95% tension mercerised, 2%				0.47	0.33	0.14	0.49	0.35	0.14	0.50	0.34	0.16	0.49	0.34	0.15
100% tension mercerised, 1%	0.61	0.32	0.29	0.58	0.31	0.27	0.46	0.30	0.16	0.47	0.30	0.17	0.47	0.30	0.17
100% tension mercerised, 2%				0.54	0.31	0.23	0.46	0.30	0.16	0.48	0.31	0.17	0.47	0.30	0.17

[a] Fabric thickness at 2 gf/cm^2 (196 Pa), T2, in mm.
[b] Fabric thickness at 100 gf/cm^2 (9.81 kPa), T100, in mm.
[c] Surface thickness, ST, in mm, ST = T2-T100.

Table V. Fabric breaking strength of untreated and enzyme treated with 1% or 2% enzyme dosage for various periods of incubation time.

	Breaking strength, N				
Fabric type	Untreated	1 h	2 h	3 h	4 h
Linen, 1%	709	673	651	652	593
Linen, 2%		627	564	504	441
Slack mercerised, 1%	803	758	757	749	733
Slack mercerised, 2%		708	665	612	537
95% tension mercerised, 1%	761	696	692	672	659
95% tension mercerised, 2%		661	655	628	616
100% tension mercerised, 1%	748	702	694	687	650
100% tension mercerised, 2%		636	638	540	450

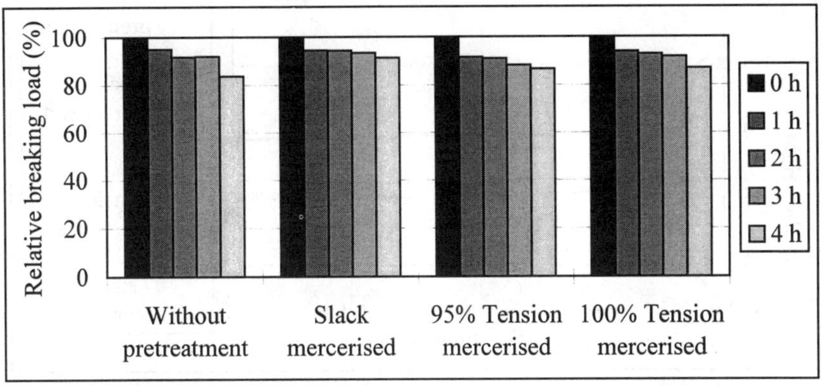

Figure 1a. Relative breaking load of unpre-treated and mercerised linen after various periods of incubation time (1% enzyme dosage).

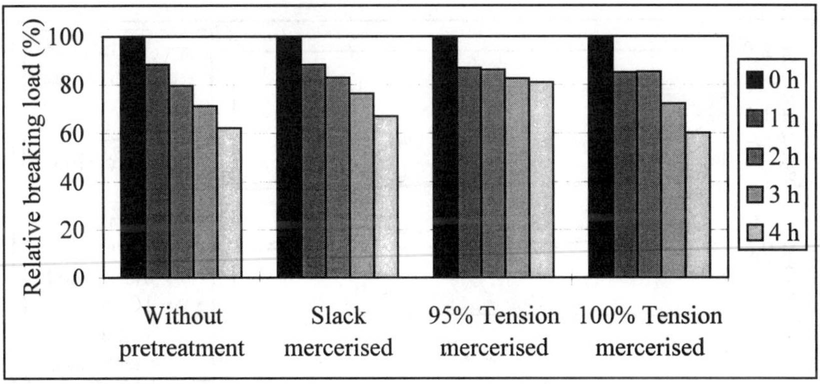

Figure 1b. Relative breaking load of unpre-treated and mercerised linen after various periods of incubation time (2% enzyme dosage).

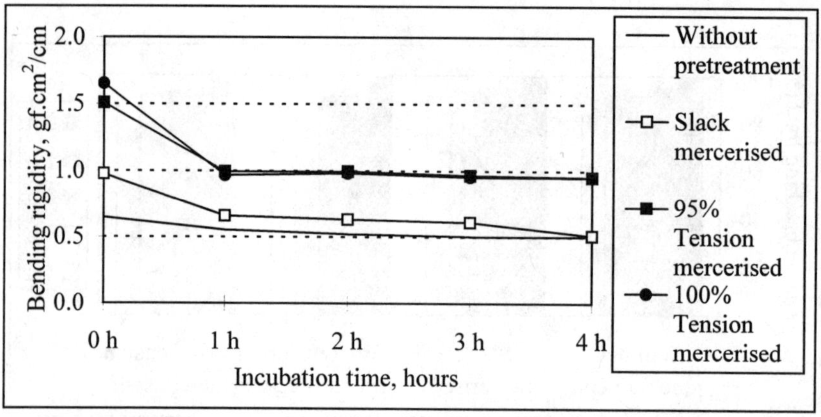

Figure 2a. Bending rigidity of unpre-treated and mercerised linen after various periods of incubation time (1% enzyme dosage).

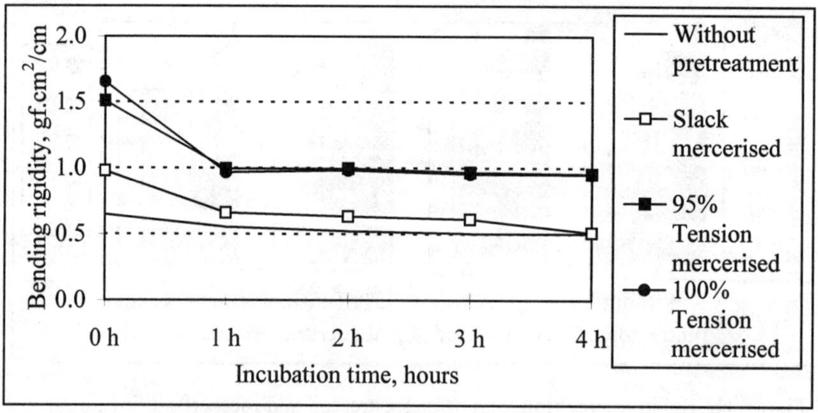

Figure 2b. Bending rigidity of unpre-treated and mercerised linen after various periods of incubation time (2% enzyme dosage).

of fabric stiffness was less predominant than the tension mercerised samples. In general, the stiffness of a fibre depends partially on the stiffness, degree of orientation and crystallinity of the polymers. As the proportion of amorphous area within a fibre becomes greater, the fibre tends to become less stiff. The more crystallinity the fibre is, the stiffer it will tend to be. Therefore, the slack mercerised sample was less stiff than the tension mercerised samples. On the other hand, fibre diameter has greater influence than the flexural rigidity of fibre. The physics involved in calculating the bending resistance as the diameter increases shows that doubling a fibre diameter will produce a 16-fold increase in resistance to bending. Likewise, a 2-fold reduction in fibre diameter decreases bending resistance by 16-fold. Furthermore, cross-sectional shape also influences the flexural rigidity of a fibre. Round fibres have high values of stiffness while flat fibres have relatively low values of stiffness *(21)*. Owing to the fact that fibre was swelled after the mercerisation process, thus the treated material would become stiffer. Work has been reported that the major disadvantage of fabric mercerisation was the large increment in fabric stiffness *(22)*. Moreover, fabric thickness and the inter-fibre or inter-yarn friction also play an important role in bending rigidity.

The results of bending rigidity of the unpre-treated and mercerised sample after various periods of incubation time are shown in Figure 2a and 2b. For the unpre-treated sample, a slight reduction of bending rigidity was found. However, up to 40% reduction of bending rigidity was found for the mercerised sample after the first hour of enzyme treatment. Even though the period of treatment was prolonged to four hours, there was no significant change in the reduction of bending rigidity. Similar extent of reduction was found even with higher enzyme dosage as shown in Figure 2b. The reduction of bending rigidity was probably due to the reduction of fibre diameter caused by the removal of surface fibrils, therefore, significant reduction of bending rigidity was found for the mercerised samples. When the bending rigidity was reduced by the application of enzyme treatment, a similar reduction of bending hysteresis could found (Figure 3a and 3b).

Fibre Surface Appearance

Appearance of the fibre surface was investigated by scanning electron microscopy (SEM). When compared to the cotton fibre, the surface of the untreated flax fibre possessed a much larger amount of loose fibrils composed of debris and adhering fragmentary tissue components as shown in Figure 4. Figure 5 showed the accumulation of the surface fibrils on the yarn surface. After four hours of enzyme treatment alone as shown in Figure 6 and 7, some of the surface fibrils still remained on the fibre surface which implied that longer incubation time will be required to remove all the surface fibrils completely.

SEM micrograph of the slack mercerised sample is shown in Figure 8. The shape of the slack mercerised fibres became rounder due to unrestricted swelling, and considerable surface fibrils still remained on the fibre surface. After four hours of enzyme treatment, a comparatively smooth and clear surface was obtained and the effect was more predominant in the case of 2% enzyme dosage as shown in Figure 9 and 10. As the mercerisation tension was increased from slack to both 95% and 100% tension, fibre cross-section was progressively changed from round to relatively flat as shown in Figure 11 and 14. Figures 12-13 and 15-16 revealed that most of the

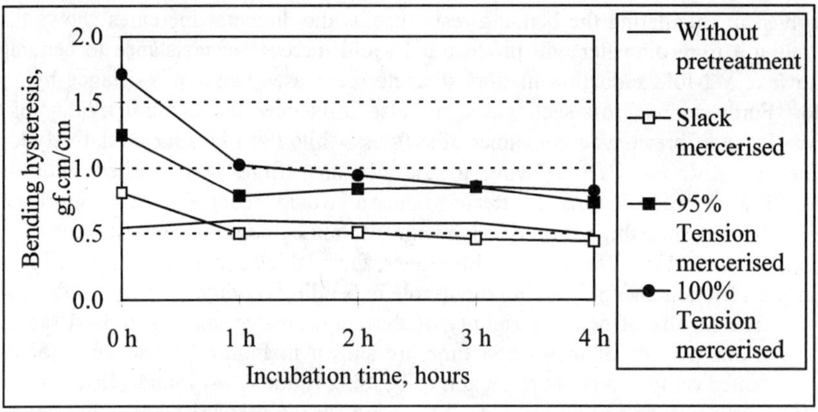

Figure 3a. Bending hysteresis of unpre-treated and mercerised linen after various periods of incubation time (1% enzyme dosage).

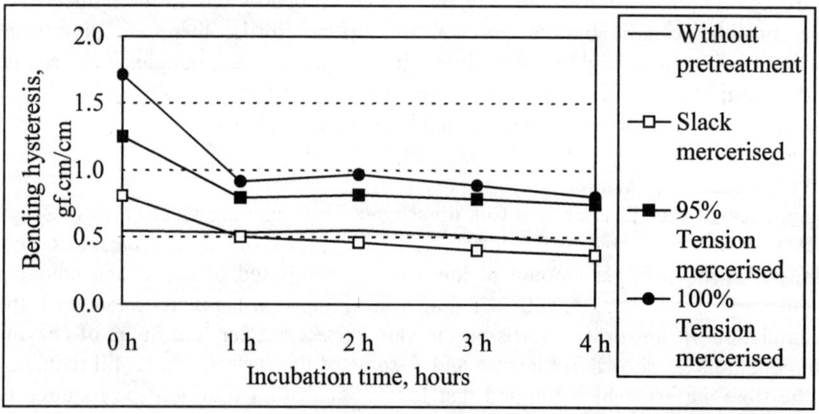

Figure 3b. Bending hysteresis of unpre-treated and mercerised linen after various periods of incubation time (2% enzyme dosage).

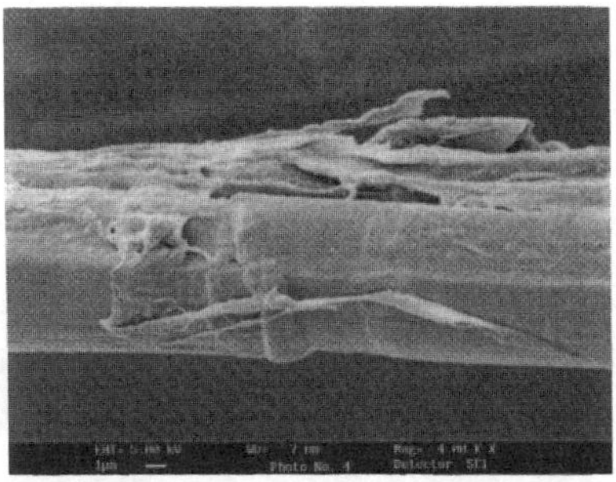

Figure 4. SEM micrograph of untreated flax fibre.

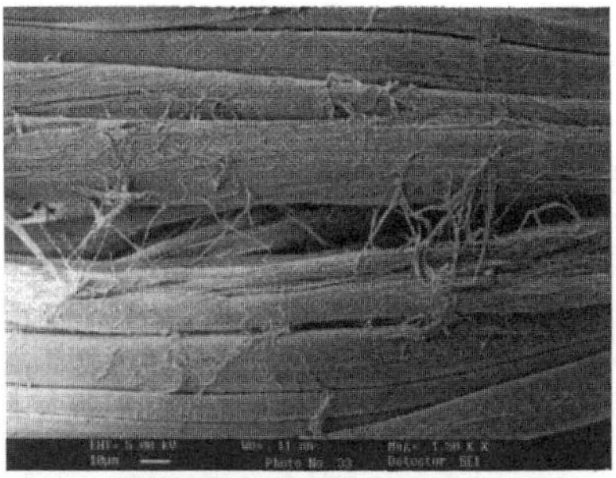

Figure 5. SEM micrograph of untreated flax yarn.

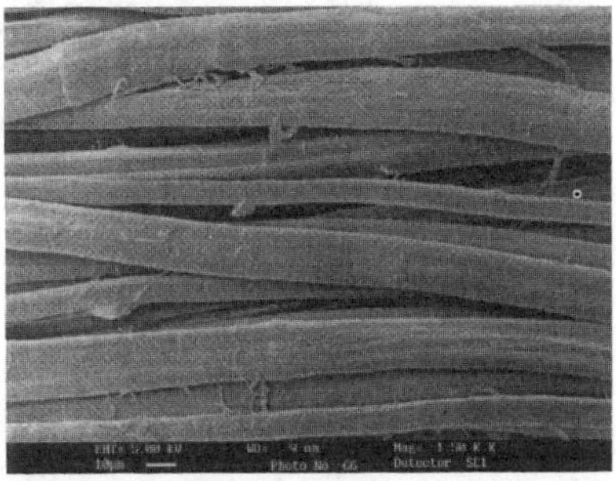

Figure 6. SEM micrograph of linen after 4 hours enzyme treatment (1% enzyme dosage).

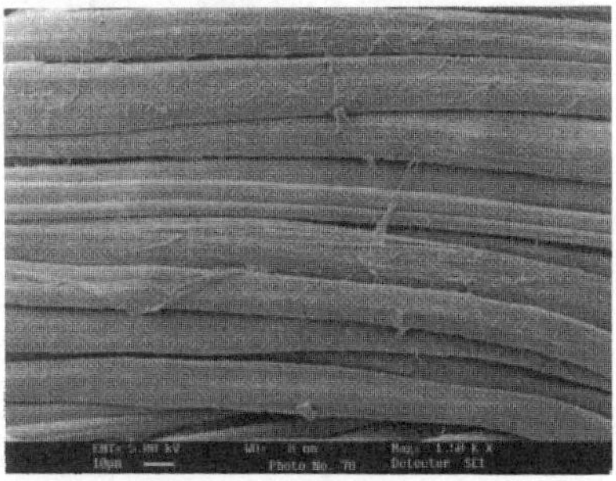

Figure 7. SEM micrograph of linen after 4 hours enzyme treatment (2% enzyme dosage).

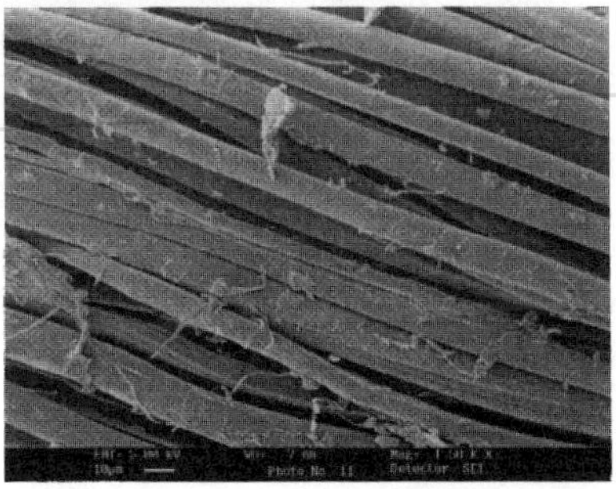

Figure 8. SEM micrograph of slack mercerised linen.

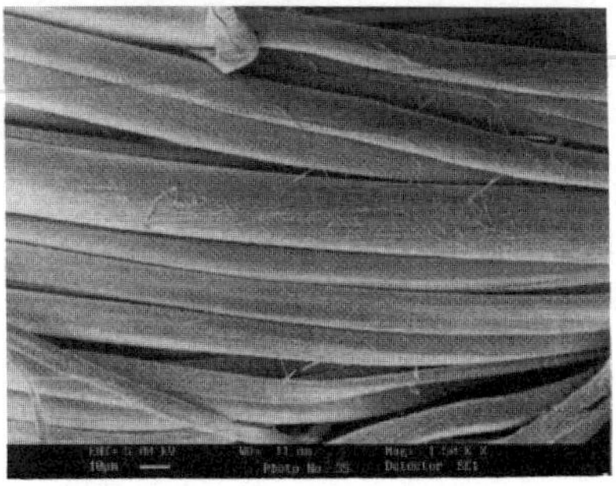

Figure 9. SEM micrograph of slack mercerised linen enzymatically treated for 4 hours (1% enzyme dosage).

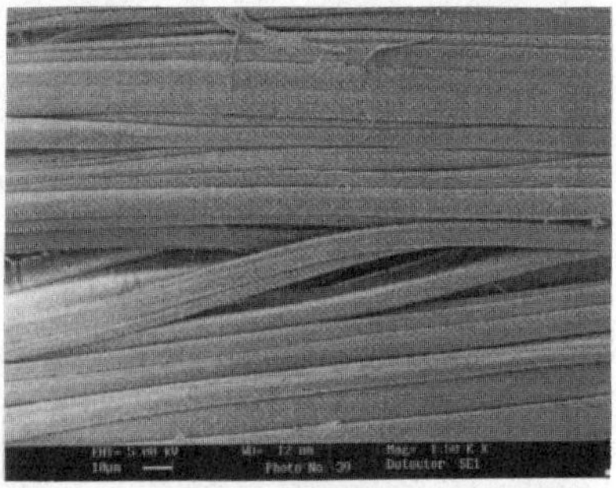

Figure 10. SEM micrograph of slack mercerised linen enzymatically treated for 4 hours (2% enzyme dosage).

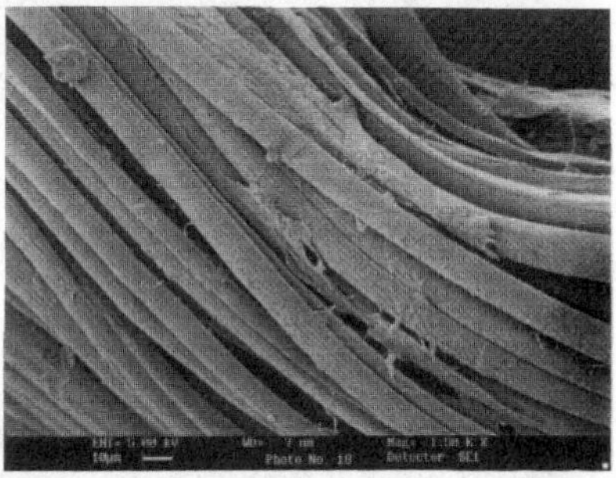

Figure 11. SEM micrograph of 95% tension mercerised linen.

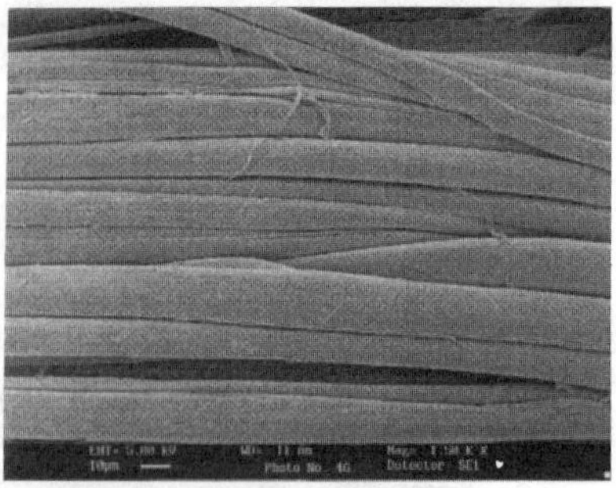

Figure 12. SEM micrograph of 95% tension mercerised linen enzymatically treated for 4 hours (1% enzyme dosage).

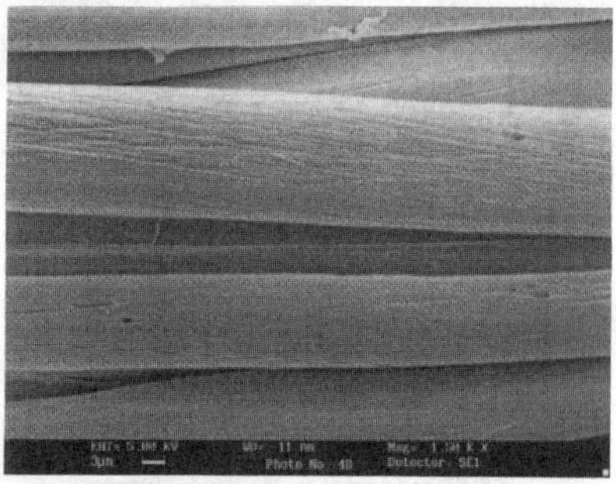

Figure 13. SEM micrograph of 95% tension mercerised linen enzymatically treated for 4 hours (2% enzyme dosage).

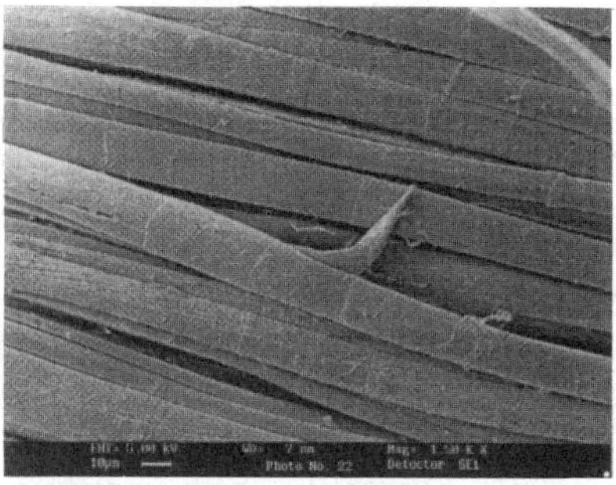

Figure 14. SEM micrograph of 100% tension mercerised linen.

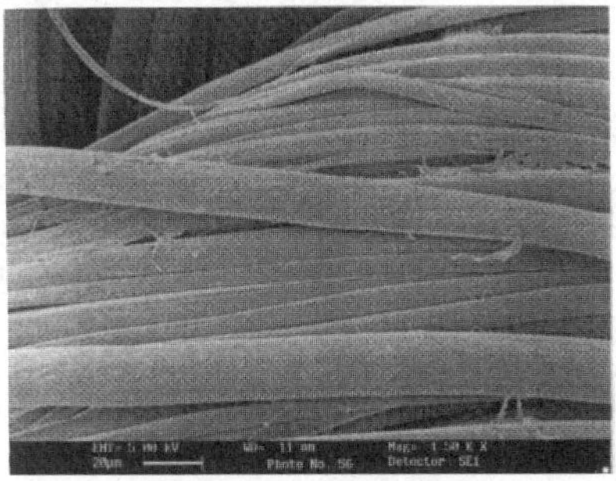

Figure 15. SEM micrograph of 100% tension mercerised linen enzymatically treated for 4 hours (1% enzyme dosage).

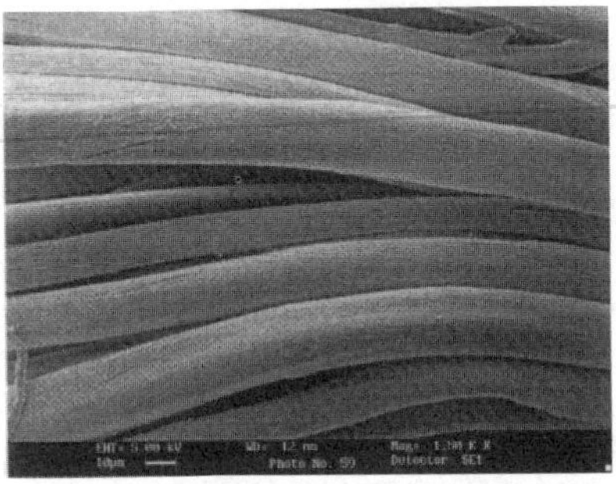

Figure 16. SEM micrograph of 100% tension mercerised linen enzymatically treated for 4 hours (2% enzyme dosage).

surface fibrillar matter was removed and fibre surfaces became smoother in the case of tension mercerised and enzymatic treated samples.

Conclusions

Reduction of crystallinity and improvement of fibre accessibility by mercerisation was employed as the pretreatment of linen material for enzymatic hydrolysis. Pretreatments under different tensions also altered the fabric sett, tensile strength and bending properties of linen. Tensile strength was increased under tension mercerisation but reduced in slack condition. In addition, large increments of bending rigidity were found especially when tension was applied during the pretreatment. Significant changes were obtained for the moisture regain and X-ray crystallinity ratio under different tensions of mercerisation. Owing to the fact that the accessible regions of the cellulosic material were often too narrow for the large enzyme molecules to penetrate effectively, both moisture absorption and the crystallinity ratio could not be further changed after the enzyme treatment. Macroscopic features such as fabric thickness and packing density also influenced the rate of enzymatic reaction. These effects were shown by the reduction of fabric weight loss and the fabric surface thickness while enzyme treatment was applied to the slack mercerised sample. Furthermore, strength retention ability during the enzymatic hydrolysis was found to be enhanced by the mercerisation pretreatment. Enzyme treatments effectively reduced the fabric stiffness of the pre-treated material which reflected that crystallinity of the material was important for the reduction of bending rigidity caused by the enzyme treatment. The fibre surface appearance was revealed by the SEM micrograph which showed that fibre surface smoothness and clearness were improved by the pretreatment. In terms of incubation time effects, significant reduction of fabric surface thickness, strength and bending rigidity were obtained after the first hour of incubation. The effect of increasing the concentration from 1% to 2% could result in higher fabric weight and strength loss, but with clearer fibre surface. However, effect was less predominant in the reduction of fabric surface thickness and bending rigidity.

Literature Cited:

1. Yamagishi, M, Kako Gijutsu, **1988,** *23,* 64.
2. Pedersen, G. L., Screws, Jr., G. A., and Cedroni, D. M., *Can. Textile J.* **1992,** *109,* 31-35.
3. Tyndall, R. M., *Textile Chem. Color.* **1992,** *24,* 23-26.
4. Rolf, S., and Petry, G., *Melliand Textilberichte*, **1995**, *76*, E253.
5. Finch, P. and Roberts, J. C. In *Cellulose Chemistry and its Applications*, Nevell, T. P. and Zeronian, S. H., Eds., John Wily & Sons, NY, 1985; 312-343.
6. Wood, T. M., In *Biosynthesis and Biodegradation of Cellulose*; Haigler C. H. and Weimer, P. J., Eds., Marcel Dekker, NY, 1991; 491-525.
7. Woodward, J., *Bioresource Technol.* **1991**, *36,* 67-75.
8. Walker, L. P. and Wilson, D. B., *Bioresource Technol.* **1991**, *36,* 3-14.
9. Focher, B., Marzetti, A., Beltrame, P. L., and Carniti P. In *Biosynthesis and*

Biodegradation of Cellulose; Haigler C. H. and Weimer, P. J., Eds., Marcel Dekker, NY, 1991; 293-310.

10. Buchle-Diller, G., Zeronian, S. H., Pan, N. and Yoon, M. Y. *Textile Res. J.* **1994**, *64,* 270-279.
11. Chikkodi, S. V., Khan, S. and Mehta, R. D., *Textile Res. J.* **1995**, *65,* 564-569.
12. Cowling E. B. In *Cellulose as Chemical and Energy Resource*, Wilke, C. R., Eds., John Wily & Sons, NY, 1975; 163-181.
13. Buchle-Diller, G., Zeronian, S. H., *Textile Chem. Color.* **1994**, *26,* 17-24.
14. Yoon, N. S. and Lim, Y. J., Tahara, M. and Takagishi, T., *Textile Res. J.* **1996**, *66,* 329-336.
15. Millett, M. A., Baker, A. J. and Satter, L. D. In *Cellulose as Chemical and Energy Resource*, Wilke, C. R., Eds., John Wily & Sons, NY, 1975; 193-219.
16. Cheek, L. and Roussel, L., *Textile Res. J.* **1989**, *59,* 478-483.
17. Kirk, T. K. In *Cellulose as Chemical and Energy Resource*, Wilke, C. R., Eds., John Wily & Sons, NY, 1975; 139-150.
18. Buchle-Diller, G. and Zeronian, S. H., *Journal of Applied Polymer Science,* **1992**, *45,* 967-979.
19. Segal, L., Creely, J. J., Martin, Jr., A. E., and Conrad, C. M., *Textile Res. J.* **1959**, *29,* 786-794.
20. Pandey, S. N. and Nair, P., *Textile Res. J.* **1987**, *57,* 532-538.
21. Hatch, K. L. In *Textile Science*, West Publishing Company: St. Paul, 1993; 113-114.
22. Murphy, A. L., Margavio, M. F., Cooper, Jr., A. S. and Welch, C. M. *Textile Res. J.* **1968**, *38,* 544-553.

Chapter 21

The Control of an Enzymatic Treatment on Linen Fabrics To Eliminate Pectic Substances

M.-H. Dézert[1,2], P. Viallier[1], and D. Wattiez[2]

[1]LPMT, Laboratoire de Physique et Mécanique Textiles, ENSITM: 11 rue Alfred Werner F-68093 Mulhouse cédex, France

[2]ITF, Institut Textile de France, Direction Régionale de Mulhouse, 185 rue de l'Illberg, F-68093 Mulhouse cédex, France

Linen fabrics are comfortable, flax fibers have good mechanical properties but linen fabrics easily crease and their dyeings have bad fastnesses. These drawbacks are related to the pectic substances, still present on fabric even after bleaching. We have applied an enzymatic treatment to partly eliminate these pectic cements instead of hydrolyzing them by a caustic soda treatment, which would damage the cellulose. The aim of our research is to continuously control the enzymatic treatment in order to stop it before a complete degradation of the mechanical properties of the fabric. After an accurate analysis of the treatment bath by chromatography, a control of the process with a specific electrode has been proposed.

The flax fiber, called technical fiber, consists of single cellulosic fibers bundles and pectic cements which are found within and around these bundles. For us pectic cements are composed of pectins and hemicelluloses. The pectic cements are amorphous materials in comparison to the crystalline cellulose of flax. Moreover, structure unevennesses can be found on the fiber surface (*1*). Finishing chemicals penetrate into the fiber thanks to these structure unevennesses and will preferably fix on the pectic cements which are more accessible than the cellulose. We have applied an enzymatic treatment because, owing to its specificity, it will not damage the cellulose. The enzymatic treatment is better too than a classical treatment from the ecological point of view. In order to specifically attack pectins and hemicelluloses, we have selected a commercial enzyme preparation which includes hemicellulase and pectinase activities. In such a non purified mixture, a low cellulase activity will be found (*2-3*). But we have verified that this low cellulase activity did not decrease the Degree of Polymerisation of cellulose (*4*) and then the single cellulosic fibers were not apparently damaged in the limits of the accuracy of the DP determination. We have chosen to control the treatment bath composition rather than the change in structure of the flax fiber. The damaged pectic substances are composed of sugars and we will be able to control the treatment by an analysis of the released sugars during the treatment.

Experimental

(Dézert, M. H.; Viallier, P.; Wattiez, D. *J. Society of Dyers and Colourists*, in press.)

According to studies of the Irish scientist Sharma on several commercial enzymatic preparations, Flaxzyme is the most suitable preparation for retting flax (*5-6*). The use of various commercial enzymes for retting flax and degumming plant bast fibers have been patented (*7-8*). Unlike the treatment we have applied on linen fabrics, all these treatments are aimed to replace the traditionnal retting or to compensate a retting which has not be carried out.

The activities of Flaxzyme, which is supplied by Novo Nordisk, have been measured thanks to the method of the reducing groups determination using the Miller reactive (*9*). We have chosen xylan from Birchwood 1% as substrate for the hemicellulase activity, sodium polypectate 0,1% as substrate for the pectinase activity and the filter paper Whatman N°1 as substrate for the cellulase activity (*10*). These substrates are dissolved in acetic acid 0,03 M / sodium acetate 0,07 M buffer. The resultant activities are 283,3 μmol xylose equivalents / min. ml enzymatic preparation for the hemicellulase activity, 25943 μmol galacturonic acid equivalents / min. ml enzymatic preparation for the pectinase activity and 0,67 units / ml enzymatic preparation for the Filter Paper Activity.

The liquor of the treatment is composed of Flaxzyme 6 ml/l diluted in acetic acid 0,04 M / sodium acetate 0,09 M buffer pH=5 40°C with 0,1% of surfactant (Tween 80). The choice of the treatment time depends on the enzyme concentration and the residual strength of the fabric. The enzymatic treatment bath must be buffered because the enzymes have optimal activities in narrow limits of pH and temperature. The temperature is regulated by the apparatus. A wetting agent is used in order to make the fabric more accessible to the enzymes and it must be a non-ionic wetting agent so as not to decrease the enzymatic activities. The apparatus is a laboratory dyeing machine with liquor circulation, supplied by Zeltex. We have worked with a liquor ratio of 1/10 and a weight of sample of 80 grams. The 100% linen fabric is pretreated as follows: desizing, boiling, bleaching, demineralisation with orthophosphoric acid (3,5 ml/l, 30 minutes at 50°C).

The Techniques Tested to Detect the released Sugars during the Treatment (*11*)

The damaged pectic cements, released in the bath during the treatment, are monosaccharides and perhaps polysaccharides. Monosaccharides, like glucose, mainly exist under two cyclic forms in solution (β-hemiacetalic form and α-hemiacetalic form) and then are found in the open aldehyde form at equilibrium only in a very weak proportion. However there is always free aldehyde enough so that the reactions with aldehyde groups can proceed.

The Classical Techniques. The determination of the refractive index with a refractometer or an interferometer, has not supplied acceptable results because of the complexity of the treatment bath. Indeed, the treatment bath consists of a mixture of different auxiliaries which have a great influence on the refractive index measurements masking the presence of the sugars.

The measures with a redox electrode were unsuccessful because these measures were disturbed by the many ions in solution due to the buffer. Indeed, these ions disturbed the equilibrium between the cyclic and linear forms, for each released sugar, which made the redox determination impossible.

The polarimetric method has not been tested because the light rotation of substances like glucose and sugars in general, is too long to stabilize, which would make a continuous control of the enzymatic treatment difficult.

A Discontinuous Control of the Enzymatic Treatment. We have determined the reducing groups of the released sugars with the Miller reactive, which has been used for the measurement of the activities in the enzymatic preparation. The time evolution which has been obtained, is shown in Figure 1.

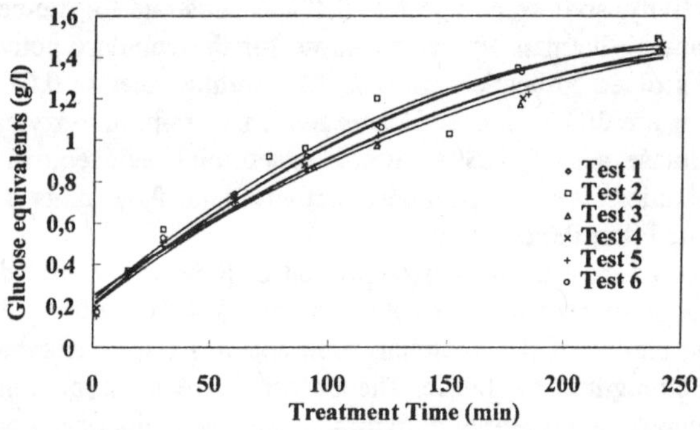

Figure 1. Time evolution of the released reducing sugars concentration during the enzymatic treatment (Reproduced with permission from ref.1 Copyright 1997 Journal of the Society of Dyers and Colourists).

This time evolution allows us to choose the treatment time as a function of the enzyme concentration and the residual strength of the fabric. However this method supposes that all released sugars have a reducing aldehyde function whereas cetoses can be released as well. Moreover, the result of the colorimetric determination does not depend on the chain length of the sugars, whose aldehyde group is measured. In addition, this method can not be used for a continuous control because each measure takes 30 minutes.

The Carbon Analyses. The enzymatic treatment, which is aiming at hydrolysing pectic cements, produces water soluble sugars. The determination of carbon in the aqueous phase should allow us to control the hydrolysis, all conditions being identical. This method is mentioned in the literature to control cellulase treatment on cotton fabric (*12*). With carbon analyses, all chemical species with carbon are detected. The carbon analyzer used is the analyzer IONICS 1555B (USA). As shown in Figure 2, the time evolution is very irregular.

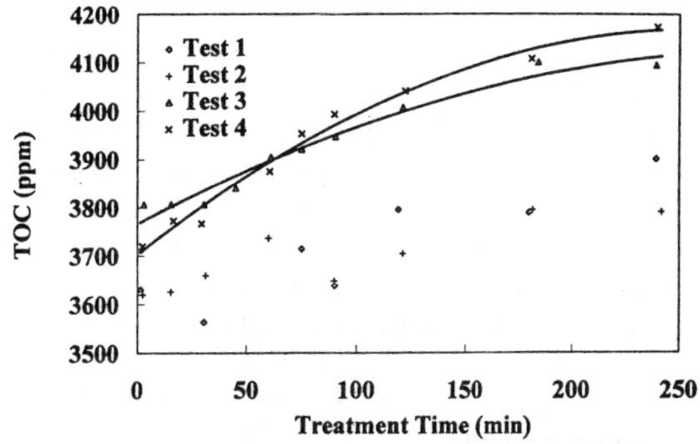

Figure 2. Time evolution of the TOC concentration during the enzymatic treatment (Reproduced with permission from ref.2 Copyright 1997 Journal of the Society of Dyers and Colourists).

The buffer, the wetting agent and the enzyme have a great influence on the carbon concentration of the bath samples which have been withdrawn during the enzymatic treatment. The measurements are particularly inaccurate during the first stages of enzymatic attack. We can attribute it to enzyme adsorption and enzyme desorption phenomenons. A sample preparation before analysis seems to be necessary. So, we have tried to separate the enzymes from the released sugars by an ultrafiltration of samples.

Carbon Analyses on Ultrafiltered Samples. We have selected ultrafiltration units fitted out with a polysulfone membrane of a 10kDa cut-off, which have enabled us to ultrafilter small volumes of treatment bath. These units have been chosen according to their efficiency to retain the enzymatic preparation. This efficiency has been tested by determining the protein concentration in the permeate (*13*). The results of the carbon measures on the ultrafiltered samples are shown in Figure 3.

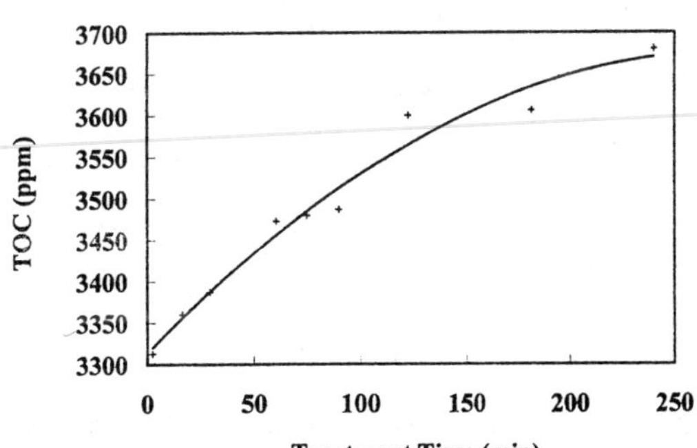

Figure 3. Time evolution of the TOC concentration during the enzymatic treatment with ultrafiltered samples (Reproduced with permission from ref.3 Copyright 1997 Journal of the Society of Dyers and Colourists).

We obtain a more regular evolution. But substances like stabilizing agents or buffer, which are present in the enzymatic preparation, are not retained by the 10 kDa cut-off membrane. Consequently the ultrafiltration does not improve the accuracy of the carbon measurements enough and sugars are still weakly involved in the measurements. However the membrane does not retain the released sugars as we have controlled it by determining the reducing sugars concentration, with the Miller reactive, in the permeate of the ultrafiltered samples. The subsequent measures by chromatography will confirm that the sugars released during the enzymatic damage are mainly monomers.

The Analyses by Chromatography

All the techniques, mentioned above, do not supplied suitable results for an accurate control of the enzymatic treatment. That is why we have decided to analyze the treatment bath composition by chromatography.

We have selected an ion-exchange column, supplied by Waters, which has allowed us to work with water as eluting solvent. This type of column can give rise to size exclusion phenomenons as well. We can predict what sort of sugars are released in the bath during the treatment thanks to the sugars composition of the pectic cements given by the literature (*14*). Thus we can calibrate the method by injecting each sugar at different concentrations (*15*). The type of chromatogram we have obtained after a 4 hour enzymatic treatment, for example, is shown in Figure 4.

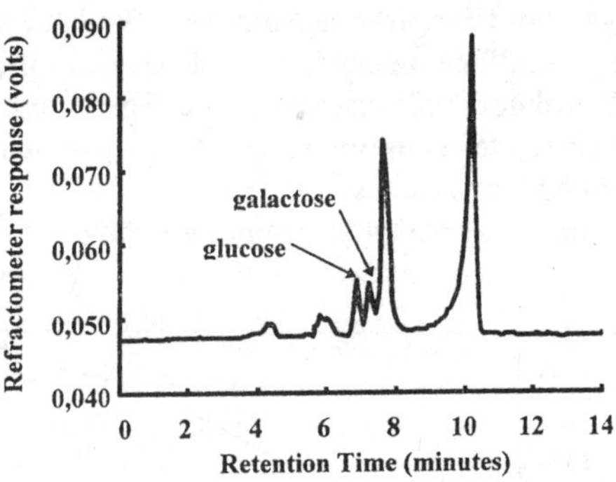

Figure 4. Chromatogram of the treatment bath after 4 hours.

We obtain the same type of chromatogram all through the enzymatic treatment. Glucose and galactose are mainly released during the treatment but the proportion of glucose/galactose changes. The two greatest peaks are due to the reference bath composed of the enzyme, the buffer and the wetting agent. The smaller peaks before the glucose one's can be due to the dimers or the trimers of glucose and/or galactose since the column implies size exclusion phenomenons. The presence of cellobiose has been only proved. This cellobiose is the result of a partial hydrolysis of cellulose, due to the cellulase activity.

Galacturonic acid and rhamnose, which can be found in the composition of pectins are not detected. Consequently only hemicelluloses remain on the fabric before the enzymatic treatment.

The residual tensile strength of the fabric with respect to the grey fabric has been studied as a function of the treatment time. The results are shown in Figure 5.

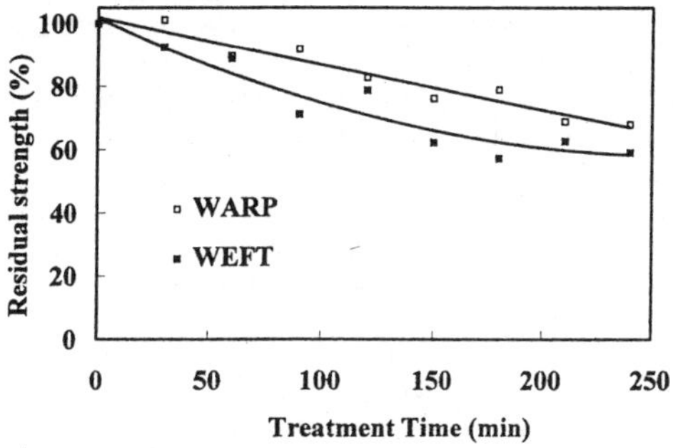

Figure 5. Evolution of the fabric tensile strength during the enzymatic treatment.

We can observe that the strength of the fabric decreases all through the treatment and has decreased by around 35% after 4 hours. It would be interesting to determine a relation between the decrease of the strength and the increase of the galactose concentration in the bath.

We have applied the same enzymatic treatment on six other linen fabrics and we have analyzed by chromatography the bath samples which have been withdrawn during the treatment. The same type of chromatogram as the one shown in Figure 4, is obtained. Therefore we have confirmed that two main sugars, glucose and galactose, are released during the treatment and that the substances we want to eliminate are hemicelluloses.

Conclusion

A simple industrial method to continuously control the enzymatic treatment does not exist. The auxiliaries of the treatment bath have a great influence on the measurements masking the presence of the released sugars. The chromatographic analyses have proved that the substances we want to eliminate are hemicelluloses and that glucose and galactose are mainly released during the treatment. The only conceivable technique is to develop a specific industrial electrode for the detection of galactose to continuously control the treatment. We prefer to control the release of galactose rather than glucose because a part of glucose can be due to a partial hydrolysis of cellulose owing to the low cellulase activity. The content of galactose on the fabric before the enzymatic treatment will have to be known as long as we do not eliminate completely the pectic cements. Indeed, a complete elimination of the pectic substances would imply insufficient mechanical properties for the fabric.

Literature Cited

1. Sotton, M.; Satta, A.; Hagege, R. *Bull. Scien. de l'ITF* **1986,** *15 (59),* pp. 3.
2. Hugo, W. B.; Mityal Eranil-Tokay, D. *Pharm. Acta Helv.* **1970,** *45 (8),* pp. 480.
3. Biely, P.; Vrsanska, M.; Kucar, S. In: *Xylans and Xylanases;* Vissers et al., Eds.; Progress in Biotechnology; Elsevier: Amsterdam, 1992, Vol. 7; pp 81-94.
4. Dézert, M. H. *Contribution à la mise au point d'une technique de contrôle en continu d'une élimination enzymatique des matières pectiques du lin sur étoffe;* thèse de doctorat, Université de Haute Alsace, Mulhouse, Fr., 1997; pp. 86.
5. Sharma, H. S. S. *The genetic engineer & biotechnologist* **1992,** *12 (2),* pp. 19.
6. Sharma, H. S. S. *International Biodeterioration* **1987,** *vol. 23,* pp. 181.
7. Novo Industri, A/S; Sumere *European Patent* **1987,** 0 220 913.
8. Helmic., Inc., Jaskowski *US Patent* **1986,** 4 617 383.
9. Miller, G. L. *Analytical Chemistry* **1959,** *31 (3),* pp. 426.
10. Ghose, T. K. *Inter. Union of Pure & Applied Chemistry* **1987,** *59 (2),* pp. 257
11. Dézert, M. H. *Contribution à la mise au point d'une technique de contrôle en continu d'une élimination enzymatique des matières pectiques du lin sur étoffe;* thèse de doctorat, Université de Haute Alsace, Mulhouse, Fr., 1997; pp. 88.
12. Ueda, M.; Yoshimura, Y.; Kumita, K.; Morita, M. In *Cellulosics: Pulp, Fibre and Environmental Aspects;* Kennedy, J. F.; Phillips, G. O.; Williams, P. A., Eds.; Ellis Horwood: Ltd, U.K., 1993, pp 187.
13. Bradford, M. M. *Analytical Biochemistry* **1976,** *72,* pp. 248.
14. Morvan, C.; Abdul-Hafez, A.; Morvan, O.; Jauneau, A.; Demarty, M. *Plant Physiol. Biochem.* **1989,** *27 (3),* pp.451.
15. Dézert, M. H. *Contribution à la mise au point d'une technique de contrôle en continu d'une élimination enzymatique des matières pectiques du lin sur étoffe;* thèse de doctorat, Université de Haute Alsace, Mulhouse, Fr., 1997; pp. 149.

Chapter 22

Enzymatic Retting of Flax

D. E. Akin[1], G. Henriksson[2], W. H. Morrison III[1], and Karl-Erik L. Eriksson[2]

[1]Russell Research Center, ARS, U.S. Department of Agiculture, P.O. Box 5677, Athens, GA 30604
[2]Department of Biochemistry and Molecular Biology, Center for Biological Resource Recovery, University of Georgia, Athens, GA 30602–7229

Industrial interest is driving a renewal of a flax and linen industry in the US. A major limitation to production of fiber from flax stems is retting, which is a microbial process involving partial degradation of the plant tissues for isolation of cellulosic fibers. Enzymatic retting of flax has been proposed as a means of replacing the current practice of dew-retting, thereby improving the quality and consistency of flax fiber for use in textiles. Since costs were too high for enzymatic retting by the previously proposed procedures, new formulations and methods were tested for improved efficiency of retting. Commercial enzyme mixtures (i.e., Flaxzyme and Ultrazym) at various concentrations and experimental supernatants from fungi isolated from dew-retted plants were tested for retting efficiency by visual (Fried test), analytical, and microscopic methods. Factors that increased retting efficiency at the laboratory level were: specific enzyme mixtures, addition of the chelators oxalic acid and ethylenediamine-tetraacetic acid to the enzymes, increased temperature, and mechanical disruption of flax stems to increase surface area for enzymes. Use of these methods and procedures should increase the efficiency of retting and reduce costs for commercial application.

Bast fibers are produced in the cortical regions of plant stems. Examples of bast fiber plants are jute, ramie, hemp, kenaf and flax *(1)*. Flax, which is the source of linen, is perhaps the oldest and best known of the bast fibers used for textiles. Linen and other bast fibers are obtained from the plant stems by a process called retting. Retting, which is usually a microbial process, is the partial degradation of the stems, resulting in the isolation of the cellulosic fibers from the other, non-cellulosic components *(2)*. In past times, flax stems were submerged in rivers and lakes, and anaerobic bacteria retted the plants. However, this practice was discontinued in western countries several decades ago because of the pollution from fermentation products and the high cost of drying. Currently, dew-retting is the accepted practice in European countries, which supply much of the linen used in textiles. In dew-retting, flax plants are pulled from the soil and laid out in fields for selective attack by indigenous fungi for several weeks. Disadvantages of dew-retting are: 1) dependence on particular geographical regions that have the appropriate moisture and temperature ranges for retting, 2) coarser and lower quality fiber than with water retting, 3) less consistency in fiber characteristics, and 4) occupation of agricultural fields for several weeks *(2)*.

Retting remains the major problem with production of flax fibers for linen. In the 1980's, research was undertaken in Europe to develop enzymatic retting as a method to replace dew-retting. Despite the production of Flaxzyme, a commercial enzyme mixture from Novo Nordisk (Denmark), and several patents *(2)*, no commercial process has been reported and dew-retting is still the practice most widely used in Europe. Cost of the enzymes, and perhaps other less obvious reasons, prevented development of a commercial enzyme retting process.

Although the US is the largest per capita user of linen, only small amounts of flax are grown in this country and only one company is currently processing (i.e., dew-retting and scutching) flax. The US Flax Initiative, which is a consortium of business and state and federal research entities, was formed in 1996 to promote a flax/linen industry in the US. Part of this effort is to research major problems related to economic production of high and consistent quality flax fiber. We have organized a research effort toward the goal of developing an enzymatic retting process for flax/linen fiber. In this paper, we review results from earlier published studies, with supplementation of unpublished results from new studies.

Structure of Flax Stems

Bast fibers are located in the cortical regions between the epidermis/cuticle and the core as shown in the scanning electron micrograph of the flax stem cross-section (Figure 1). Bundles are comprised of several ultimate fibers, which have thick secondary walls and small lumens at maturity. Thin-walled parenchyma cells surround the bundles, occupying the inner area (i.e., towards the core), outer area, and regions between bundles. Cambium is located between the bast fibers and the core. Histochemical examination of cross-sections of flax stems stained with ruthenium red, which has been used to locate pectin in plants *(3)*, showed intense positive reactions in epidermis, parenchyma, and cambium cells. A less intense reaction occurred in the secondary walls of the fibers, and middle lamellae, which bind the ultimate fibers together, gave stronger reactions. Transmission electron micrographs show structural characteristics of the epidermis and parenchyma cells (Figure 2) and the fibers (Figure 3), which are closely associated and connected by an electron dense middle lamella.

Modifications in Structure and Chemistry after Retting

Effective retting separates fiber bundles from the epidermis/cuticle and the core (Figure 4). In addition, fiber bundles are divided into smaller bundles and at times into ultimate fibers. Transmission electron micrographs show attack by indigenous fungi, which primarily carry out dew-retting, on the middle lamella of fiber cells walls (Figure 5). The fibers are released from the non-fiber components, and the middle lamellae are degraded to produce ultimate fibers.

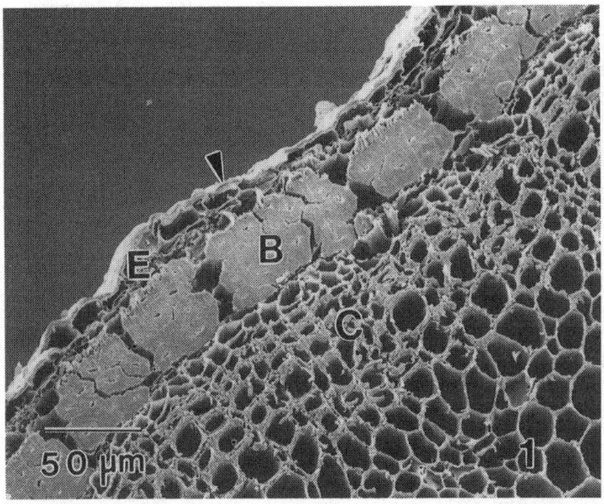

Figure 1. Scanning electron micrograph of unretted flax stem showing bast fiber bundles (B), epidermis (E) with cuticle (arrow) on the outer side, and core (C). Bar = 50 µm. From Akin et al. *(6)*.

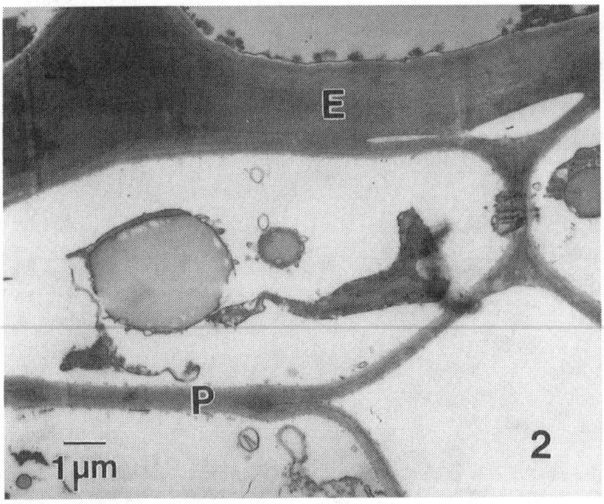

Figure 2. Transmission electron micrograph of unretted flax stem showing thick-walled epidermis (E) and thin-walled parenchyma (P) cells. Bar = 1 µm. From Akin et al. *(6)*.

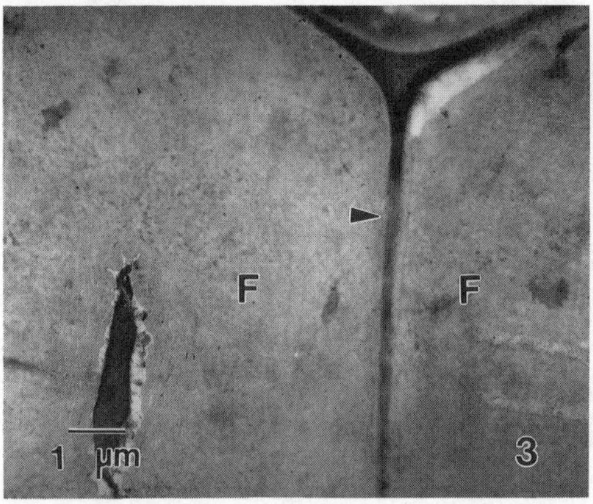

Figure 3. Transmission electron micrograph of unretted flax stem showing thick-walled fiber cells (F) with electron dense middle lamella (arrow). The cell lumen is small in mature fibers. Bar = 1 μm. From Akin et al. *(6)*.

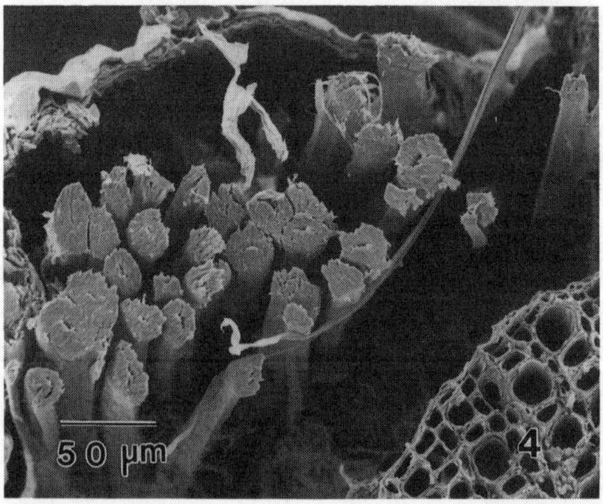

Figure 4. Scanning electron micrograph of dew-retted flax stem showing separation of fiber bundles from epidermis and cuticle and the core and separation into smaller groups and to ultimate fibers. Bar = 50 μm. From Akin et al. *(6)*.

The analysis of structural carbohydrates that comprise plant cell walls in flax stem and modifications after retting are shown in Table I. Components representative of pectins (e.g., uronic acids), arabinose, and xylose were reduced in amounts, while levels of glucose (indicative of cellulose), mannose, and galactose increased after retting. These data, as well as those from other workers *(4, 5)*, suggest that the cellulosic fibers have non-glucose components within the fiber structure. Aromatic compounds are low in amounts in the bast fibers but were localized using microspectroscopy in the middle lamellae and particularly in the cell corners of bast fibers before and after dew-retting *(6)*. The nature of these aromatic components are not well defined *(7*; Gamble, personal communication), and some appear to have tannin-like properties.

Enzymatic Retting

We tested commercial enzymes with high levels of pectinase and which have been previously used (e.g., Ultrazym), or expressly developed (e.g., Flaxzyme), for enzymatic retting of flax *(8)*. We used scanning electron and light microscopies and the Fried test to assess fiber separation from the stems *(9)*. The Fried test, which scores retting efficiency based on standard images, has been used to evaluate enzyme retting *(2)*. For these initial tests, twelve 9 to10 cm long sections of flax stems were incubated in screw-cap vials with 13 ml sodium acetate buffer, pH 5.0, in an end-over-end fashion. Various enzymes and levels, chelators, temperature, and mechanical pretreatments were among the variables tested.

The commercial enzyme mixtures are high in pectinolytic activity and also have xylanase and cellulase activities *(2,8)*. We found that experimental cultures with high pectinase activity but without substantial xylanase and cellulase activities were very efficient in retting flax. It is clear from our work as well as others *(2)* that the presence of pectinases in enzyme mixtures is of paramount importance for effective retting. Scanning electron microscopy had indicated that Ultrazym and Flaxzyme were effective in separating fibers where the plant site was exposed to enzymes but did not ret effectively when stems were more intact. The structural modifications effected by Flaxzyme are shown in Figure 6. Flaxzyme appeared to isolate fibers more effectively than Ultrazym in these studies *(9)*, although other work had indicated that Ultrazym outperformed Flaxzyme with some flax samples *(8)*. Our results documented, as expected, that mechanical pretreatment to fracture and open the stem surface increased the efficiency of enzymatic retting with Flaxzyme also *(10)*. Increased temperature also enhanced the activity of cell wall-degrading enzymes, with retting efficiency at incubation temperatures of 40 to 50 °C about 2 times faster than that at 22 °C *(10)*.

A major disadvantage of enzymatic retting is cost of the enzyme required to effectively ret flax stems *(2)*. Therefore, methods to reduce the level of enzyme required would enhance the chances of success of a commercial process. We evaluated a series of chelators added to pectinase mixtures for the increased effectiveness of retting. We are not aware of other work in which chelators have been added to enzyme mixtures for flax retting, although other work *(11, 12)* indicated that pretreatment with chelators increased retting by fungi. Table II indicates that oxalic acid at 50 mM concentrations substantially increased the effectiveness of retting by Flaxzyme or Ultrazym. This level

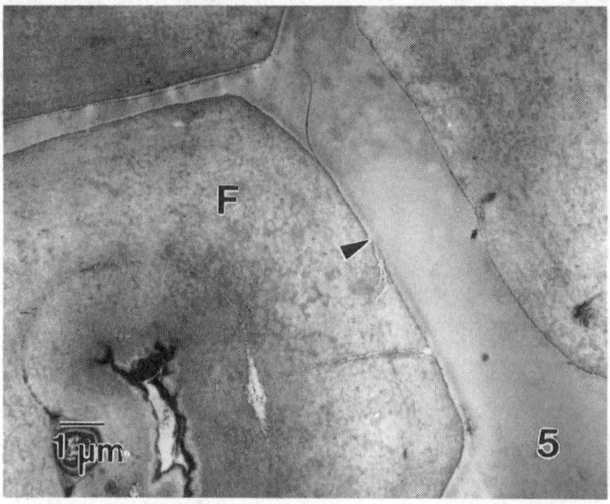

Figure 5. Transmission electron micrograph of dew-retted flax stem showing loss of middle lamella (arrow) and separation into ultimate fibers (F). Bar = 1 µm. From Akin et al. *(6)*.

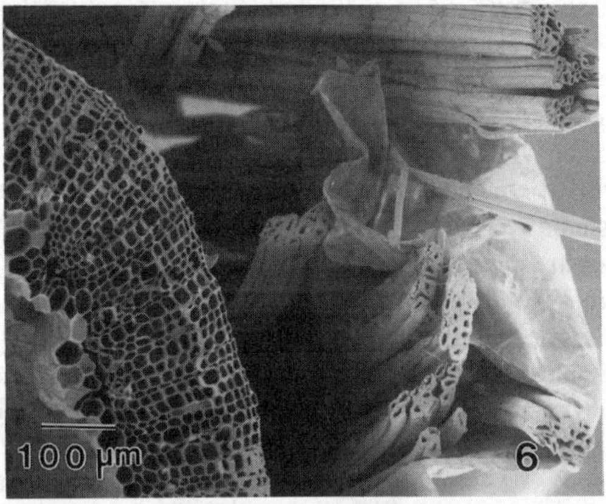

Figure 6. Scanning electron micrograph of enzymatically retted flax stem with 4% (dilution of commercial product as supplied by Novo Nordisk) showing separation of fiber bundles from core and cuticle. Bar = 50 µm. From Akin et al. *(9)*.

chelators would facilitate the action of pectinases within the epidermal wall structure, enhance disorganization of the stem tissues, and increase retting efficiency. Further work is required to confirm this theory and, if true, to maximize the influence of enzyme/chelator formulations.

Preliminary results confirm that retting of mechanically disrupted flax stems with 0.05 % Flaxzyme (% of undiluted commercial product as supplied by Novo) plus 50 mM EDTA produces fibers of similar fineness (based on relative air-flow values) to that with 0.3 % Flaxzyme without chelators. The strength of fibers derived by the former method were about 38% higher than those of Flaxzyme without chelators based on stelometer measurements (i.e., g/tex). These results support the contention that chelators enhance the activity of Flaxzyme, thus facilitating flax retting and reducing amounts of enzymes required.

Conclusions

Enzymatic retting in larger scale is required to assess the commercial feasibility of these procedures discussed above. Further, research on fungal filtrates of higher activity than Flaxzyme in combination with other chelating agents should be continued to acquire enzyme mixtures of maximum retting efficiency. Our preliminary results indicate that the following aspects warrant consideration in any commercial process for enzymatic retting of flax:

1) mechanical disruption of the surface (epidermis/cuticle) of the flax stem to increase surface action of the enzyme mixtures *(10)* is required to obtain optimal retting,
2) the addition of chelators (i.e., oxalic acid and EDTA) to commercial and experimental pectinase mixtures substantially increases the efficiency of enzymatic retting, and
3) increasing the temperature to around 40 °C increases the retting efficiency of Flaxzyme/oxalic acid over that at ambient temperature *(10)*.

Literature Cited

1. van Dam, J.E.G.; van Vilsteren, G.E.T.; Zomers, F.H.A.; Shannon, W.B.; Hamilton, I.T. *Industrial Fibre Crops. Increased Application of Domestically Produced Plant Fibres in Textiles, Pulp and Paper Production, and Composite Materials;* ATO-DLO: Wageningen, The Netherlands, 1994; 249pp.
2. Van Sumere, C.F. In *The Biology and Processing of Flax*; Sharma, H.S.S., Van Sumere, C.F., Eds.; M Publications: Belfast, Northern Ireland, UK, 1992; pp 157-198.
3. Jensen, W.A. *Botanical Histochemistry;* W.H. Freeman and Co.: San Francisco, CA, 1962; pp 201-205.
4. Gorshkova, T.A.; Wyatt, S.E.; Salnikov, V.V.; Gibeaut, D.M.; Ibragimov, M.R.; Lozovaya, V.V.; Carpita, N.C. *Plant Physiol.* **1996,** 110, 721-729.
5. Stewart, D.; McDougall, G.J.; Baty, A. *J. Agric. Food Chem.* **1995,** 43, 1853-1858.
6. Akin, D.E.; Gamble, G.R.; Morrison, W.H., III; Rigsby, L.L.; Dodd, R. B. *J. Sci. Food Agric.* **1996,** 72, 155-165.
7. Love, G.D.; Snape, C.E.; Jarvis, M. C.; Morrison, I.M. *Phytochemistry* **1994,** 35, 489-491.

8. Sharma, H.S.S.; Van Sumere, C.F. *Genetic Engin. & Biotechnol.* **1992,** 12, 19-23.
9. Akin, D.E.; Morrison, W.H., III; Gamble, G.R.; Rigsby, L.L.; Henriksson, G.; Eriksson, K.-E.L. *Textile Res. J.* **1997,** 67, 279-287.
10. Henriksson, G.; Akin, D.E.; Rigsby, L.L.; Patel, L.; Eriksson, K.-E.L.; *Textile Res. J.* **1997,** In press.
11. Brown, A.E.; Black, D.L.R. *Trans. Br. Mycol. Soc.* **1985,** 85, 625-630.
12. Bratt, R.P.; Brown, A.E.; Black, D.L.R. *Records Agric. Res.* **1986,** 34, 9-15.
13. Jauneau, A.; Cabin-Flaman, A.; Verdus, M.-C.; Ripoll, C; Thellier, M. *Plant Physiol. Biochem.* **1994,** 32, 839-846.
14. Rihouey, C.; Jauneau, A.; Cabin-Flaman, A.; Demarty, M.; Lefebvre, F.; Morvan, C. *Plant Physiol. Biochem.* **1995,** 33, 497-508.

Chapter 23

Enzymes for Wool Fiber Modification

E. Heine, B. Hollfelder, W. Lorenz, H. Thomas, G.Wortmann, and H. Höcker

German Wool Research Institute, Aachen University of Technology, Veltmanplatz 8, 52062 Aachen, Germany

With the aid of enzymes and combined processes handle, dyeability and degree of whiteness of wool were improved and the feltability was reduced. The molecular origin of these effects was studied by monitoring the wool protein pattern using gel electrophoresis. It was deduced that among others also peptide bonds in the intermediate filaments are cleaved by the enzyme catalytic action. Changes in wool morphology were documented by using TEM. By the protease catalytic effect the cell membrane complex is modified and the endocuticle of wool is removed.

Wool as a natural, animal derived product is an environmentally friendly, renewable textile material. There is a broad range of finishing processes aiming at converting wool into different kinds of clothing material and home textiles (*1*). Some of these processes still include the use of chlorine as a fiber-modifying agent leading to environmentally unfriendly adsorbable organic halogen (AOX) by-products in the effluent of these plants (*2*). Therefore research work to establish alternative processes for wool fiber modification is performed in the field of physical treatments like plasma and corona (*3*) but also in the field of biochemical treatments like enzymatic processes as forward-looking technologies. The application of enzymes for textile finishing processes was reviewed (*4*). In the present paper results concerning wool fiber modification by the use of enzymes and combined processes are given.

Wool morphology. Wool as a biological composite material is built up of two types of cells: cortex cells and surrounding cuticle cells. Each cuticle cell consists of three layers with different amounts of cystine and isodipeptides: the outer a-layer, the exocuticle, and the endocuticle. A cell membrane complex (cmc) is found between overlapping cuticle cells, between cuticle and cortex and between cortex cells. The surface of the cuticle cells is highly hydrophobic due to covalently bound fatty acids.

Furthermore the a-layer and the exocuticle contain high amounts of cystine. A cortex cell is composed of 5 - 8 macrofibrils. Compared to the macrofibrils the intermacrofibrillar material (cytoplasmic and nuclear remnants) swells in water to a larger extent. One macrofibril is composed of 500 - 800 Keratin intermediate filaments (KIF). Keratin-associated proteins (KAP) surround the intermediate filaments and their aggregates. The central rod-like domains of the Keratin intermediate filaments are microcrystalline (α-Keratin). The other wool components like cuticle, cmc, intermacrofibrillar material, interfilament material and parts of the intermediate filaments build up the matrix phase (*1*).

Experimental

Materials. Wool top (merino 20.9 µm), woven (test fabric, plain weave, 127 g/m^2) and knitted fabrics (tex 25x1, 250 g/m^2) were used. Proteases (Subtilisin-type) were supplied by NOVO Nordisk A/S, Bagsvaerd Denmark. Lipase preparation from Pseudomonas spec. was obtained from SIGMA.

Enzyme treatment. The enzyme treatment of the wool material was performed in an Ahiba Turbocolor dyeing machine at a liquor ratio of 1:20 and 1:40.

Photobleaching. Wool samples were bleached for 8 h with blue light in the dry and in the wet state (*5*).

Pretreatment of wool. Plasma treatment and chemical oxidation as pretreatment procedures prior to enzyme treatment were performed as described previously (*6*).

Gel electrophoresis. For electrophoresis the wool samples were extracted according to ref. (*7*). The electrophoretic protein separation was performed as described in ref. (*8*). The visualization of the protein bands in the gel was carried out with Coomassie Brilliant Blue R-250 as described in ref. (*9*).

HPDSC. Thermoanalytical measurements were performed on a differential scanning calorimeter (Perkin-Elmer DSC 7) with a heating rate of 10 °C per min in a temperature range of 0 - 200 °C in sealed pans in the presence of water (8 - 10 mg wool and 50 µl distilled water).

TEM. Transmission electron microscopy was performed on an EM 109 (Zeiss) at an acceleration voltage of 50 kV. As a contrasting method, wool cross-sections were stained with uranyl acetate and lead citrate according to ref. (*10*).

Fiber diameter. Wool fiber diameters were determined by using an optical fiber diameter analyser (OFDA). 4000 fibers were measured per sample.

Felting test. The felting test was performed according to IWS test method 31 (TM 31).

Dyeing test. The dyeing test was performed with Lanasol Blue 8G (CIBA) at a liquor ratio of 1:30 in an Ahiba Turbocolor dyeing machine.

Colour measurements. The colour depth of the dyed wool samples was measured at a Datacolor System (Marl).

Handle evaluation. The handle of wool fabrics was measured by using a KAWABATA Evaluation System Fabrics (KES-FB).

Results

Wool was treated with proteases concerning different aspects of wool finishing like handle improvement, enhancement of the degree of whiteness and dyeability, and reduction of feltability. To achieve these aims enzymes were not only applied to untreated wool but also to chlorinated, plasma treated and photochemically bleached wool.

Handle Improvement. Chlorinated wool was used not ignoring the environmental aspects but solely for fundamental studies. It had been observed that the application of proteases especially to chlorinated wool led to a handle improvement (*11*). It was one aim of the study to determine the origin of this handle improvement. One approach was the recording of the protein patterns of the differently treated wool samples by one-dimensional electrophoresis. This method allows up to 50 distinct wool proteins to be separated. Comparing electrophoretic patterns of digests of differently treated wool textiles gives information on the involvement of morphological components and chemical constituents of wool fibres (*12-15*). In Table I the extractable proteins from an untreated wool are listed and their relation to morphological wool components is given.

Table I. Soluble wool proteins extracted from untreated merino wool and their relation to morphological wool components[a]

Protein fraction	*Molecular weight in kD*	*Origin*
Low Sulfur (58 wt%)	45 - 50	KIF
High Sulfur (18 wt%)	14 - 28	KAP
Ultra High Sulfur (8 wt%)	28 37	Cuticle
High Glycine High Tyrosine (6 wt%)	9 - 13	KAP

[a]Adapted from ref. (*16*)

The protein pattern of chlorinated wool treated with different amounts of protease was analysed via one-dimensional electrophoresis. In Figure 1 the densitograms of protein separation patterns of the differently treated wool samples in comparison to an untreated wool (0 %) are given. The protein separation pattern of a solely chlorinated wool is not depicted because it does not differ significantly from that of the untreated wool. The protein separation patterns of the chlorinated samples treated with different amounts of enzymes show a decrease in the intensity of the bands in the region of the low sulfur proteins. This decrease is more noticeable as the enzyme concentration increases. Already the protein pattern of the wool sample treated with an enzyme concentration of 0.5 % owf shows a drastic decrease of the low sulfur proteins. On the other hand in the low molecular weight region new protein bands most probably originating from the degradation products are recorded. In the case of the application of higher enzyme dosages (2.5 and 5 % owf) these degradation products are further decreased. Figure 2 shows a comparison between densitograms of untreated and chlorinated wool samples all treated with 1 % owf protease. In the case of the untreated wool weak protein bands in the region of the low sulfur proteins were still recorded whereas in the case of the chlorinated wool samples they were no longer detected. Thus, chlorinating and partial oxidative degradation of the wool cuticle accelerates the enzyme catalytic effect on the bulk of the fiber.

Considering that low sulfur proteins of wool build up the intermediate filaments that belong to the α-keratin, the crystalline region of the wool fiber, degradation of these parts of the fibers should impair the tenacity of the fibers. Therefore, we measured the wet yarn strength of the samples. After a treatment with 5 % owf protease the yarn tenacity only decreased by about 16 % compared to the chlorinated sample. The sample treated with 1 % owf protease showed a loss of 6 % tenacity and the sample treated with 2.5 % a reduction of the yarn strength of 12.5 %.

The protease used in these trials has a slight preference for the cleavage of peptide bonds adjacent to aromatic amino acids. Taking this into account when interpreting the above mentioned results it was deduced that only some peptide bonds within the intermediate filaments were cleaved by the enzyme catalytic action. Figure 3 shows the amino acid sequence of a polypeptide chain of an intermediate filament and a schematic drawing of a molecular unit of an intermediate filament. Due to steric hindrance only the peptide bonds at exposed tyrosine residues at the end of the linker regions (L1, L2) might be cleaved.

Thermoanalysis. Thermoanalytical measurements on the enzyme treated chlorinated samples show that neither the absolute value of the melting peak (as a measure for the cystine content of wool) nor the melting peak area (as a measure of the α-helix content) differ significantly from the reference sample. This supports the hypothesis that the protease only cleaves exposed peptide bonds in the wool leaving the overall wool morphology with the keratin and the keratin-associated proteins intact but increasing the extractability of the wool proteins during electrophoretic analysis.

Fiber Diameter. The fiber diameter of the chlorinated wool samples that were treated with different amounts of enzymes is listed in Table II.

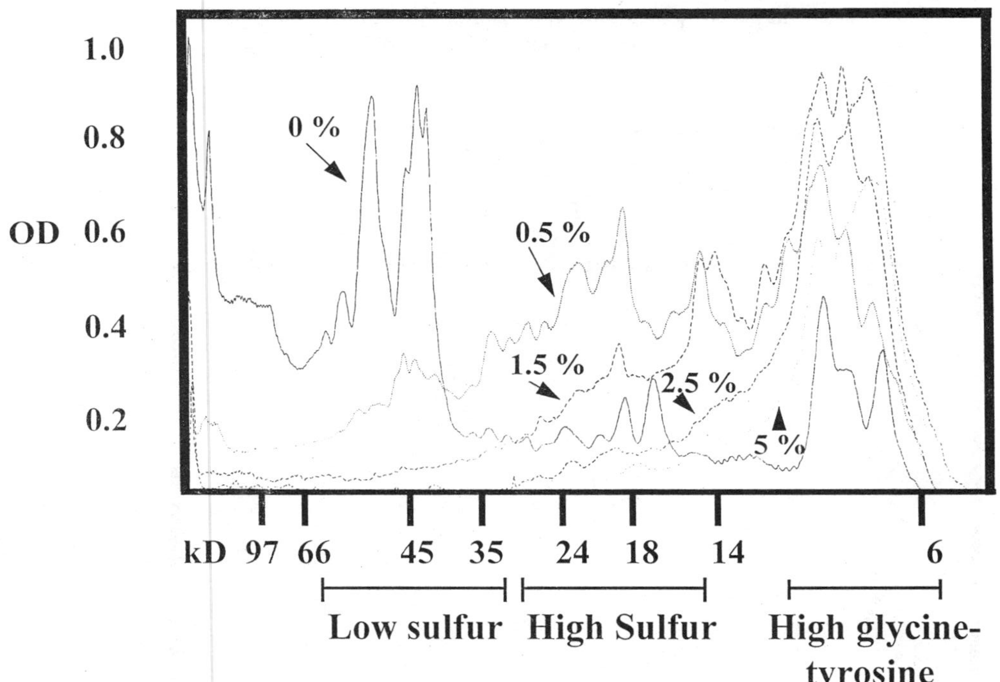

Figure 1. Densitograms of protein separation patterns of the extracts of untreated and chlorinated wool treated with different amounts of protease.

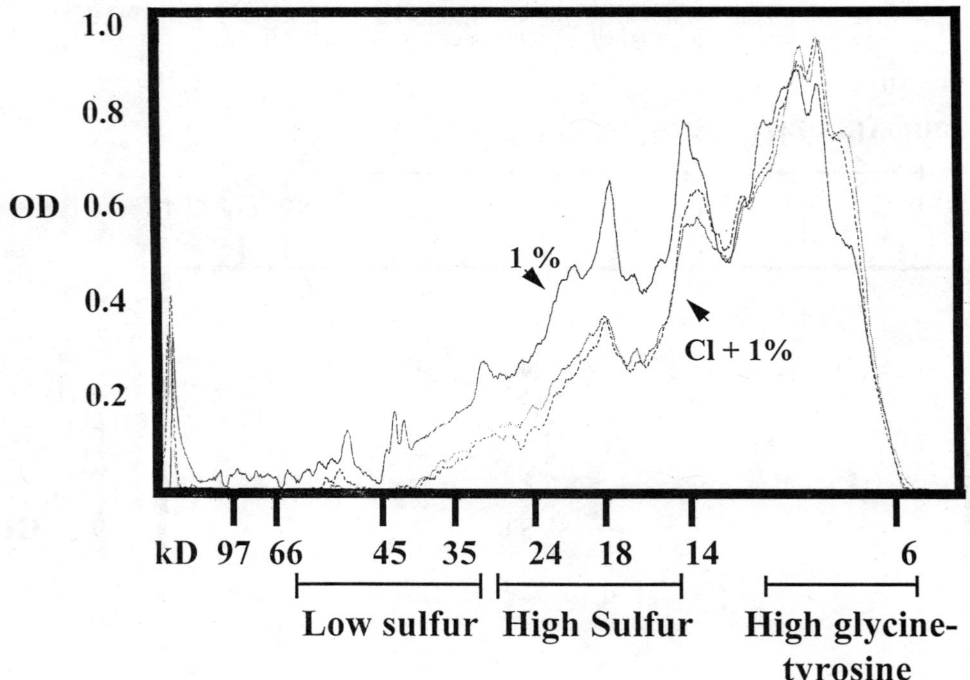

Figure 2. Densitograms of protein separation patterns of the extracts of untreated and chlorinated wool samples all treated with 1 % owf protease.

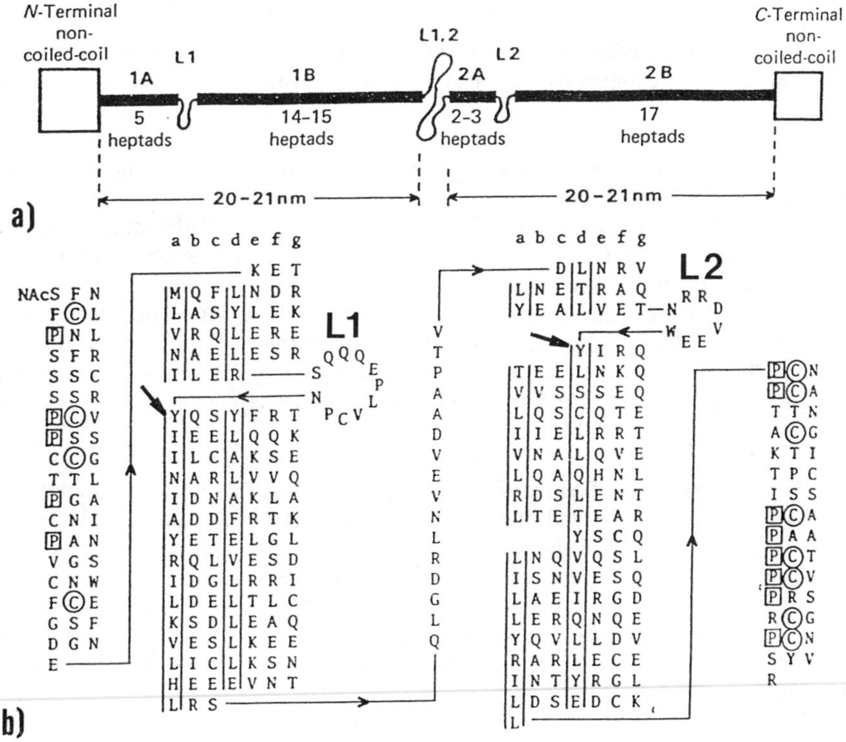

Figure 3. Schematic drawing of a polypeptide chain of a KIF (a) and its amino acid sequence (b); arrows indicate exposed tyrosine molecules; adapted from ref. (*16*)

Table II. Fiber diameter of chlorinated wool samples treated with different amounts of protease

Samples	*Fiber diameter in μm*	*Δ Fiber diameter in μm*
Chlorinated Reference	21.2	-
1.0 % owf Protease	21.0	0.2
2.5 % owf Protease	20.2	1.0
5.0 % owf Protease	20.2	1.0

The enzyme catalytic effect reduces the fiber diameter by 0.2 μm in the case of the 1 % owf treatment. A treatment with 2.5 and 5 % owf protease reduces the fiber diameter by 1 μm.

Morphology of the Enzyme Treated Samples. The morphology of the wool fibers was investigated by using TEM (Figure 4).
Treating wool with proteases results in a complete removal of the endocuticle and parts of the cell membrane complex (cmc). The modification of the cmc is visible after contrasting the fiber cross sections with uranyl acetate and lead citrate. Partly material between the cortical cells has been removed completely. Furthermore, the enzyme catalytic effect leads to a degradation of the nuclear remnants.
The handle improvement has to be a composed effect of cmc modification and reduction of fiber diameter and of partial cleavage of single peptide bonds within the KIF.

Handle after Protease Treatment Alone. Wool fabric was treated with protease respectively lipase alone and the handle was evaluated by using KAWABATA evaluation system. Figure 5 shows the bending stiffness of the enzyme treated samples and the corresponding references as one parameter within the KAWABATA evaluation system. Treating wool with buffer (pH 7 or pH 8) already has an influence on the handle. After protease treatment the bending stiffness is reduced to a larger extent compared to a lipase treatment, the difference in the values for the bending stiffness amounting to 12.5 %.

Antifelting Finish. The majority of enzymatic processes published in the last few years are combined processes (*4*). We applied proteases on physically and chemically pretreated wool. Both plasma treated and chemically oxidized samples were treated with 0.17 % owf protease for 120 min at 50 °C. In both cases the enzyme post-treatment led to a further improvement of the antifelting effect (*6*). The felting test was performed according to IWS TM 31. Table III summarizes the results of the felting test of the plasma pretreated wool and Table IV of the chemically preoxidized wool.
After plasma treatment the feltability of knitted wool fabric is reduced approximately by a factor of 2 (2 wash cycles) respectively by a factor of 2.4. After combined plasma and enzyme treatment the feltability is reduced by a factor of 3 (after 2 wash cycles) respectively a factor of 6 (after 5 wash cycles).

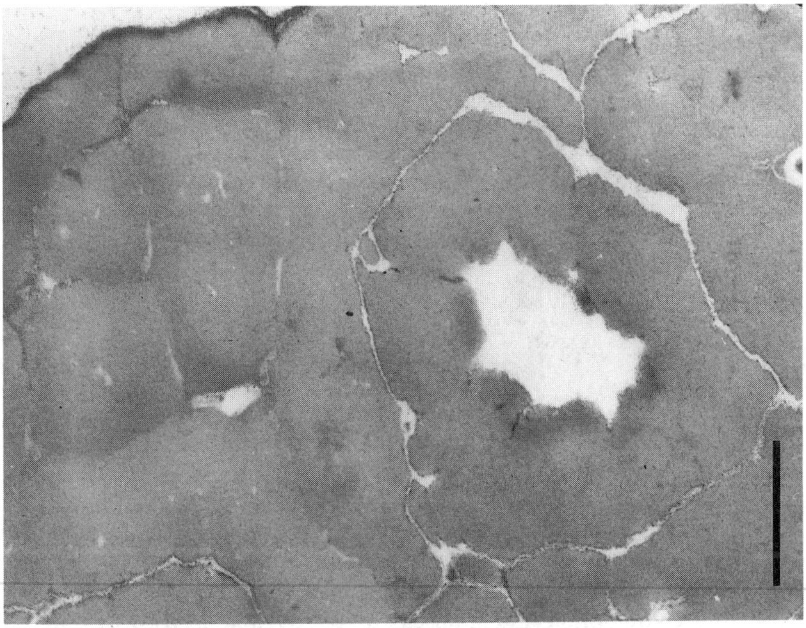

Figure 4.. TEM picture of a cross section of chlorinated wool treated with 2.5 % owf protease (— 500 nm).

In the case of the chemically oxidized woven wool fabric the feltability is reduced by a factor of 3 for the oxidation step and by a factor of 19 after the additional protease treatment (2 wash cycles). After 5 washings the felting is reduced by a factor of 2 for the oxidative treatment respectively by a factor of 9 after the enzyme post-treatment. The feltability of the untreated reference material was also decreased by the enzyme treatment (Table V). After treating knitted wool fabric with 0.83 % owf protease for 120 min the feltabilty was reduced by a factor of 1.3 after 2 wash cycles respectively 5 wash cycles. After a treatment of 60 min the feltability was reduced by a factor of 1.5 (2 and 5 wash cycles). Using higher enzyme concentrations the treatement time is decisive for the antifelting effect.

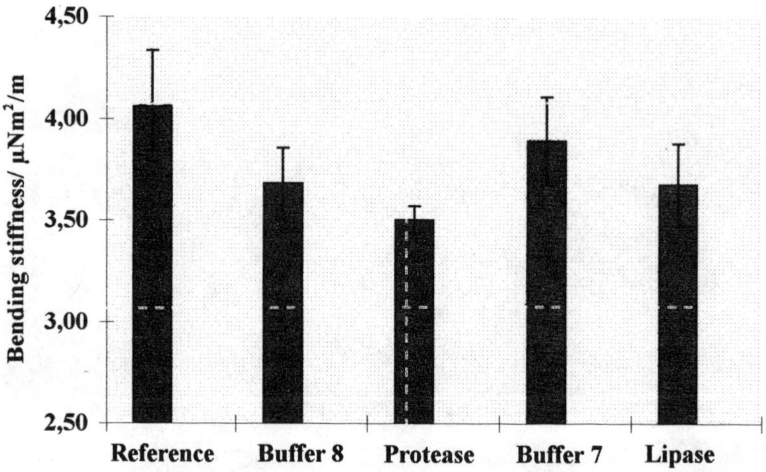

Figure 5 **Bending stiffness of enzyme treated wool fabrics and the corresponding references (KAWABATA evaluation system)**

Table III. **Felting properties of plasma pretreated and protease treated knitted wool fabric (according to IWS TM 31)**[b]

Samples (Knitted fabric) (size 225 mm x 300 mm)	*Total shrinkage in % (2 wash cycles)*	*Total shrinkage in % (5 wash cycles)*
Reference	-22.27	-50.20
Plasma treated	-11.60	-20.82
Plasma + buffer	-8.95	-12.30
Plasma + protease	-6.95	-8.51

[b] Adapted from ref. (*6*)

Table IV. Felting properties of chemically oxidized and protease treated woven wool fabric (according to IWS TM 31)[c]

Samples (Woven fabrics) (size 280 mm x 280 mm)	*Total shrinkage in % (2 wash cycles)*	*Total shrinkage in % (5 wash cycles)*
Reference	-26.91	-71.90
Chemically oxidized	-8.75	-31.30
Chemically oxidized + buffer	-6.07	-24.60
Chemically oxidized + protease	-1.45	-8.00

[c] Adapted from ref. (*6*)

Table V. Feltability of the non-pretreated reference material after enzymatic treatment

Sample (Knitted wool fabric) (size 300 mm x 400 mm)	*Total shrinkage in % (2 wash cycles)*	*Total shrinkage in % (5 wash cycles)*
120 min		
Reference buffer treated	-20.63	-46.01
Reference protease treated	-15.36	-35.20
60 min		
Reference buffer treated	-19.78	-48.77
Reference enzyme treated	-13.42	-32.60

Dyeability. Both the enzyme treatment alone and the combined processes enhance the dyeability of wool. Table VI summarizes the colour differences between the pretreated respectively enzyme treated samples and their corresponding references. The enzyme catalytic effect enhances the dye uptake of the plasma pretreated respectively chemically oxidized wool. A more negative DL value refers to a darker shade. The highest difference in the colour depth between reference and sample is obtained after chemical oxidation. The enzyme treatment leads to an additional increase of the colour shade. After treating the chemically oxidized samples with a buffer solution the effect of an enzyme post-treatment is more clearly visible than after chemical oxidation alone. Possibly, the buffer treatment removes surplus chemicals that impair the enzyme catalytic action. Already the enzyme treatment alone enhances the dye uptake of knitted wool fabrics. But even after a plasma treatment an additional buffer and also the enzyme treatment lead to deeper colour shades.

The dyeing of untreated wool top is also enhanced by the protease treatment. Figure 6 shows the exhaustion curves of enzyme treated wool top and the corresponding references.

Table VI. Colour differences of the samples and references dyed in competition (DL values for the clour depth are given)

Samples[a] *(Reference / Sample)*	*DL*
Woven	
untreated / chemically oxidized	-10.2
chemically oxidized buffer treated / chemically oxidized, enzyme treated	-7.1
chemically oxidized / chemically oxidized, enzyme treated	-4.0
Knitted	
untreated / enzyme treated	-6.1
plasma / plasma, buffer treated	-3.6
plasma, buffer treated / plasma, enzyme treated	-0.5

[a]Samples treated according to the samples in Table III and Table IV

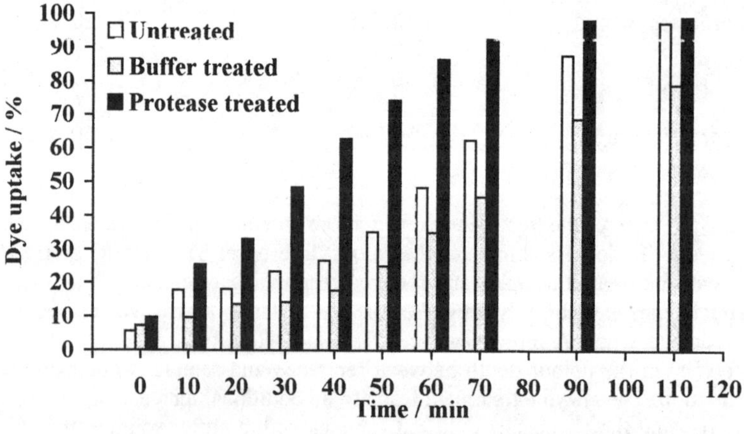

Figure 6 Dye exhaustion curves of protease treated wool top and the corresponding references

Using the dyestuff Lanasol Blue 8G without pH adjustment the dyeability of enzyme treated wool is higher compared to the corresponding untreated or buffer treated references. In the case of the enzyme treated sample the dyeing rate is enhanced, e.g., after 40 min the dye uptake is 63 % in the case of the enzyme treated sample compared to 27 % for the untreated sample. Untreated and protease treated wool end up at nearly the same value (about 99 %) after 110 min dyeing time but the total dye uptake of the buffer treated sample is lower (about 80 %).

Degree of Whiteness. The degree of whiteness was enhanced by treating wool fabric with blue light in the dry and in the wet state. After an enzyme post-treatment the degree of whiteness was further enhanced (Figure 7).

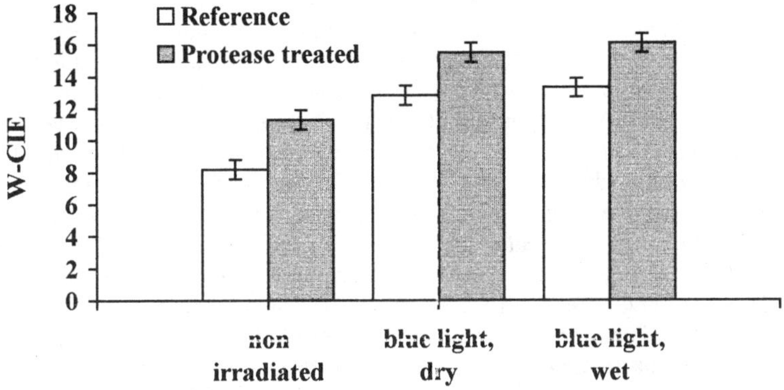

Figure 7 **Degree of whiteness (W-CIE) of the photobleached and enzyme treated wool fabric**

The degree of whiteness achieved by the protease treatment alone is lower than the whiteness achieved by photobleaching. However, treating photobleached wool samples with proteases leads to an additional enhancement of the degree of whiteness.

Conclusion

Handle improvement and shrinkproofing treatment are quality enhancing steps in wool finishing increasing the value of the resulting product. It was one aim of the study to clarify why the enzyme treatment of wool leads, e.g., to handle improvement and shrinkproofing especially, when enzymes are applied after pretreatment methods like plasma treatment or chlorination.

All measurable effects like enhancement of the degree of whiteness and dyeability, reduction of felting and handle improvement reported here seem to be based upon a combination of surface and bulk effects. Both modification of the surface by chemical or physical methods on the one hand and modification of the bulk by the enzyme catalytic effect on the other hand result in improved properties of the treated wool. However, the combined application of both procedures leads to much higher effects. The origin of these synergistic effect was investigated by using different analysis methods. On the molecular level the protein patterns obtained via one-dimensional electrophoresis of the differently treated wool samples were compared. In the protein pattern of enzyme treated wool the low sulfur proteins were no longer detected as distinct bands indicating that the extractability of these proteins was enhanced after enzyme treatment. As the helical segments of the keratin intermediate filaments, that are responsible for the tenacity of wool fibres, are composed of low sulfur proteins it was expected that the strength of these samples should have been drastically decreased. However, the yarn tenacity of the sample treated with the highest enzyme concentration applied in these tests was only 20 % lower than the corresponding reference. These findings led to the conclusion that the protease catalyses the cleavage of only some peptide bonds being located in the KIF leaving the overall structure and organization within the filaments intact. Interaction of the residual bondings and junctions seems to be sufficient to account for the residual strength. Additionally, by applying TEM on enzyme treated chlorinated wool samples removal of endocuticle and modification of the cmc were reported. On these samples a decrease in the fiber diameter was measured. The effect of chlorination on the wool fibers was also investigated by using TEM. Only the outer layers of the cuticle, the a-layer and the exocuticle, were partly degraded by the chlorination step. Thus, the handle improvement measured on the enzyme treated chlorinated samples most probably is the result of a multicomponent modification of wool.

The enhanced shrinkproofing effect achieved after plasma and enzyme treatment among other factors might be due to a smoothing of the surface derived from the enzyme catalyzed modification of the cuticle reducing the fiber friction.

Chlorination and plasma treatment both attack the hydrophobic surface of the fibers creating a more hydrophilic wool sample. On this substrate the diffusion of enzymes to the surface and into the fibers (via the cmc) is enhanced. Therefore the enzyme catalytic effect is higher on pretreated than on untreated wool. Conclusively it may be stated that the enzyme catalytic effect on wool is able to improve wool properties but this effect is additionally enhanced when applying enzymes to pretreated wool.

Acknowledgments

The authors gratefully thank the Forschungskuratorium Gesamttextil for the financial support of the research project (AiF-No. 9703) which was effected by funds of the Bundeswirtschaftsministerium via a grant of the Arbeitsgemeinschaft industrieller Forschungsvereinigungen (AiF). We gratefully acknowledge the financial support of NOVO Nordisk for the research work of W. Lorenz and the excellent cooperation with L. Dybdal and M. B. Andersen from NOVO Nordisk, Bagsvaerd, Denmark. The authors thank K.-H. Phan for TEM investigations, K. Schäfer for the photobleaching, D. Ganssauge for KAWABATA measurements and Beate Tetzlaff for excellent technical assistance. The participation of E. Heine at the ACS Symposium was sponsored by the Deutsche Forschungsgemeinschaft (DFG).

Literature Cited

1 Zahn, H., Wortmann, F.-J., Wortmann, G., Hoffmann, R. In: *Ullmann's Encyclopedia of Ind. Chem.* Ed.; VCH Verlagsgesellschaft, Weinheim, Germany, **1996**, Vol. 28A, pp 395-421

2 Müller, B., *Rev. Prog. Coloration,* **1992**, *22*, 14-21

3 Thomas, H., Herrling, J., Rakowski, W., Höcker, H., *Int. Text. Bull. Veredlung*, **1993**, *39*, 42-49

4 Heine, E., Höcker, H., *Rev. Prog. Coloration*, **1995**, *25*, 57-63

5 Goddinger, D., Heine, E., Müller, B., Schäfer, K., Thomas, H., Höcker, H., *Report 7 of the European Workshop on „Technologies for Environmental Protection"* 1995; Bilbao, Spain; Büsing, J. H., Pippich, B. Eds, 419-431

6 NOVO Nordisk A/S, Dybdal, L. Heine, E., Höcker, H., Appl. No. PCT/DK95/00517

7 Marshall, R. C., *Text. Res. J.* **1981**, *51*, 106-108

8 Schägger, H., Jagow, G. v., *Anal. Biochem.* **1987**, *166*, 368-379

9 Weber, K., Osborne, M., *J. Biol. Chem.* **1969**, *244*, 4406-4412

10 Lickfeld, K. G., *Elektronenspektroskopie*, Ulmer Verlag Stuttgart, **1990**, 254

11 Smith, S. M., Jackson, J., Ramazio, F., Dybdal, L., Andersen, M. B., *Int. Dyer*, **1995**, 9, 10-15

12 Marshall, R. C., Zahn, H., Blankenburg, G., *Text. Res. J.*, **1984**, *54*, 126-128

13 Zahn, H., Souren, I., Altenhofen, U., *Melliand Textilber.* **1984**, *65*, 467-471

14 Wilrich, C., Wortmann, G., Wortmann, F.-J., Höcker, H., *Proc. 9th Int. Wool Text. Res. Conf.* Biella Italy, Bona, M., Grupallo, G., Eds. **1995**, vol II, 289-295

15 Seiler, P., Wilrich, C., Wortmann, G., Wortmann, F.-J., DWI Reports **1996**, *117*, 637-648

16 Powell, B. C., Rogers, G. E., In: Cellular and Molecular Biology of the Intermediate Filaments, Goldman R. D., Steinert, P. M., Eds. Plenum Press, New York, **1990**, Chapter 10, 267

17 Dowling, L. M., Crewther, W. G., Parry, D. A. D., Biochem. J. **1986**, *236*, 705-712

Chapter 24

Interaction of Subtilisin-Type Protease with Merino Wool Fibers

Y. M. Galante[1], D. Foglietti[2], C. Tonin[2], R. Innocenti[2], F. Ferrero[3], and R. Monteverdi[1]

[1]Central R&D, Lamberti spa, 21041 Albizzate (VA), Italy
[2]Ist."O.Rivetti", Consiglio Nazionale delle Ricerche, 13051 Biella, Italy
[3]Politecnico di Torino, Dip. Ingegneria Chimica, 10100 Torino, Italy

In order to understand the structural modifications caused by proteolytic enzymes on wool fibers, the interaction of subtilisin-type protease with Merino wool was studied by physical and physico-chemical techniques. Enhanced wool dye affinity and optical microscopy examination indicate increasing fiber damage as a function of enzyme action. Scanning electron microscopy suggests a preferential proteolytic attack on the Cellular Membrane Complex (CMC), a non-cellular and non-keratinic material, while no significant degradation of cortical and cuticular cells was observed. SDS-PAGE (after fiber dissolution in urea) shows progressive proteolysis of low-sulfur containing proteins, which are considered to be part of non-keratinic structures. High ionic strength in the incubation medium provides protection of the fiber interior, by limiting proteolysis to the epicuticule. FTIR-ATR indicates that no change in the redox state of -S-S- cystine bonds takes place, which occurs with currently used oxidative treatments (e.g., Basolan DC). Accordingly, wool top treated with an oxidatively stable serine protease acquires minimal non-felting properties, as proven by the Aachen felting test, even in the presence of hydrogen peroxide.
It is concluded that commercially available subtilisin-type proteases are useful tools in the study of wool / enzyme interactions and can be used to enhance hand feel or lustre of wool articles, but might be of limited use in shrink-resistance treatments.

One of the intrinsic properties of wool is its tendency to felt and to shrink in the wet state under certain conditions of mechanical agitation.
A few theories have been put forward to explain this property (1). There is a general consensus that the differential frictional effect (DFE), which is related to the "scaly" structure of the fibers causing them to move unidirectionally towards their roots' ends,

plays a fundamental role in the felting phenomenon, in combination with the fibers' hydrophobic character.

Shrink-resistance technologies modify the fiber surface with relatively harsh oxidative treatments, eventually followed by applications of cationic polymer resins to preserve wool dimensional stability. Most of these processes use either gaseous chlorine or chlorine salts. However, the industrial use of chlorine as wool fiber-modifying agent produces environmentally unfriendly absorbable halogen by-products (AOX). Therefore, there is an increasing worldwide interest in the development of alternative environmentally safe approaches to the current shrink-resistant processes used by the wool industry (2).

Wool can be considered an almost ideal substrate for various enzymes, such as esterases, proteases, lipases and enzymes acting on disulfide bonds. This fiber is mainly composed of proteins and lipids, and it consists of two major morphological components: the cuticle and the cortex. The cuticle is made of overlapping cells, each surrounded by a thin outermost membrane, called the epicuticle. By weight, the latter is composed of about 75% proteins and 25% fatty acids (3), is largely hydrophobic and acts as a diffusion barrier, also thanks to a large content of disulfide bonds of cystine residues. The cortex is composed of the ortho- and the para-cortex with different composition and properties.

A hydrophilic compound can enter into the wool fiber, by penetrating between cuticle cells, through the Cell Membrane Complex (CMC), which is a material mainly composed of non-keratinous proteins and of some lipids.

Applications of enzymes on wool, mostly of proteases, have been studied by numerous Authors. Several reports in the literature mention the use of proteases and describe some of the changes that occur in wool (4-7), but no correlations have been established between structural , morphological and functional modifications.

Although a few enzyme-based products have been introduced into the market, claiming to give shrink-resistance properties to wool, no enzyme technology has so far withstood the test of industrial trials or, to our knowledge, has been adopted by the industry.

The availability of new recombinant and bioengineered subtilisin-type proteases with novel biochemical features, including enhanced oxidation resistance, allows the study of enzyme/fiber interactions under conditions commonly found in the wool industry (e.g., in the presence of hydrogen peroxide) and to evaluate their practical application potential.

Experimental

Australian wool with a 19 micron mean diameter was kindly provided by the Filatura di Grignasco (Grignasco, Italy). Scoured and washed wool was extracted with petroleum ether and ethanol (8).

Protease treatments were done in an Ahiba equipment, at pH 8.0-9.0, 45°C and with 0.1 to 4.0% of enzyme at 1:50 liquor ratio, for variable incubation times. SDS-PAGE was carried out according to Laemmli (9) on a 12% or 15% acrylamide mini-gel slab. Before electrophoresis, wool fiber was dissolved in urea and dithiotreitol essentially following the method of Marshall (10) with few modifications. Protein thiol groups were blocked with iodoacetamide.

Scanning electron microscopy (SEM) was performed with a Cambridge Stereoscan 240 Scanning Electron Microscope, at an acceleration voltage of 15 KV and a 16 mm working distance. Samples were prepared by cutting 3-5 mm fiber snippets from the wool slivers, which were mounted on aluminum specimen stubs with double-sided adhesive tape.

Samples were finally coated with a 50 nm-thick gold layer in a rarefied argon atmosphere at 0.1-0.2 mbar, using an Emitech K 550 Sputter Coater Unit with an application voltage of 0.6 KV and plasma current of 18 mA for 5 min.

A Nicolet 510P FTIR spectrophotometer was used for infrared spectroscopy. The spectra of the wool samples were measured in the ATR mode with a thallium iodide crystal to determine fiber surface modifications. The spectra recorded represented an average of 200 scans at 4 cm^{-1}. The baseline was corrected and normalized on the amide III band.

Dyeing affinity was determined with Shirlastain A, following the IWS recommendation. In every set of analysis, a non treated and a control sample without enzyme were included.

Felting was evaluated by the Aachener Filztest (IWTO-20-69).

Proteases used in these studies were oxidative resistance bioengineered subtilisin-type, serine proteases of commercial origin either from Genencor International or from Novo Nordisk, with comparable proteolytic activity per gram of product.

Dye Affinity and Optical Microscopy

Dye uptake by wool depends on several factors, mostly the hydrophylicity of the fiber and chemical nature of the dye. It appears that a dyeing solution penetrates the wool fiber interior through the tiny channels of the CMC between adjacent cuticular cells. Therefore, any modification of the fibers able to widen this access, should result in an increased affinity for the dye, which is in turn a useful parameter to detect structural modifications of the wool.

The dye mixture chosen was Shirlastain A, which actually is a combination of three dyes (i.e., picric acid, chlorazil blue, crocein scarlet) and is recommended by the International Wool Secretariat (IWS).

While untreated wool top gives a light yellow color, oxidative treatments, eventually followed by cationic resins application, generates a red/orange or green/brown coloring, respectively.

Figure 1 shows the increase in dyeing affinity of wool top due to protease modification, either as a function of time or of enzyme concentration, suggesting facilitated dye penetration into the fiber interior.

Optical microscopy at 200x magnification of the same Shirlastain A-dyed samples confirms that the enzyme causes substantial fiber damage as shown in Figure 2 (non treated wool in Figure 2a and treated with 1g/l of protease in Figure 2b).

From Figure 2b, it is quite obvious that: a) the proteolytic attack is not uniform either on different fibers or even on a single fiber, as one can see an extensively damaged fiber (in the forefront) next to an apparently unaltered fiber (in the background); b) breakage of the fiber is the consequence of loss of the ordered cortex structure and of

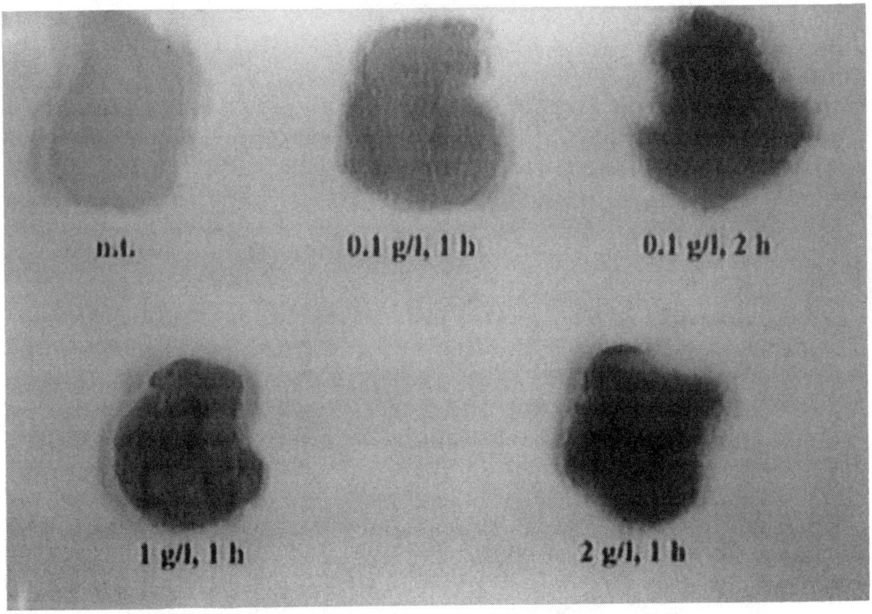

Figure 1. Shirlastain A dyeing of wool top with increasing enzyme treatment.

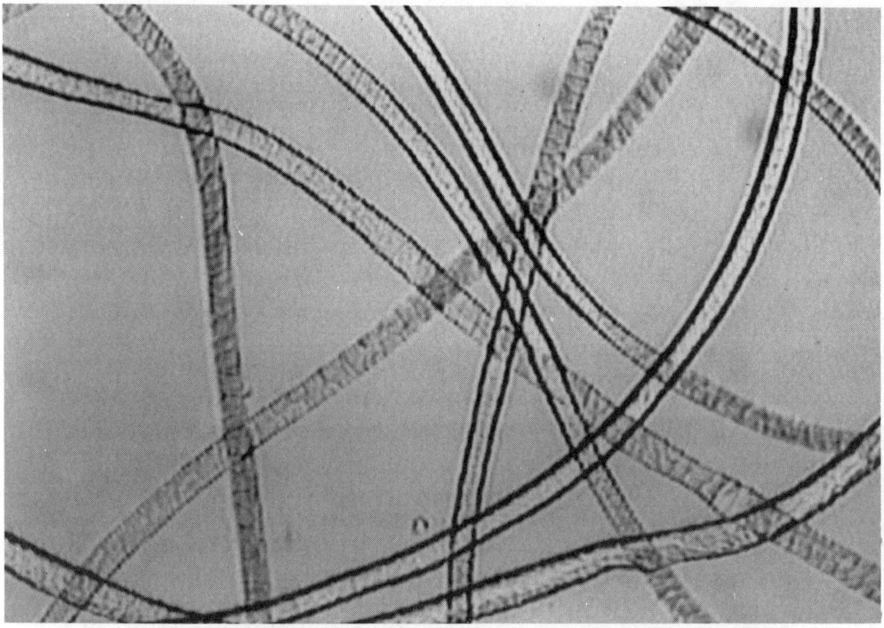

a

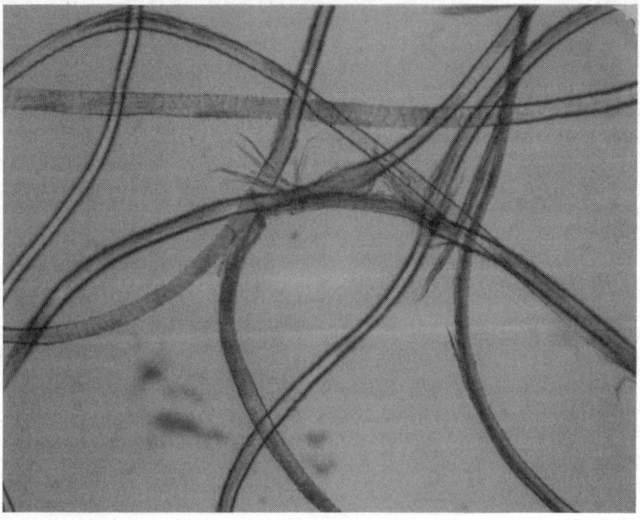

b

Figure 2. a) Optical microscopy (x200) of wool fibers treated with 1 g/l of protease and dyed with Shirlastain A; b) Optical microscopy (x200) of wool fibers treated with 1 g/l of protease and dyed with Shirlastain A.

leakage of cortical cell material. Figure 3 is the electronic elaboration of an optical microscopy picture at 400x magnification of the cross-section of a wool fiber bundle treated and dyed as in Figure 2b.
It shows that enzyme modification, as detected by uneven brown dyeing throughout the fiber section, is not only limited to the outer region, but also occurs in the orto-cortical structure.

Gel electrophoresis and HPLC analysis

In order to have a better understanding of the wool proteins affected by proteolytic attack, electrophoresis on polyacrylamide gel slab in the presence of sodium dodecylsulfate was used (SDS-PAGE).
Before electrophoresis, wool proteins were dissolved in 8 M urea, as described by Marshall (10), reduced with 1,4-dithiothreitol (DTT) and finally carboxymethylated with iodoacetamide (IOA). It should be mentioned, however, that with such an approach it is possible to solubilize between 60% and 95% of the total wool proteins. Therefore, the results of a SDS-PAGE analysis should be considered in qualitative, rather than quantitative terms.
Figure 4 is the picture of a 15% acrylamide gel slab of wool fiber proteins, treated with protease as detailed in the figure legend, fixed and dyed with Coomassie blue. It is clear that, compared to untreated wool (lane 1), there is an extensive loss of staining intensity of the proteins in the mol.wt range between 43 and 100 Kda (see lanes 4 and 5). These proteins are considered to be non-keratinic in nature and to have a low cysteinic sulfur content, in comparison with the high sulfur-containing proteins in the range of mol.wt between 17 and 20 Kda (10).
The above results confirm a preferential attack of the protease on the proteins of the CMC and the endocuticle, although a less extensive degradation of keratinic proteins cannot be ruled out from the gel profile, also considering that those proteins might not be fully solubilized in urea. Presence of hydrogen peroxide along with the enzyme in the incubation medium gave the same SDS-PAGE profile.
Amino acid analysis by reverse-phase HPLC was used to investigate whether any significant change in amino acid composition was detectable after proteolysis and fiber dissolution in 6 N HCl. At 1 g/l enzyme, no significant change (greater than 0.15% moles) was noticed, compared to non-treated wool (not shown). At higher enzyme concentration (2 and 4 g/l), there is a decrease in amino acids typical of low sulfur proteins (e.g., Asp and Glu) and an increase in amino acids of high sulfur proteins (e.g., Thr and Pro), although surprisingly Cys and Ser decreased. This apparently contradictory result can be explained assuming that, upon enzyme treatment, detachment and net loss of keratinic cuticular material from the fiber occurs, rather actual proteolysis (see also SEM results below). If this is so, SDS-PAGE would not show significant high-sulfur proteins degradation, while some inconsistency could be found by HPLC analysis of total amino acids after extensive enzyme treatment.
Indeed, the cuticular fraction of wool fiber contains high amounts of Cys and Ser, and is more easily lost under stress conditions (11).

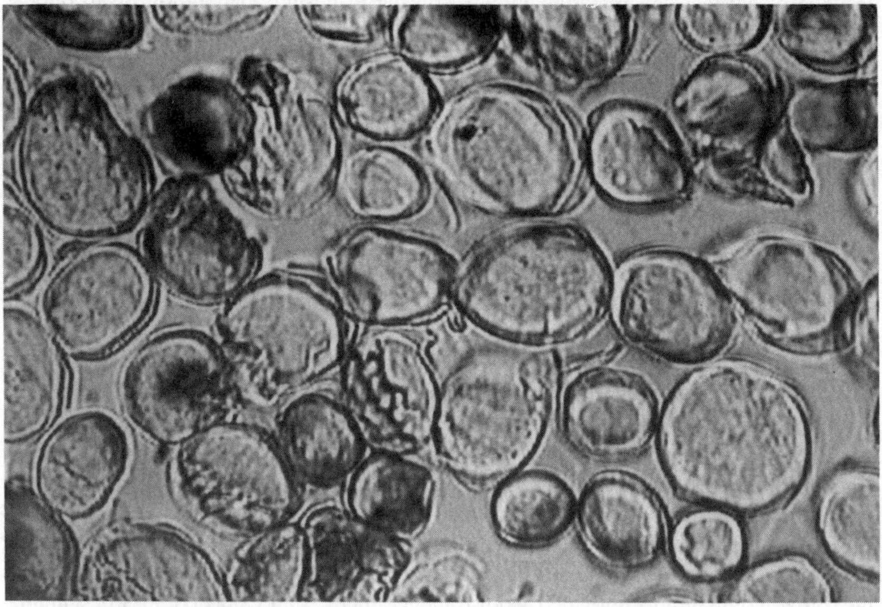

Figure 3. Optical microscopy (x400) of a cross section of wool fibers treated and dyed as in Figure 2b.

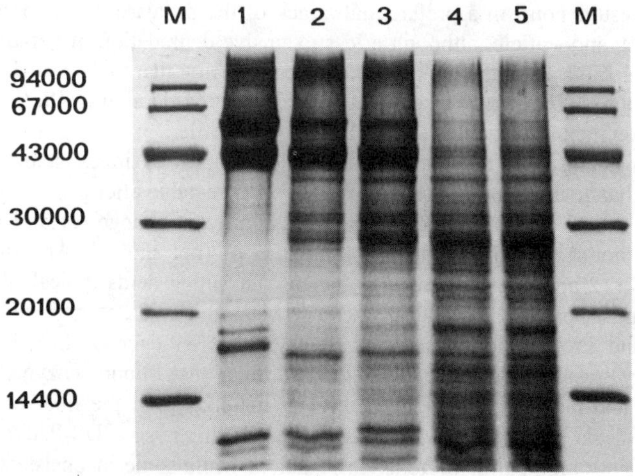

Figure 4. SDS-PAGE of wool proteins after solubilization in urea under reducing conditions. Lane 1: untreated wool; Lane 2: wool treated with 0.1 g/l of protease for 1 hr; Lane 3: wool treated with 0.1 g/l of protease for 2 hr; Lane 4: wool treated with 1 g/l of protease for 1 hr; Lane 5: wool treated with 2 g/l of protease for 1 hr.

Scanning electron microscopy at low and high ionic strength

Scanning electron microscopy (SEM) is a powerful technique for studying wool fiber morphology. Figure 5 shows a "typical" SEM picture of untreated wool, where it can be noticed that the cuticular cells have a well defined "scaly" structure with a sharp profile and closely adhere to each other. Figure 6 is a SEM picture after treatment with 1 g/l of protease at low ionic strength. Clearly, under such conditions enzyme attack causes massive, although uneven, fiber damage with detachment of cortical cells from the amorphous CMC. The same result can be obtained at lower enzyme concentration for longer incubation times.

Preferential proteolysis of CMC proteins and of the non-keratinic material of the endocuticule is likely to cause separation of cortical cells, as observed by SEM, and is in agreement with the results of SDS-PAGE analysis (see Figure 4).

In the presence of high ionic strength in the incubation medium (i.e., from 5 to 20 g/l of sodium sulfate), it was noticed that the Shirlastain A dye affinity increases much less than in the absence of salt (not shown), suggesting a "protective effect" of the fibers. SDS-PAGE analysis also indicates less extensive degradation of the low sulfur proteins in the presence of high ionic strength (not shown).

That at high ionic strength protease attack is mostly limited to the cuticle with limited damage to the fiber interior is confirmed by the SEM picture in Figure 7 of wool treated with 1 g/l protease in the presence of 20 g/l of sodium sulfate. Rather than the massive damage observed in Figure 6, one can notice a filing or de-scaling of the structure.

This altered morphology, as detected by SEM, closely resembles the one obtained by the industrial anti-felting Basolan DC treatment of wool based on dichloroisocyanuric acid (DCCD). Indeed , Figure 8 is an example of a 2.7% Basolan DC-treated wool, showing a very similar morphology of the enzyme-treated wool at high sodium sulfate concentration, as seen in Figure 7.

In order to check whether there is a correlation between morphology and non-felting properties, various samples of protease-treated wool were subjected to the Aachener filztest, according to the official method IWTO-20-69 ("Method for the Determination of the Felting Properties of Loose Wool and Top"9.

It turned out that at any enzyme concentration used, also at high ionic strength, no change in the felting properties of wool was obtained. Under standard conditions and at constant shaking time, non-treated wool top gave a ball diameter of 2.4 cm, while wool treated with protease (in a range of 0.1-1.0 g/lt of enzyme) gave a ball diameter of 2.35-2.4 cm, in the presence or absence of sodium sulfate. This should be compared to a ball diameter greater than 3.0 cm after 60 min shaking time, as specified by the IWTO-20-69 method, for Basolan DC-treated wool top.

Therefore, these results strongly indicate that filing and smoothening of the cuticle scales do not represent the structural basis of wool anti-felting properties.

Fourier transform infra-red spectroscopy analysis (FTIR-ATR)

It was elegantly demonstrated by Douthwaite and Lewis by FTIR-ATR (12) that shrink

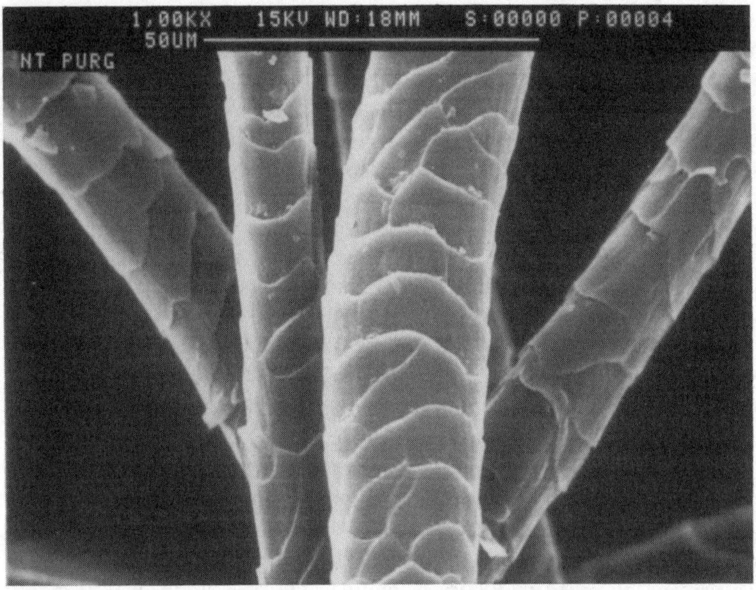

Figure 5. SEM (x1000) of untreated wool.

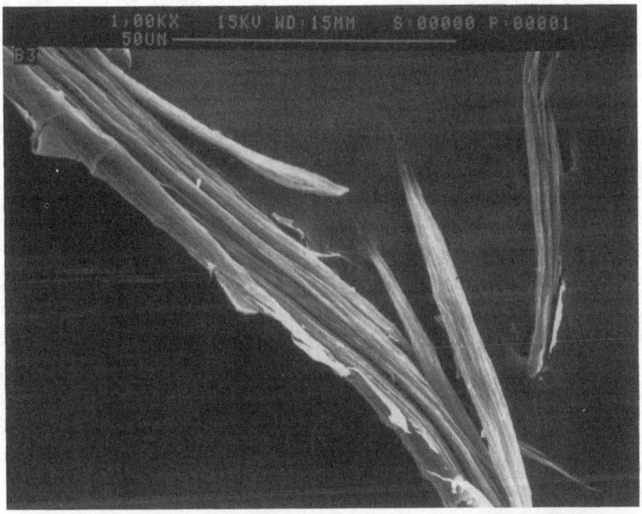

Figure 6. SEM (x1000) of wool treated with 1 g/l of protease for 1 hr at low ionic strength.

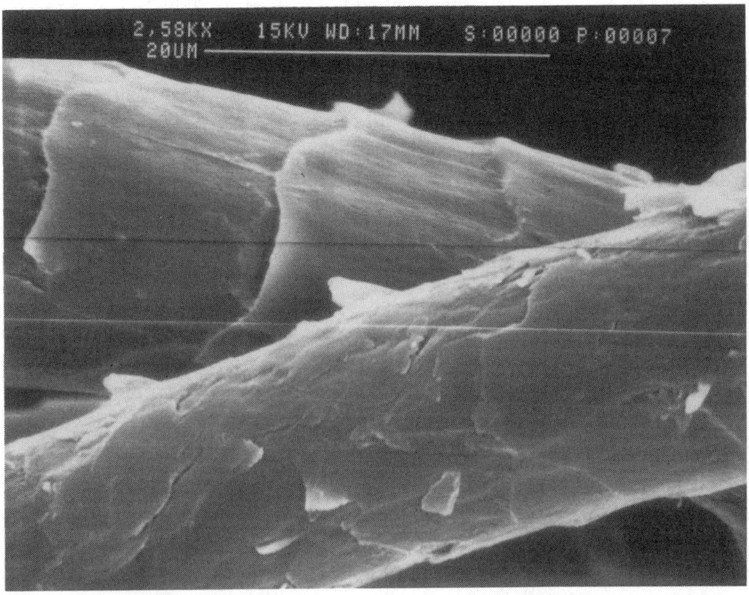

Figure 7. SEM (x2500) of wool treated with 1 g/l of protease for 1 hr in the presence of 20 g/l of sodium sulfate.

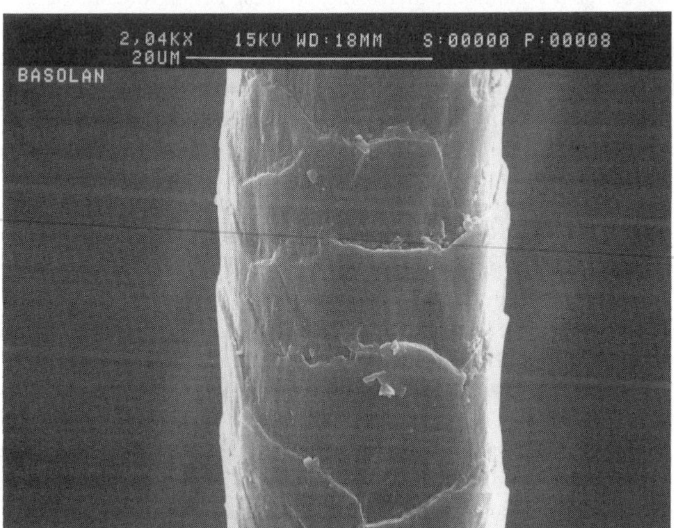

Figure 8. SEM (x2000) of Basolan DC-treated wool.

resistance induced in wool by an oxidative treatment can be correlated to the appearance of an IR band at 1040 cm^{-1}, indicative of cystine oxidation to cysteic acid, and to the appearance of cysteine-S-monoxide (at 1060-1070 cm^{-1}) and of cysteine-S-dioxide (at 1124 cm^{-1}). Furthermore, upon sulphitolysis with bisulfite, a new IR band appears in the IR spectrum spectrum at 1023 cm^{-1}, suggesting Bunte salt formation (cysteine-S-sulphonate). Conversely, when these residues are reconverted to cysteine disulfide bonds, the IR band around 1023 cm^{-1} disappears and the non felting characteristics are lost. From this work it appears crucial that cysteine -S-S- bonds of keratins be oxidatively modified in order to induce shrink-resistance features.
We compared by FTIR-ATR our enzyme-treated wool samples to untreated and Basolan DC-treated wool, which is shown in Figure 9. The spectra were taken and recorded at different times and, for the purpose of comparison, normalized on the amide III peak at 1230 cm^{-1} and superimposed in the figure. No significant difference is apparent in the spectra of untreated or enzyme-treated wool, even in the presence of high ionic strength. On the other hand, the Basolan DC-treated sample (same as in the SEM picture of Figure 8) shows a typical IR band of cysteic acid at 1040 cm^{-1} and has a clearly different overall profile than untreated wool.

Discussion

We have studied the structural and morphological modifications caused to Merino wool top by microbial subtilisin-type bioengineered proteases, under different conditions of incubation medium.
Whenever a significant structural modification was obtained, the corresponding wool top was subjected to the Aachen felting test to assess its shrinkage properties under standardized treatment conditions. Increasing protease action, obtained either by long contact time at constant enzyme or at fixed incubation time with increasing enzyme, causes progressive fiber damage mostly of the CMC, rather than of the keratin structure. SDS-PAGE analysis, after dissolution of the wool fiber, confirms a preferential attack on the low sulfur-containing proteins of the CMC in the mol.wt range of 45-60 Kda.
Higher ionic strength appears to limit diffusion of the protease within the fiber, thereby limiting the proteolytic activity on the epicuticle. In the latter case, SEM shows "filing" of the fiber scales, a modified morphology that closely resembles that obtained by oxidative treatment with DCCD. Nevertheless, no significant felt-resistance property is obtained.
No striking difference in the FTIR spectra was observed after protease treatment, in the wavenumber interval from 950 to 1280 cm^{-1}, as after chlorine oxidation. This suggests that the redox state of cystine residues remains unchanged.
Taken together, the above and our previous results (13) indicate that the use of proteases can, under carefully controlled conditions, be exploited in processes aimed to obtain special effects on wool top or fabrics, that might be required by the market (e.g., softening, hand modification, "aged" look, enhanced lustre, etc.), but that felt-resistance requires specific modifications of the cystine disulfide bonds.

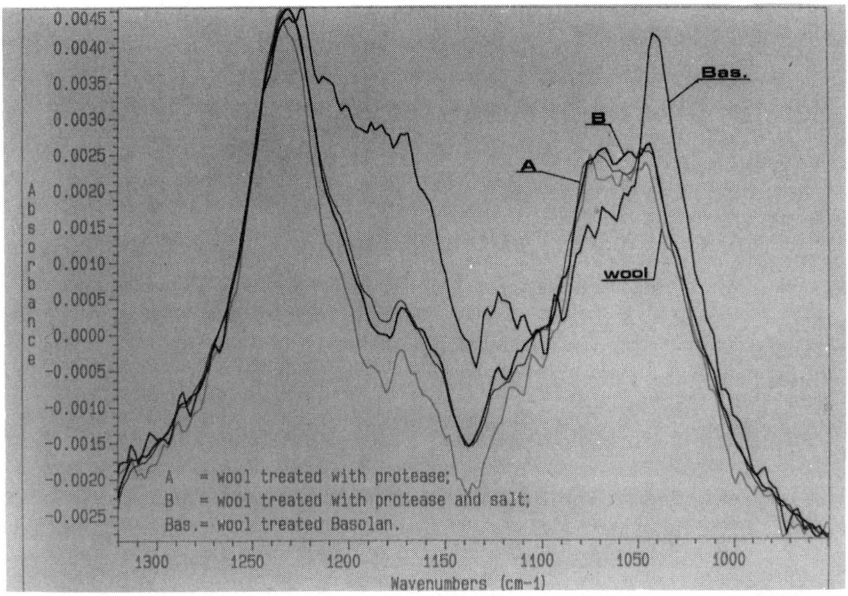

Figure 9. FTIR-ATR of wool samples: untreated (wool); treated with 1 g/l protease without (A) or with 20 g/l sodium sulfate (B); Basolan DC-treated (Bas).

Literature Cited

1. Makinson, K.R., "Surface Properties of Wool Fibers" in *Surface Characteristics of Fibers and Textiles*, Schick, M.J. ed., Marcell Dekker, New York, **1975**, 109-156.
2. Heine, E.; Hocker, H., *Rev.Prog.Coloration* **1995**, *25*, 57-63.
3. Negri, A.P.; Cornell, H.J.; Rivett, D.E., *Tex. Res. J.* **1993**, 63(2), 109-115
4. Haefely, H.R., *Textilveredlung* **1993**, *7/8*, 271-276.
5. Riva, A.; Cegarra, J.; Prieto, R., *JSDC* **1993**, *109*, 210-213.
6. Levene, R.; Cohen, Y.; Barkai, D., *JSDC* **1996**, *112*, 6-10.
7. Nolte, H.; Bishop, D.P.; Hocker, H., *J.Text.Inst.* **1996**, *87(1)*, 212-226.
8. MacLaren, J.A.; Milligan, B., *Wool Science. The Chemical Reactivity of Wool Fiber*, Science Press, Australia, 1981.
9. Laemmli, U.K., Nature, **1970**, *227*, 680-685.
10. Marshall, R.C., *Text.Res.J.*, **1981**, *2*, 87-91.
11. Innocenti, R.; Zoccola, M.; Freddi, G.; Motta, A.; Sudano, C., *Proc. 9th Int. Wool Text. Res. Conf.* **1995**, Biella, Italy 128-137.
12. Douthwaite, F.J.; Lewis, D.M., *J.S.D.C.* **1994**, *110*, 304-307.
13. Monteverdi, R.; Lavelli, V.; Galante, Y.M.; Tonin, C.; Innocenti, R.; Ramazio, F., *Proc. 9th Int. Wool Text. Res. Conf.* **1995**, Biella, Italy, 307-314.

Chapter 25

Perspectives of Enzymes for Processing Cellulose for New Chemical Fibers

H. Struszczyk and D. Ciechanska

Institute of Chemical Fibres, M. Sklodowskiej-Curie 19, PL-90570 Lodz, Poland

Progress for the manufacture of cellulose fibres, without use of toxic carbon disulfide as in a viscose method, requires applications of the new alternative methods. Biotechnological processing of cellulose pulp using selected types of enzymes, especially cellulases, allows to produce a new cellulosic material which is directly soluble in the aqueous alkali or selected organic solvents as well as a high reactive toward specific chemicals. Biotransformation of cellulose pulps using a cellulase complex originating from *Aspergillus niger* strain seems to be a solution for perspective processing cellulose for cellulosic fibers.

The conversion of wood into pulp, mainly for paper production, comprises one of the larger industries (*1*). Another large area of pulp utilization is its processing into the man-made cellulosic fibers. However, demand for more environmentally friendly processes has been caused because of tightening the environmental regulations (*2*).
The aim of this paper is to present the perspective of enzyme applications for processing pulps, by their biotransformation, to directly soluble or highly reactive cellulose which should be useful in the chemical fiber industries.

Biotransformation of Cellulose

Cellulolytic enzymes have been used, for varying reasons, to treat pulps or cellulose fibres (*1,3-5*). However, degradation and saccharification were the more important actions of enzymes during such cellulose treatments (*3-5*). Discovery of a new enzymatic treatment of cellulose, by the scientists from Institute of Chemical Fibres, Lodz, Poland and Tampere University of Technology, Finland in cooperation of Technical University of Lodz, Institute of Technical Biochemistry, Poland, using selected cellulolytic complexis has become the basis of a novel biotransformation process (*6-9*).

Biotransformation of pulps, using the cellulolytic enzyme complexe containing mainly: β-1,4- endoglucanase, β-1,4 - exoglucanase, cellobiohydrolase and glucosidase using several phenomena such as:

- controlled degradation
- controlled activation
- controlled saccharification
- reduction of hydrogen bonds energy

occurs on the molecular, super-molecular and morphological level of cellulose structure (*8-10*,).

The biotransformation process based on the applications the selected groups of these cellulolytic enzymes, catalysed the decomposition of the β-1,4-glucoside bonds. Two major modified strains of *Aspergillus niger* (*11*) and *Trichoderma reesei* (*12*) were selected as being the best for biotransformation of cellulose to achieve the following solutions such as:

- direct soluble pulps CELSOL, with solubility in an aqueous sodium hydroxide solution up to 99.5 wt % (*6-12*),
- highly soluble pulps, with improved solubility in NMMO (*9,13*),
- highly reactive pulps with improved reactivity toward reactants such as acetic anhydride (*9,14*).

The enzymatic transformation of the cellulose pulps occurs in an heterogenic system. However, the biotransformation rate and final results depend on the cellulose structure parameters such as average molecular weight and its distribution, crystallinity, capillary system or swelling behaviour as well as the enzyme activity. The application of special pretreatment of pulps, both mechanical or chemical, improves their intrinsic surface accessibility to enzyme action with a higher effectivity of biotransformation process (*7*). The enzymatic transformation of pulps into direct soluble cellulose CELSOL is shown schematically as Figure 1.

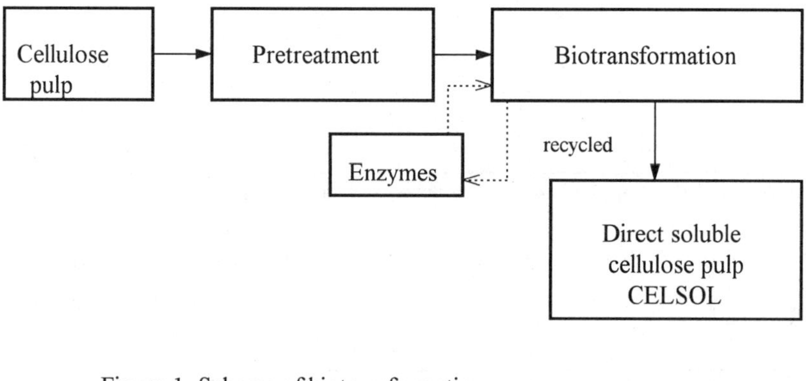

Figure 1. Scheme of biotransformation process

This biotransformation, especially using the cellulolytic complex of *Aspergillus niger* strain, is characterized by several advantages such as:

- soft reaction conditions
- high action specifity toward cellulose
- minimum amounts of by - products
- options for continous processing
- recirculation of enzymes
- high effectivity of process.

Several factors involved in the biotransformation process, such as enzymatic complex type and its composition, enzyme concentration, process conditions, including pretreatment, or structure of used pulps, have an affect on the properties of direct soluble cellulose. Some results of studies, on the perspective of cellulolytic enzymes for pulp biotransformation to the direct soluble cellulose suitable for preparation the new cellulose fibres and other products from regenerated cellulose, are presented. The cellulases originated from a modified *Aspergillus niger* strain with a CMC activity of 5.7 U/cm^3, an FPA activity of 0.28 U/cm^3, a CB activity of 5.9 U/cm^3 and ratio of CMC/FPA = 20 was produced by the Institute of Technical Biochemistry, Technical University of Lodz, Poland. These were used in the investigations discussed. Several commercial pulps were also tested. The biotransformation of pulp was carried out according to the original method (*6*). The analytical methods used in these studies have been described previously (*6-12*)

Pretreatment of Pulp Before Biotransformation. The necessity of pretreatment of pulps before biotransformation is shown in the results presented in Tables I-II.
Some additives for improving the mechanical pretreatment of a Fibrenier pulp were also used (Table I). Biotransformation was carried out for 6h at temperature of 50°C with an enzyme activity of 2U/cm^3.

Table I. Influence of mechanical pretreatment on the properties of biotransformed Fibrenier pulp

Type and amount of additive	*Properties of pulp*						
	Before pretreatment		*After pretreatment*		*After biotransformation*		
	$\overline{DPw}^*$	*WRV* %	$\overline{DPw}^*$	*WRV* %	$\overline{DPw}^*$	*WRV* %	S_A %
No additive	756	62	688	90	431	75	88
Berol V-4044	756	62	702	89	376	71	90
Berol BV-32	756	62	697	90	416	76	97

S_A - solubility degree (*14*); *- viscometric method

Table II. Influence of pretreatment type on the properties of biotransformed Fibrenier pulp

Type of pretreatment	*Properties of biotransformed pulp*		
	$\bar{DPw}$*	*WRV* %	S_A %
No	756	62	17.0
Mechanical	376	71	90.0
Mechanical + hydrothermic	307	70	99.5

Pretreatment of pulp seems to be the key factor for its effective processing using cellulases to prepare a directly soluble cellulose (Tables I-II).
The molecular structure parameters of cellulose, during its processing to biotransformed pulp, are presented in Table III.

Table III. Some molecular parametres of cellulose [a]

Type of treatment	*Cellulose properties*						
	$\bar{Mn}$ kDa	$\bar{Mw}$ kDa	$\bar{DPw}$	*Pd*	*Percentage content of DPw fraction*		
					<200	*200-550*	*>550*
None (initial)	47	119	740	2.6	17	39	44
Mechanical (with Berol BV-32)	42	118	738	2.9	19	42	39
Enzymatic	39	111	682	2.8	23	39	38
Mechanical + enzymatic	25	64	395	2.5	40	42	18

a - by GPC method

Influence of Biotransformation Conditions. The effect of the enzyme concentration on biotransformed Fibrenier cellulose molecular properties such as the average polymerization degree, its solubility in an alkaline solution as well as a weight loss during process is shown in Figures 2 - 4.

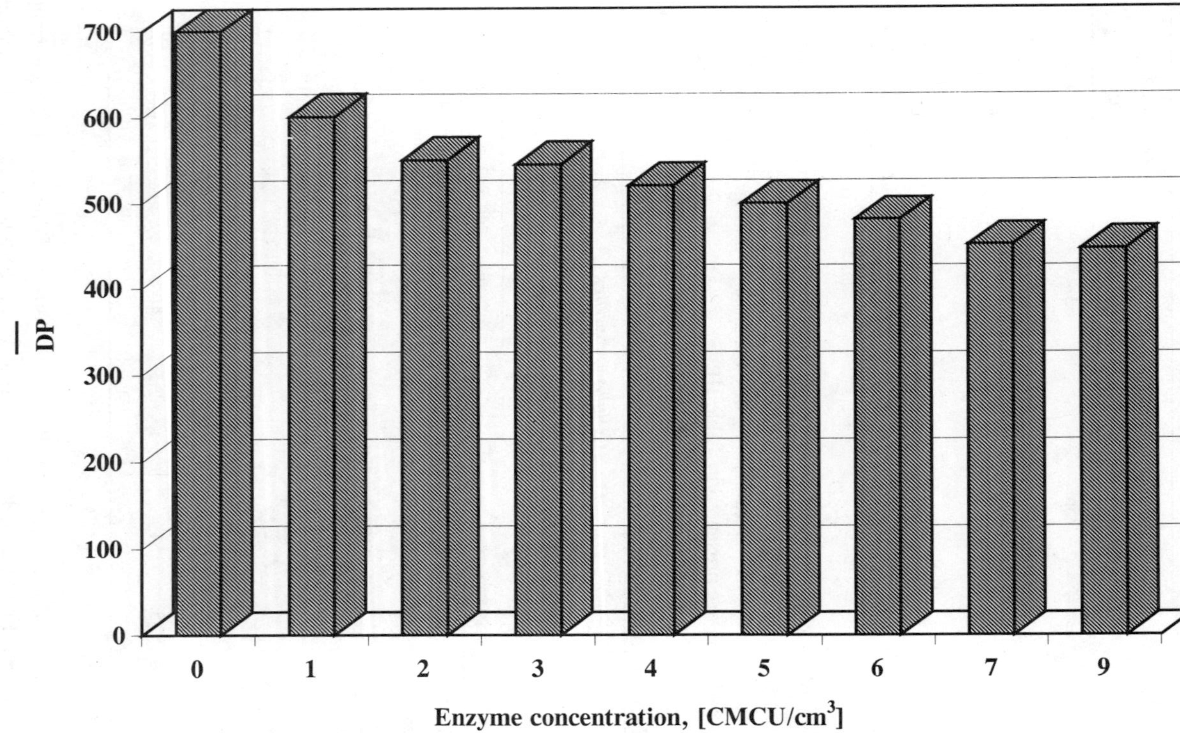

Figure 2. Effect of cellulose complex concentration on the $\overline{DP}$ of biotransformed cellulose

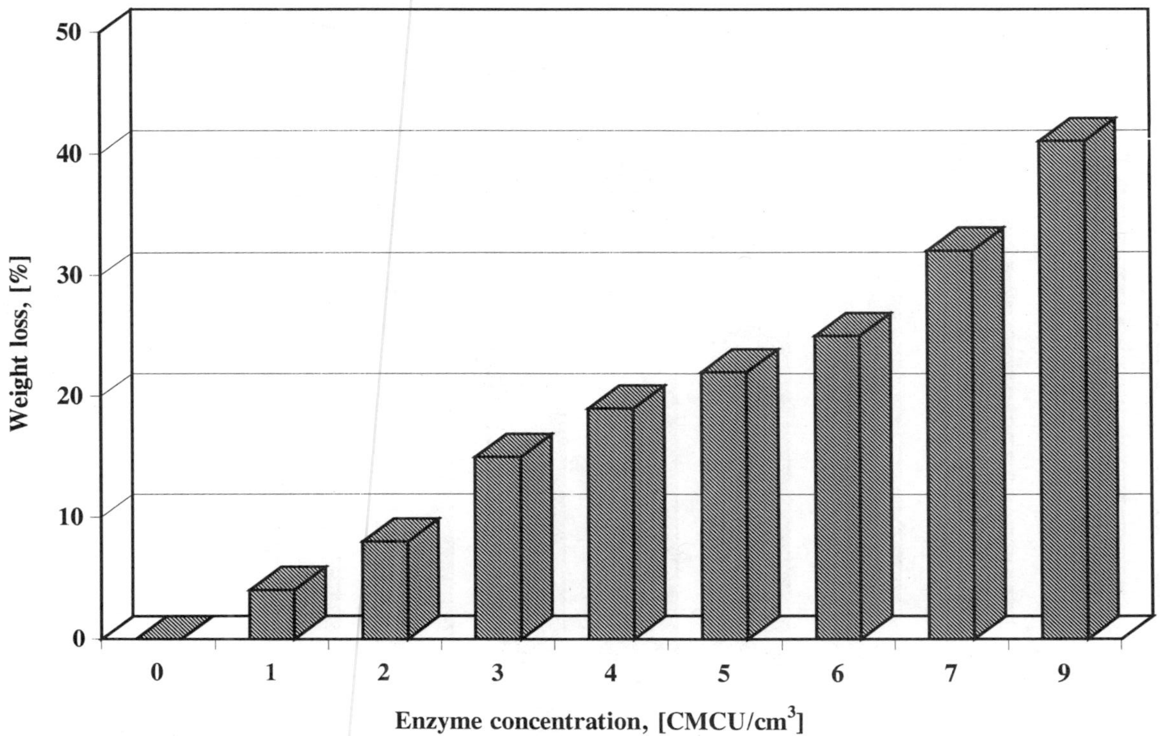

Figure 3. Effect of cellulose complex concentration on the weight loss of cellulose

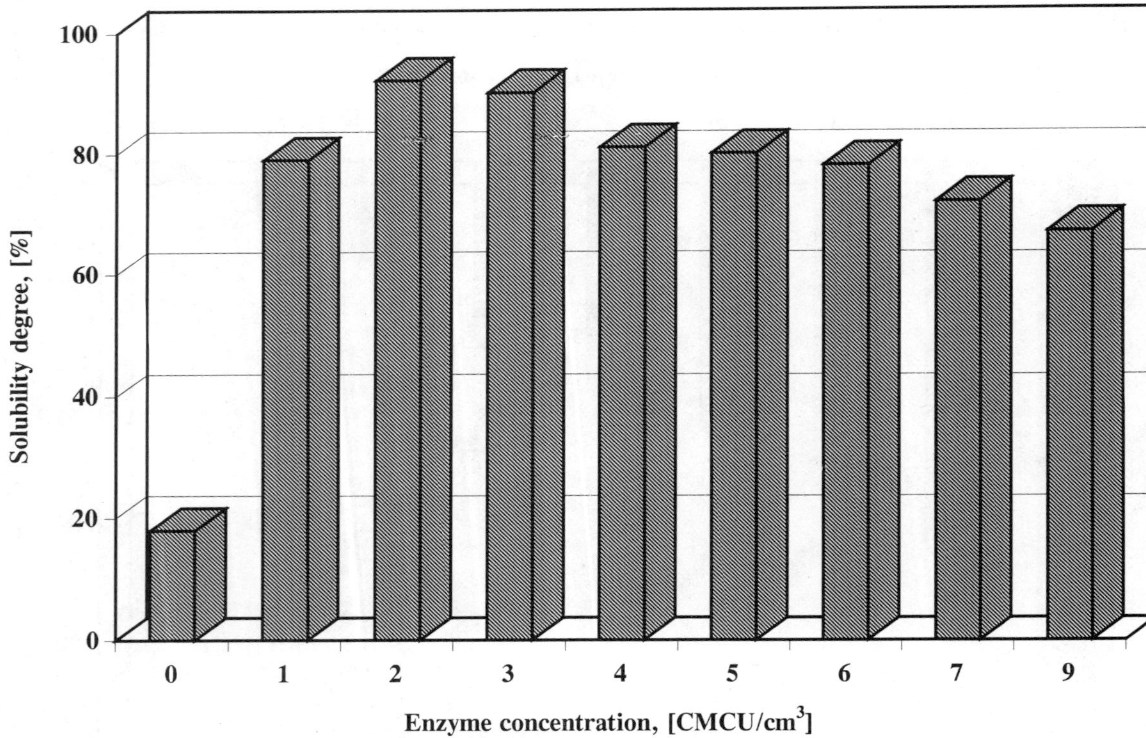

Figure 4. Effect of cellulose complex concentration on the solubility degree of cellulose sodium hydroxide lye

Based on the above results (Figures 2 - 4), it can be pointed out that the cellulase, originated from modified *Aspergillus niger* strain, affects the solubility of biotransformed Fibrenier cellulose in the aqueous hydroxide solution with concentration of approx. 2 CMC U/cm^3. The Fibrenier pulp was also subjected to mechanical pretreatment with cellulase treatment with its concentration of 1.8 CMC U/ cm^3 for 6h at different temperatures (Table IV).

Table IV. Some molecular parameter of Fibrenier pulp biotransformed at different temperatures.

Temperature of biotransformation	*Biotransformed pulp properties*					
	$\bar{D}Pw$	*Pd*	*Percentage content of DPw fractions*			S_A
°C			*<200*	*200-550*	*>550*	%
30	433	2.4	31	42	27	74
40	395	2.3	34	43	23	80
50	389	2.4	36	40	24	91
60	426	2.4	33	41	26	81

Generally, it can be pointed out that the biotransformed cellulose properties, including solubility, depend on the temperature of process. However, the biotransformation process can be shown to be the most sensitive in a temperature range of 40 - 60 °C for 6h (Table IV, Figures 2 - 4).

The type of pulps, characterized by individual parameters of molecular, super-molecular and morphological structure, affects also the biotransformation specifity (Table V).

Table V. Some properties of different pulps

Type of pulp	*Properties of pulp*						
	Initial			*Biotransformed*			
	CrI %	*WRV* %	*DPw**	*CrI* %	*WRV* %	*DPw**	S_A %
Fibrenier	74	62	756	74	77	485	92
Riocell	75	60	654	66	79	475	85
Excell	62	58	686	65	66	502	79
Sappi saicor	67	63	726	65	77	474	80

The structural parameters of cellulose were subjected to changes during its processing. Some results of these studies using Fibrenier pulp, with biotransformation carried out for 6h at temperature of 50 °C and enzyme concentration of 2 CMC U/cm^3, are presented in Table VI.

Table VI. Some structural parameters of Fibrenier pulp

Type of treatment	*Crystalline degree*	*Total volume of pores* mm^3/g	*Total porosity* %	*Intrinsic surface* m^2/mm^3
No (initial)	0.52	253.8	6.09	4.9
Mechanical	0.46	458.3	24.74	8.3
Enzymatic	0.50	359.7	6.47	5.5
Mechanical + enzymatic	0.48	316.4	9.17	9.6

The results of studies on structure of Fibrenier pulp during its processing show specific changes in the values of investigated parameters with increase of intrinsic surface, total porosity and total volumes of pores as well as with decreasing of crystallinity degree, in comparison to the initial cellulose (Table VI).
The cellulolytic enzymes originated from modified *Aspergillus niger* strain appeared to be effective in the biotransformation of pulps for its solution in aqueous sodium hydroxide and some other processes but the treatment conditions may be differed according to destination. The effectivity of biotransformation ranged from 93 to 98 % depends among others to the process conditions and pulp type and properties of pulp.

Practical Utilization of Biotransformed Cellulose

Biotransformed cellulose which is soluble in aqueous sodium hydroxide solution seems to be one of the more useful raw materials for the chemical fibers industry. It will have application in the manufacture of regenerated cellulose products as well as in the production of high reactive pulps for special applications. The biotransformed cellulose, both never-dried or dried, seems to be a suitable raw material for the preparation cellulose films, cellulose beads and cellulose fibres.
The research carried out on a laboratory scale showed the possibility of developing this process in the pilot scale as the next step to obtain the industrially acceptable technology.

Preparation of Cellulose Film. The regenerated cellulose film based on its mechanical, higienic and biodegradation properties is a continously required product, especially for the packing industry. Biotransformed cellulose seems to be a suitable, ecologically safe raw material to be used for preparation the film, in comparison to the existing viscose method (*15*). Some mechanical properties of film made on a laboratory scale, on the basis of biotransformed cellulose, according to the scheme (Figure 5), are presented in Table VII.

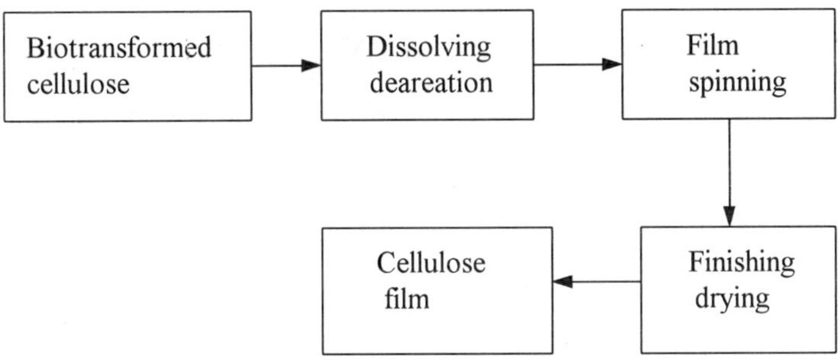

Fig. 5. Scheme of cellulose film preparation on the base of biotransformed pulp

Table VII. Some mechanical properties of cellulose film

Parameter	*Value*	*Unit*
Thickness	0.035 - 0.04	mm
Breaking power	16 - 23	N
Tenacity	30 - 50	MPa
Breaking elongation	4 - 15	%

The tenacity of cellulose film, obtained without stretching on the base of biotransformed pulp (Table 7) is much higher than a value of the same parameter characteristic for a viscose film prepared under similar conditions.

Preparation of Cellulose Beads. The biotransformed pulp can also be a useful raw material for preparation of cellulose beads useful as a carrier for slow release agricultural products. This method (*16*) seems to be the most advanced technologically for practical applications.
Some properties of cellulose beads, prepared on the basis of biotransformed pulp, are shown in Table VIII.

Table VIII. Some properties of cellulose beads

Parameter	Value	Unit
Colour	white	
Moisture content		
- wet	90 - 95	wt %
- dry	5 - 10	wt %
Diameter		
- wet	4 - 10	mm
- dry	1 - 2	mm
Volume density		
- wet	0.50	g/cm^3
- dry	0.74	g/cm^3
DPw*	300 - 350	
WRV		
- wet	800 - 1000	%
- dry	70 - 300	%

Preparation of Cellulose Fibers. Cellulose fiber preparation seems to be the most exciting, but at the same time difficult, technique of biotransformed pulp processing. The biotransformed pulp is easily soluble in an aqueous sodium hydroxide, with cellulose solutions as characterized by their average properties obtained presently on a laboratory scale as shown in Table IX (*11*).

Table IX. Some properties of alkaline solution of biotransformed pulp (11).

Parameter	*Value*	*Unit*
α - cellulose content	5 - 7	wt%
alkali content	7 - 8	wt%
viscosity	40 - 100	s
ripeness degree	8 - 10	°H
stability of solution at 5 °C	48	h

The spining of regenerated cellulose fibres (CELSOL) (*16*) using an alkaline solution of biotransformed pulp and acidic regeneration bath, to produce the fibres with high purity and sorption capacity, produces the fibres with a tenacity value of 11 - 13 G/den. There are great opportunities to improve this property during development stage of discussed method. The effectivity of fiber and film production using a

biotransformation process is generally on a level of 93 - 95 % to the weight of initial pulp.
Biotransformed pulps seems to be useful as a source of modified cellulose having a high ability to exchange the present viscose technology towards a more environmently friendly method.

Acknowledgment

A part of presented investigation was carried out within the COPERNICUS project No CIPA - CT94 - 0141 sponsored by the European Community.

Literature Cited

1. Jeffries, T.W, in „ Materials and Chemical from Biomass", ACS Symposium Series 476, American Chemical Society Publ., Washington D.C., 1992, chapter 18, pp.313 - 329.
2. Struszczyk, H., Chemical Fibres International, 1996, 46, 265 - 267.
3. Symposium on „Natural Textile Processing", Novo Nordisk (Danmark), 1995, Lodz.
4. Struszczyk, H., Wrześniewska-Tosik, K., Ciechańska, E., Wesołowska, E., Fibres & Textiles in Eastern Europe, 1994, vol.2, 4(7), 46 - 48.
5. Noé, P., Chevalier, J., Mora, F., Comtat, J., J. Wood Chem. Techn. 1986, 6(2), 167-184.
6. Polish Pat. 167519 (1992) (Struszczyk, H., Wawro, D., Ciechańska, D.).
7. Struszczyk, H., Ciechańska, D., Wawro, D., Fibres & Textiles in Eastern Europe, 1995, 1, 47 - 49.
8. Struszczyk, H., Ciechańska, D., Wawro, D., Nousiainen, P., Matero, M., in „Cellulose Derivatives: Physical-Chemical Aspects and Industrial Applications", Kennedy J.F., Phillips G.O., Williams P.O., ed., Woodhead Publ., Cambridge, 1995, chapter 4, 29-35.
9. Schleicher, H., Struszczyk, H., Wetzel, H., Das Papier, 1996, 12, 674-681.
10. Ciechańska D., „Biotransformation of cellulose", Ph.D. Dissertation 1996, Lodz, Poland.
11. Struszczyk, H., „Alternative Wet Spinning Technologies for the Manufacture of Cellulose Fibres", in „Cellulosic Men-Made Fibres", Akzo Nobel Co., Singapore, 1997, chapter 18.
12. Vehviläinen, M., Nousiainen, P., „CELSOL - Modification of Pine Sulphate Paper Grade with Trichoderma reesei Cellulases for Fibre Spinning", in „Cellulosic Man-Made Fibres", Akzo Nobel Co., Singapore, 1997, chapter 17.
13. Ger. Pat. Appl., 19624866.3. (1996), „Verfahren zur Verbesserung der Reaktionsaktivität von Cellulose zur Herstellung von Celluloseacetat" (Schleicher, H., Wetzel, H., Struszczyk, H., at al.).
14. Ger. Pat. Appl. 19624867.1. (1996), „Verfahren zur Verbesserung der Löslichkeit von Cellulose in wasserhaltigem Aminoxid" (Schleicher, H., Weigel, P., Wetzel, H., at al.).
15. Kamide, K., Okajima, K., Matsui, K., Kowsaka, K., Polymer J., 1984, 16, 857.
16. Polish Pat. 167776 (1992) (Struszczyk, H., Wawro, D., Ciechańska, D.).

Chapter 26

Modification of Dissolving Pulp by Hydrolysis with Cellulase Enzymes

L. Rahkamo[1], M. Vehviläinen[2], L. Viikari[1], P. Nousiainen[2], and J. Buchert[1]

[1]VTT Biotechnology and Food Research, P.O. Box 1501, 02044 VTT, Finland
[2]Tampere University of Technology, Fiber, Textile and Clothing Science, Box 589, 33101 Tampere, Finland

The environmental risks caused by the use of carbon disulfide in the viscose process have lead to the search for new types of processes for fiber production from cellulose. As a novel approach, cellulose degrading enzymes have been investigated in the manufacture of a directly soluble cellulose. In this paper, the potential of monocomponent cellulases from *Trichoderma reesei* in the transformation of dissolving pulp into directly alkaline soluble cellulose is discussed.

Cellulose can be processed to textile fibers by the viscose process. The use of carbon disulfide in the viscose process causes several environmental problems. Carbon disulfide is the main reactant in converting cellulose into alkaline soluble cellulose xanthate. The problems associated with carbon disulfide have lead to the search for more environmentally friendly techniques. These alternative processes for fiber production from cellulose include direct dissolution of cellulose pulp in organic solvents (*1*), or the use of cellulose derivatives such as cellulose carbamate (*2,3,4*). Also, cellulose degrading enzymes, cellulases, have been investigated in the manufacture of directly alkaline soluble cellulose (*5,6*).

Cellulases are produced by many fungi. Due to the complex structure of cellulose, several different enzymes are required for its complete degradation. The enzymes involved in cellulose degradation typically include several endoglucanases (EG), two or more cellobiohydrolases (CBH) and at least one ß-glucosidase (*7,8*). Cellulases act in synergy in the hydrolysis of crystalline cellulose. Endoglucanases randomly attack the amorphous regions in native cellulosic substrates, resulting in a rapid decrease in cellulose chain length (*9*). Cellobiohydrolases cleave cellobiose units either from the reducing or non-reducing ends of cellulose chains and they have been shown to hydrolyze crystalline cellulose without the aid of endoglucanases (*10*).

Vehviläinen *et al.* (*5*) have studied the activation of cellulose with mixtures of cellulases produced by *Trichoderma reesei* and *Aspergillus niger*. They concluded that cellulose can be converted into directly alkaline soluble form by the use of cellulase enzymes. Films and fibers can be spun from this cellulose solution.

In order to optimize the composition of the cellulase mixture it is necessary to understand the effects of individual cellulase enzymes. In our studies we have used purified *Trichoderma reesei* cellulases to investigate their individual effects in the activation of cellulose into alkaline soluble form. In our studies we have found that of the purified *Trichoderma reesei* cellulases investigated, endoglucanase II was most effective in reducing the viscosity and improving the alkaline solubility of hardwood dissolving pulp (*6*). The effect of EG II was not clearly improved by the action of hemicellulases or other cellulases (unpublished data). Also, a two-step solubilization of hardwood dissolving pulp has been investigated. The readily alkaline soluble fraction was first solubilized in alkali and the resulting insoluble residue was treated with cellulases.

Materials and Methods

Dissolving Pulps. Dissolving pulps were obtained from Saiccor (SA) and Borregaard. The Saiccor pulp was hardwood dissolving pulp (eucalyptus and acacia with a ratio of 3:1) and the Borregaard pulp was softwood dissolving pulp. The characteristics of the pulps are presented in Table 1. Prior to the enzymatic treatments the pulps were disintegrated in an L&W SE 003 apparatus at 1.5 % concentration and with 30 000 revolutions.

Table I. Characteristics of Saiccor and Borregaard Pulps

	Saiccor	Borregaard
Origin	Hardwood	Softwood
Viscosity (dm^3/kg)	555	510
Carbohydrate composition (%)		
glucose	97.9	96.0
xylose	2.1	2.0
mannose	<0.3	2.0

Enzymes. *Trichoderma reesei* cellulases were purified from culture filtrate by chromatographic methods as described earlier by Pere *et al.* (*11*) and Rahkamo *et al.* (*6*). Protein concentrations of the purified enzyme preparations were assayed by the method of Lowry *et al.* (*12*) using bovine serum albumin as standard.

Xylanase (pI 9) was purified from *T. reesei* culture filtrate as described by Tenkanen *et al.* (*13*), omitting the final gel filtration step. Mannanase (pI 5.4) was purified from a genetically modified strain of *T. reesei* with an inactivated EGII gene, kindly provided by Primalco Ltd. Biotec, according to Rättö *et al.* (*14*) but omitting the last polishing step by gel filtration. Xylanase activity was measured by the method described by Bailey *et al.* (*15*) and mannanase activity as described by Rättö and Poutanen (*16*).

Enzymatic Treatments. Enzymatic treatments were performed in deionized water at 5 % concentration at 50°C and pH 5 for 2 hours. The dosage for the cellulases (EG II and CBH I) was 0.1-2.5 mg/g (dw) pulp. The xylanase and mannanase were dosed by enzyme activities. In the 2 hour hydrolysis experiments the dosage for both xylanase and mannanase was 1000 nkat/g (dw) pulp. In the extensive hemicellulase treatments, the dosage for xylanase and mannanase was 5000 nkat/g (dw) pulp and the hydrolysis time 24 hours. The control samples were treated similarly as the enzyme-treated samples, without addition of enzyme.

Analyses. The solubilized carbohydrates liberated in the enzyme treatments were analyzed by HPLC after a secondary enzymatic hydrolysis to monomers as described earlier (*17*). The viscosity of the pulp was measured according to SCAN C15:1988. The carbohydrate composition of the pulps was analyzed by HPLC after acid hydrolysis (*18*). The alkaline solubilities of the reference and enzyme-treated pulps were determined by the method described previously (*6*). The crystallinity of the samples was determined by the x-ray powder diffraction method (*19*).

Results and Discussion

Comparison of the Effects of Individual Purified *Trichoderma Reesei* Cellulases in the Hydrolysis of Hardwood Dissolving Pulp. Mechanically pretreated hardwood dissolving pulp (*5*) was treated with the purified endoglucanases and cellobiohydrolases. The enzyme dosage of endoglucanases was 0.1-2.5 mg/g (dw) pulp and of cellobiohydrolases 0.5-5.0 mg/g (dw) pulp.

As expected, endoglucanases were found to be more efficient in hydrolyzing pulp carbohydrates than were the cellobiohydrolases at the same protein dosage (Table II). Endoglucanases also lowered the viscosity and improved the alkaline solubility more dramatically. A clear correlation between the decrease in viscosity and alkaline solubility after enzymatic treatment was observed (Figure I). Thus, the improvement in the alkaline solubility was caused by the hydrolysis and subsequent decrease in viscosity. When compared at the same degree of cellulose degradation, endoglucanase II was found to be most effective in reducing the viscosity and thus improving the solubility. Cellobiohydrolases had a less pronounced effect on the viscosity or solubility.

Table II. Composition of Solubilized Carbohydrates of the Enzyme-treated Pulps

Enzyme[1]	Sugar composition of solubilized carbohydrates (% of pulp dw)			Viscosity (dm^3/kg)	Alkaline solubility (%)
	Glucose	Xylose	Total		
EG I	1.33	0.27	1.60	419	55
EG II	1.44	0.00	1.44	379	66
CBH I	0.49	0.00	0.49	498	43
CBH II	0.77	0.00	0.81	478	48
Control	0.03	0.08	0.11	509	41

[1] Dosage of the enzymes: 2.5 mg/g pulp (dw)

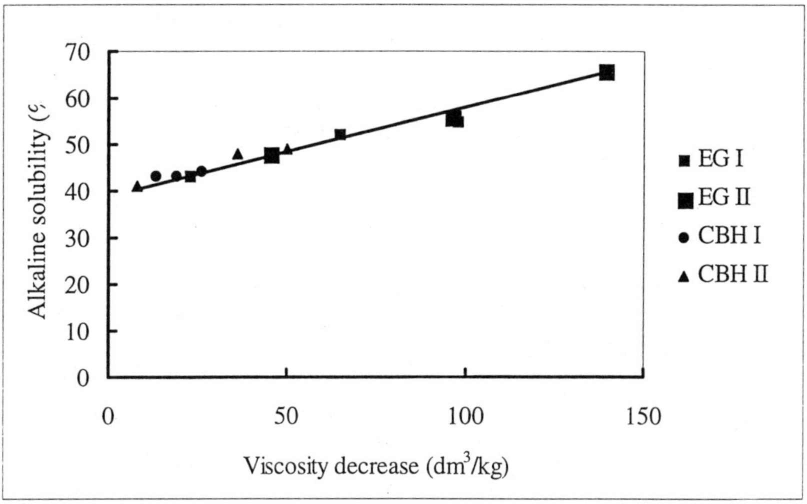

Figure 1. Alkaline solubilities as a function of viscosity decrease.

Effect of the Removal of the Residual Hemicelluloses of Hardwood and Softwood Dissolving Pulps. The effect of xylanases and mannanases on the solubility of softwood and hardwood dissolving pulps was investigated by combined treatments with endoglucanase II and xylanase and mannanase. The dosage for EG II was 2.5 mg/g (dw) pulp and the dosages for xylanase and mannanase were 1000 nkat/g (dw) pulp.

Less than 20 % of original xylanase and less than 15 % of glucomannan was hydrolyzed although a high dosage of hemicellulases was used (5000 nkat/g pulp, 24 h). The limited removal of the hemicellulose had no effect on the alkaline solubility or viscosity of the pulps.

The effect of EG II could not be clearly improved by the action of hemicellulases. The partial enzymatic removal of the residual hemicellulose did not increase the hydrolysis of cellulose by EG II (Table III). No synergism in decreasing the viscosity or improving the alkaline solubility was observed between the cellulases and hemicellulases.

Table III. Solubilized Sugars of the Cellulase and Hemicellulase Treated Dissolving Pulps

Enzymes[1]	Composition of solubilized carbohydrates (% of pulp dw)			
	Glucose	Xylose	Mannose	Total
Softwood dissolving pulp				
Xylanase	0.09	0.13	0.03	0.25
Mannanase	0.06	0.10	0.10	0.26
Xylanase + Mannanase	0.09	0.12	0.11	0.32
EG II	1.13	0.10	0.10	1.33
EG II + Xylanase	1.07	0.13	0.10	1.30
EG II + Mannanase	1.18	0.09	0.16	1.43
EG II + Xylanase + Mannanase	1.18	0.13	0.16	1.47
Control	0.04	0.11	0.03	0.18
Hardwood dissolving pulp				
Xylanase	0.04	0.14	0.01	0.19
EG II	1.05	0.07	0.05	1.17
EG II + Xylanase	1.22	0.22	0.05	1.49
Control	0.05	0.11	0.01	0.17

[1] Dosages for the enzymes: EG II 2.5 mg/g pulp dw, Xylanase and Mannanase 1000 nkat/g pulp dw

Synergism of Cellulases in Increasing the Alkaline Solubility of Hardwood Dissolving Pulps. Possible synergism between hemicellulases and cellulases was investigated by combining endoglucanase II with both xylanase and mannanase. The EG II dosage in 2 h hydrolyses was 2.5 mg/g (dw) pulp, and the xylanase and mannanase dosages were 1000 nkat/g (dw) pulp.

The combined treatment of hardwood dissolving pulp with endoglucanases I and II showed no synergy and had no effect on the viscosity or the alkaline solubility (results not shown). On the other hand, synergism was observed in the hydrolysis of hardwood dissolving pulp with endoglucanases and cellobiohydrolases as judged by solubilization of oligomeric sugars at the same total protein dosage (2.5 mg/g pulp dw) (Figure II). However, this was neither reflected in viscosity nor alkaline solubility.

Two-step Solubilization of Hardwood Dissolving Pulp in Alkali. In order to improve the treatment a two-step approach was studied. The readily alkaline soluble portion of the hardwood dissolving pulp was first solubilized in 9 % NaOH at -5°C. In the treatment 32 % of the pulp was dissolved. This solubility is lower than in our earlier studies (*6*). This is assumingly due to differences in the scale and thus less efficient mixing of the solution.

From the monosaccharide content of the insoluble residue (Table IV) it can be seen that most of the pulp hemicelluloses were solubilized in alkali. This was expected since hemicelluloses are commonly isolated from wood pulp by alkali extraction (*20*). By the x-ray measurements it was found that the structure of the insoluble residue was close to the structure of cellulose II (Table IV).

Table IV. Characteristics of the original hardwood dissolving pulp and the alkaline insoluble residue

Pulp	Carbohydrate composition (% of dw)			Viscosity	Crystalline structure
	Glucose	Xylose	Mannose	dm^3/kg	
Original pulp	97.6	2.1	0.3	470	Cell I
Insoluble residue	99.6	0.4	-	526	Cell II

The effect of cellulases on the alkaline solubility of the insoluble residue from the first step was further investigated in hydrolysis experiments with EG II and CBH I. The hydrolysis yields in the cellulase treatments of the insoluble residue were remarkably high as compared to treatments of the original pulp (Table V). It is clear from the results that the cellulose fraction after the alkaline extraction is a better substrate for the cellulases. These findings support the higher susceptibility of cellulose crystalline form II to hydrolytic attack reported by Atalla (*21*).

The effects of cellulases on the viscosity of the insoluble pulp residue were more pronounced than on that of the original pulp (Table V). From the results it can be seen, that there is a slight increase in the total alkaline solubility by the two-step method compared to the one-step method (Table VI).

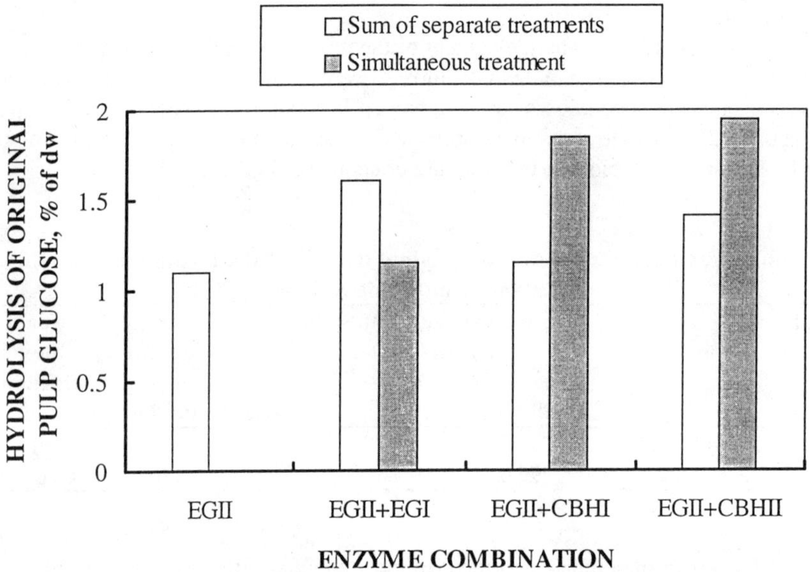

Figure 2. Cellulose hydrolysis of hardwood dissolving pulp treated with combinations of *T. reesei* cellulases.

Table V. Composition of Solubilized Carbohydrates and the Viscosities of the Enzyme-treated Hardwood Dissolving Pulps and Alkaline Insoluble Residues

Enzyme[1]	Sugar composition of solubilized carbohydrates (% of pulp dw)				Viscosity
	Glucose	Xylose	Mannose	Total	dm^3/kg
Original pulp					
EG II	1.44	0.00	0.00	1.44	379
CBH I	0.49	0.00	0.00	0.49	498
Control	0.03	0.08	0.00	0.11	509
Insoluble residue					
EG II	8.51	0.04	0.12	8.67	143
CBH I	0.66	0.00	0.02	0.68	429
Control	0.07	0.01	0.02	0.10	518

[1]Enzyme dosages: 2.5 mg/g pulp dw

Table VI. Alkaline Solubilities in the Single and Two-Step Solubilization Methods

Enzyme[1]	Single-step method (%)	Two-step method (%)
EG II	66	71
CBH I	43	38
Control	40	37

[1]Dosages for the enzymes: 2.5 mg/g pulp dw

Conclusions

Enzymatic solubilization of dissolving pulp offers an environmentally more friendly way of producing regenerated cellulosic fibers. Dissolving pulp can be modified into a directly alkaline soluble form by cellulases. According to preliminary fiber spinning trials from enzyme-treated hardwood dissolving pulp it has been shown that it is possible to spin fibers from enzyme-treated cellulose using an advanced wet spinning line (*5*).

Using two hour enzymatic treatments of dissolving pulp with pure cellulases, alkaline solubilities as high as 66 % have been obtained (*6*). Similar alkaline solubilities of dissolving pulps treated with a mixture of cellulases were earlier reported by Vehviläinen *et al.* (*5*). The alkaline solubility could not be improved by simultaneous hydrolysis with hemicellulases or with different purified cellulases. In the two-step alkaline solubilization the overall alkaline solubility was 71 %, which was slightly higher than in the one-step enzymatic solubilization. According to our results, alkaline solubilities considerably higher than 70 % seem to be difficult to obtain with purely enzymatic methods. This is apparently due to the limited accessibility of the substrate and the limited capability of the cellulases to open the cellulose structure and to break hydrogen bonds between the cellulose chains. In order to produce regenerated cellulosic fiber with acceptable quality, higher alkaline solubilities are required. To achieve this goal, different pretreatments of the cellulose and further optimization of the dissolving and spinning stages must be investigated.

Acknowledgments

The financial support of Neste Oy Foundation (Finland) is gratefully acknowledged.

References

1. Chanzy, H. and Paillet, M. *Polymer* **1990,** *32*, 400

2. Struszczyk, H. *Wlókna Chemiczne* **1988**, *14*, 365-379

3. Struszczyk, H. *Wlókna Chemiczne* **1989,** *15*, 7-22

4. Struszczyk, H. *Wlókna Chemiczne* **1989,** *15*, 169-179

5. Vehviläinen, M., Nousiainen, P., Struszczyk, H. and East, G. In *Proc. Cellucon International Cellulose Conference*, September 19-23, Wales, Ellis Horwood Series in Polymer Science and Technology (ed. J. F. Kennedy), 1994, In press

6. Rahkamo, L., Siika-aho, M., Vehviläinen, M., Dolk, M., Viikari, L., Nousiainen, P. and Buchert, J. *Cellulose* **1996**, *3*, 153-163

7. Ladish, M. R., Lin, K. W., Voloch, M. and Tsao, G. T. *Enzyme Microb. Technol.* **1983** , *5*, 82-102

8. Nevalainen, H. and Penttilä, M. In *The Mycota II: Genetics and Biotechnology* Kück, ed.; Berlin/Heidelberg; Springer-Verlag, 1995, pp. 304-319

9. Wood, T. M. and Garcia-Campayo, V. *Biodegradation* **1990**, *1*, 147-161

10. Chanzy, H. and Henrissat, B. *FEBS Lett.* **1985**, *184*, 285-288

11. Pere, J., Siika-aho, M., Buchert, J. and Viikari, L. *Tappi Journal* **1995**, *78*, 71-78

12. Lowry, O. H., Rosenbrough, N. H., Farr, A. R. and Randall, R. J. *J. Biol. Chem.* **1951**, *193*, 265-275

13. Tenkanen, M., Puls, J. and Poutanen, K. *Enzyme Microb. Technol.* **1992**, *14*, 566-574

14. Rättö, M., Siika-aho, M., Buchert, J., Valkeajärvi, A. and Viikari, L. *Appl. Microb. Biotechnol.* **1993**, *40*, 449-454

15. Bailey, M. J., Biely, P. and Poutanen, K. *J. Biotechnol.* **1992,** *23*, 257-270

16. Rättö, M. and Poutanen, K. *Biotechnol. Lett.* **1988**, *10*, 661-664

17. Buchert, J., Siika-aho, M. and Bailey, M. *Biotechnol. Techniques* **1993**, *7*, 785

18. Hausalo, T. In *Proc. 8th ISWPC*, June 6-9, Helsinki, Finland, 1995, vol. III, pp. 131-136

19. Paakkari, T., Blomberg, M., Serimaa, R., and Järvinen, M. *J. Appl. Cryst.* **1988**, *21*, 393-397

20. Timell, T. E. *Wood Sci. Technol.* **1967**, *1*, 45-70

21. Atalla, R. H. *Adv. Chem. Ser.* **1977**, *181*, 55

INDEXES

Author Index

Subject Index

C

D

E

M

R

X